Subsurface Contamination Remediation

ACS SYMPOSIUM SERIES **904**

Subsurface Contamination Remediation

Accomplishments of the Environmental Management Science Program

Edgar Berkey, Editor
Concurrent Technologies Corporation

Tiffany Zachry, Editor
Concurrent Technologies Corporation

Sponsored by the
ACS Divisions of Environmental Chemistry, Inc.,
Geochemistry, Inc., and Analytical Chemistry

American Chemical Society, Washington, DC

Library of Congress Cataloging-in-Publication Data

Subsurface contamination remediation : accomplishments of the Environmental Management Science Program / Edgar Berkey, editor, Tiffany Zachry, editor.

p. cm.—(ACS symposium series ; 904)

Developed from a symposium sponsored by the Divisions of Environmental Chemistry, Inc., Geochemistry, Inc., and Analytical Chemistry at the 225th National Meeting of the American Chemical Society in New Orleans, Louisiana, Mar. 23–27, 2003.

Includes bibliographical references and index.

ISBN 978-0-8412-3906-7

1. Hazardous waste site remediation—United States—Congresses. 2. Radioactive waste sites—Cleanup—United States—Congresses. 3. United States. Dept. of Energy. Environmental Management Science Program—Congresses.

I. Berkey, Edgar. II. Zachry, Tiffany. III. American Chemical Society. Division of Environmental Chemistry, Inc. IV. American Chemical Society. Division of Geochemistry, Inc. V. American Chemical Society. Division of Analytical Chemistry. VI. American Chemical Society. Meeting (225th : 2003 : New Orleans, La.) VII. Series.

TD1040.S84   2005
628.5—dc22

The paper used in this publication meets the minimum requirements of American National Standard for Information Sciences—Permanence of Paper for Printed Library Materials, ANSI Z39.48–1984.

PRINTED IN THE UNITED STATES OF AMERICA

Foreword

The ACS Symposium Series was first published in 1974 to provide a mechanism for publishing symposia quickly in book form. The purpose of the series is to publish timely, comprehensive books developed from ACS sponsored symposia based on current scientific research. Occasionally, books are developed from symposia sponsored by other organizations when the topic is of keen interest to the chemistry audience.

Before agreeing to publish a book, the proposed table of contents is reviewed for appropriate and comprehensive coverage and for interest to the audience. Some papers may be excluded to better focus the book; others may be added to provide comprehensiveness. When appropriate, overview or introductory chapters are added. Drafts of chapters are peer-reviewed prior to final acceptance or rejection, and manuscripts are prepared in camera-ready format.

As a rule, only original research papers and original review papers are included in the volumes. Verbatim reproductions of previously published papers are not accepted.

ACS Books Department

Contents

Overview

Remedial Science and Technology
for Subsurface Contamination

Characterization, Fate, and Transport
of Subsurface Contamination

Environmental Sensing and Monitoring

Indexes

Preface

Research on remediation of subsurface contamination is helping to solve important environmental problems associated with hazardous and radioactive contaminants in soil and groundwater. In the United States, significant research in this area has been sponsored by the U.S. Department of Energy (DOE) Environmental Management Science Program (EMSP). Since 1989, the DOE has been responsible for managing the environmental cleanup of sites and facilities involved in the nation's program for development and production of nuclear weapons. In support of this mission, the EMSP since 1996 has funded hundreds of scientific research projects intended to address and support the longer-term needs of the cleanup program.

Following a successful symposium in 1999 that focused on the first accomplishments of the EMSP, which was sponsored by the American Chemical Society (ACS) Division of Nuclear Chemistry and Technology, a second ACS symposium was organized to highlight EMSP achievements specifically in the area of subsurface contamination remediation. This symposium was held March 26–27, 2003, at the 225th ACS National Meeting in New Orleans, Louisiana, with primary sponsorship by the division of Environmental Chemistry. Based on the leading-edge research presented by the 20 invited speakers, this symposium proceedings volume was proposed and approved by the ACS.

The chapters in this book provide a cross-section of some of the most significant projects in the subsurface science component of the EMSP. They are organized into three sections that focus on the core problems of remediation: c haracterization, m odeling o f f ate a nd t ransport, and monitoring.

We appreciate the opportunity to participate in this book project, made possible by the EMSP. Funding from the EMSP supported the research projects presented in this book as well as the administrative effort to organize the symposium and the editorial work required to prepare this proceedings volume. We also thank the EMSP principal investigators and their colleagues, who diligently worked to provide clear, concise, and thorough explanations of their work for this publication; in so doing, these authors have made this book a successful representation of their scientific accomplishments, and thus, we hope, a useful resource for the scientific community.

Edgar Berkey
Vice President, Research and Development, and Chief Quality Officer
Concurrent Technologies Corporation
425 Sixth Avenue
Pittsburgh, PA 15219

Tiffany Zachry
Senior Technical Editor
Concurrent Technologies Corporation
P.O. Box 350
Millville, UT 84326

Overview

Chapter 1

Subsurface Contamination Remediation: Accomplishments of the Environmental Management Science Program

Roland F. Hirsch

Office of Biological and Environmental Research, Office of Science,
U.S. Department of Energy, Washington, DC 20585

The U.S. Department of Energy is responsible for the nation's nuclear weapons and for the complex of sites and facilities that developed and produced the weapons. In 1989 DOE established the Office of Environmental Management (EM) to manage all aspects of the environmental cleanup. In 1996, Congress mandated establishment of the Environmental Management Science Program (EMSP) to support scientific research to address the longer-term needs of the EM cleanup program. From the outset, the projects supported by the EMSP have been in a wide range of scientific disciplines and have carried out research relating to many of the components of the cleanup program. The majority of the projects, however, involve research relating to high-level radioactive waste or to subsurface contamination. Remediation of subsurface contamination at DOE sites poses numerous challenges, because of the size of the problem and because of some unique contaminants and environments. The EMSP is responsible for basic research to help overcome these challenges. The projects described in this volume are addressing modeling, immobilization and extraction of contaminants, characterization, and monitoring. Thus, they represent a good cross section of the EMSP program in subsurface science. The progress reported in each chapter gives confidence that scientific research will contribute to the solution of many of the major problems facing the DOE cleanup program.

Overview

Research and development of nuclear weapons in the United States began around sixty-five years ago and was followed in the early 1940s by development of large-scale facilities for production of these weapons. The wastes generated in the research and production activities generally were disposed of according to the best practices of the time, although the pressures of wartime (during World War II as well as the Cold War that followed) did not always allow adequate time for determination of optimal disposal procedures.

These wastes contained a great variety of components, including, for example, radionuclides of the actinide elements and the fission products, heavy metals such as mercury, and organic compounds used in the processing of materials used in the weapons. Much of the highly radioactive wastes was stored in underground tanks, which have in many cases developed leaks as they have remained in use well beyond their design lifetimes. Leakage from these tanks has released not only radionuclides but also hazardous chemicals into the subsurface environment underneath the tanks. Other, less highly radioactive wastes were placed in containers such as metal barrels that were buried in landfills at the weapons production sites. These, too, have in some cases leaked, releasing contaminants into the subsurface. Significant amounts of contaminants also were released by discharge of untreated wastewater directly into the soil.

The U.S. Department of Energy (DOE) inherited from the Atomic Energy Commission (through the Energy Research and Development Administration) the responsibility for the nation's nuclear weapons and for the complex of sites and facilities that developed and produced the weapons. In 1989 DOE established the Office of Environmental Management (EM) to manage all aspects of the environmental cleanup, including existing contamination and prevention of further contamination. In 1996, Congress mandated establishment of the Environmental Management Science Program (EMSP) to support scientific research to address the longer-term needs of the EM cleanup program.

The EMSP is a program of basic research focused on a specific mission area. It was started as a collaboration between EM and the DOE Office of Energy Research (now the Office of Science, SC), in which SC's expertise in managing basic research programs in many relevant areas of science was allied with EM's knowledge of the cleanup problems. The budget for the EMSP was in the EM Office of Science and Technology from 1996 until 2003, when it was transferred to the Office of Biological and Environmental Research in SC.

From the outset, the projects supported by the EMSP have been in a wide range of scientific disciplines and have carried out research relating to many of the components of the cleanup program. The majority of the projects, however, involve research relating to high-level radioactive waste or to subsurface contamination. In Fiscal Year 2005, about 40 projects address the former area and 60 the latter, with another 10 projects in the other EM areas.

Program notices inviting research proposals for the EMSP have stressed collaboration among investigators across disciplinary lines and at different institutions. Thus, while the lead institution for 65 of the current projects is a DOE National Laboratory and for 40 of them a university (the rest are at companies, research institutes, or other Federal agency laboratories), many of the projects, in fact, are joint efforts among two or more institutions.

The Subsurface Contamination Problem

The scope of the national DOE environmental remediation problem can be gathered from these estimates:

- More than 5,000 individual plumes of contaminants in soil and groundwater
- Spread of some plumes exceeding 5 km^2
- Millions of cubic meters of contaminated soil
- Trillions of liters of contaminated subsurface water

The initial sources of this contamination include leakage from storage tanks and buried containers, seepage from disposal and storage ponds, spills during transfer of materials, and mixing into ordinary trash of substances not considered hazardous at the time of use decades ago.

As cleanup and closure of the facilities used to produce nuclear weapons proceed at an accelerated pace, attention is increasingly being focused on dealing with this uncontained subsurface contamination. Several of the major radioactive contaminants have very long half-lives exceeding hundreds of thousands of years (notably isotopes of technetium, neptunium, and plutonium) and must be immobilized to ensure that they do not migrate into the water supply. The same is true for some of the heavy metals, such as mercury, and the halogenated organic compounds such as trichloroethylene. The latter is quite stable under the typical underground conditions. The quantities that are known to be underground are sufficient to contaminate billions of gallons of water at levels considered unsafe in Federal drinking water standards.

EMSP Research Addressing Subsurface Contamination

Research in the EMSP addresses several of the needs in the subsurface contamination remediation program. New techniques for characterizing the subsurface environments are being developed, both for identification of the chemical forms of various contaminants and for understanding the geological structure of the subsurface formations on a fine enough scale to identify the likely paths through which the contaminants would be transported into rivers and aquifers. It is now realized, for example, that measurements of bulk porosities of the components of a formation are not sufficient to identify transport patterns, as a few pores or channels in an otherwise solid rock formation would be sufficient for more rapid than predicted migration of contaminants.

Characterization of the subsurface provides data for modeling transport of chemical contaminants. Research into modeling of fate and transport is itself an important aspect of dealing with the subsurface remediation problems. Improvements in the models will enable better understanding of how to deal with contamination at specific locations within the DOE complex.

Finally, new technologies are needed to immobilize substances that cannot be decomposed in place, or to extract them from the ground using cost-effective technologies. For example, radionuclides in soluble chemical forms could be converted to insoluble forms that would have much less risk of migration into water supplies. Once immobilized, however, long-term monitoring will be required to detect possible remobilization. Techniques will need to be developed for this purpose that are robust, remotely addressable, and highly automated.

Organization of this Volume

The chapters in this book provide a cross section of the projects in the subsurface science component of the EMSP, discussing the core problems of remediation, characterization, modeling of fate and transport, and monitoring.

The first section, on remediation strategies and technologies, demonstrates the extent to which EMSP complements its partner program, Natural and Accelerated Bioremediation Research (NABIR). Five of the six chapters discuss either remediation of organic compounds or remediation by other than microbial processes. The organics to be remediated include not only the halogenated compounds already mentioned, such as trichloroethylene (the dense nonaqueous phase liquids or DNAPLS), but also phenolic compounds and their enzymatic coupling with lignin and other humic substances. Remediation techniques include sonication and vapor stripping for removal of semivolatile organic compounds from groundwater, oxidation of these compounds by permanganate (with particular attention to the manganese-containing mineral products), and phytoremediation through accumulation of arsenic or mercury in *Arabidopsis thalania* varieties that contain genes that adjust the chemical form of the target metal.

The second section of this volume contains chapters that discuss aspects of the fate and transport of contaminants in subsurface environments. These papers demonstrate the growing recognition that modeling of subsurface systems requires knowledge of the composition of more than just the contaminants. How do the other components that may have been released along with the contaminants from the source, such as a leaking tank, impact how the contaminant interacts with the subsurface? What geochemical processes must be taken into account when modeling transport by subsurface flow? A final chapter in this section is devoted to modeling the composition of the major components of the high-level radioactive waste tanks at Hanford. This ties in closely with the other chapters, for accurate knowledge of the chemical speciation of the major components of the tanks is needed to model the interaction of leaked contaminants with the subsurface.

The final section of this volume focuses on advances in instrumentation for monitoring different types of contaminants. Each chapter describes new ways of achieving high sensitivity and high specificity in measuring and monitoring for contaminants in the subsurface. Two of the chapters discuss analytical techniques for measurement of multiple components, microcantilever arrays, and automated analyzers for radioisotopes. Two other chapters cover ingenious techniques for meeting requirements for highly specific measurement of specific contaminants, in one case a spectroelectrochemical sensor for technetium and in the other a reliable means of measuring size distributions of radioactive atmospheric particulate matter over long periods of time. A final chapter discusses a new single-molecule technique for visualizing breaks in DNA caused by exposure to ionizing radiation.

The Future of EMSP in its New Setting

As previously indicated, the EMSP was initiated in the Office of Environmental Management (EM) of the DOE. In 2003, the budget was transferred to the Office of Biological and Environmental Research (BER) in the DOE Office of Science (SC). At that time, a new Division was formed in BER, the Environmental Remediation Sciences Division (ERSD). Responsibility for the EMSP was placed in this Division, along with the NABIR Program and two environmental research laboratories, the Savannah River Ecology Laboratory (SREL) and the William R. Wiley Environmental Molecular Sciences Laboratory (the EMSL) at the Pacific Northwest National Laboratory (PNNL).

The Division also collaborates with the National Science Foundation in co-funding several Environmental Molecular Sciences Institutes (EMSI).

The EMSP fits well in the new ERSD. This is especially true for the projects in the EMSP that address subsurface contamination issues, the focus of this volume. Several EMSP-funded scientists also direct projects in the NABIR and EMSI program, while others are located at the SREL or the EMSL. Examples include projects that seek new techniques for characterizing subsurface contamination, projects developing improved models of subsurface transport to enable estimation of future risks from movement of contaminants, and projects studying ways to remediate or immobilize contaminants in situ. These areas are, in fact, the focus of the chapters in this volume.

As stated on the ERSD website, "the mission of the NABIR program is to provide the fundamental science that will serve as the basis for development of cost-effective bioremediation and long-term stewardship of radionuclides and metals in the subsurface at DOE sites. The focus of the program is on strategies leading to long-term immobilization of contaminants in place to reduce the risk to humans and the environment. The NABIR program encompasses both intrinsic bioremediation by naturally occurring microbial communities, as well as accelerated bioremediation through the use of biostimulation (addition of inorganic or organic nutrients). The program focuses on in situ bioremediation of heavy metals and radionuclides. Scientific understanding will be gained from fundamental laboratory and field research on biotransformation processes, community dynamics and microbial ecology, biomolecular science and engineering, biogeochemical dynamics, and innovative methods for accelerating and assessing in situ biogeochemical processes. The societal implications and concerns of NABIR are also being explored."

The EMSP complements the NABIR program by including research that addresses non-metal contamination that is a major concern at DOE sites, such as trichloroethylene and other DNAPLs. The EMSP also supports research into phytoremediation—the use of plants for bioremediation—while NABIR research is focused on bioremediation by microbes.

The EMSL is one of the DOE's National User Facilities. It is located near the Hanford Site, which is the focus of many EMSP projects, and serves as a resource for scientists in the EMSP program. The EMSL offers a suite of major instrumentation and capabilities within its 200,000 ft^2 building, in contrast to most of the other DOE facilities, which are accelerators producing beams of particles or photons. One of the component facilities of the EMSL is devoted to Environmental Spectroscopy and Biogeochemistry and is used by EMSP researchers focusing on subsurface contamination. The Interfacial and Nanoscale Facility also supports EMSP research into subsurface transport. Among the noteworthy instruments in the EMSL are nuclear magnetic resonance (NMR) instruments (including 800 and 900 MHz machines) with special capabilities for solid-state and microimaging experiments, unique

Fourier Transform Ion Cyclotron Resonance (FTICR) mass spectrometers that are pioneering the study of proteins in cells (proteomics), and a state-of-the-art Molecular Sciences Computing Facility with a supercomputer having a peak performance of 11 teraflops. More than a thousand scientists from outside PNNL use the EMSL facilities annually, many of them accessing its capabilities remotely, and many of them studying environmental problems.

The SREL was founded in 1951 by Eugene Odum of the University of Georgia as a pioneering field ecology laboratory. Initial funding was from DOE's predecessor, the Atomic Energy Commission. During the period 1985 through 2002, funding for the SREL came from EM through the DOE Savannah River Site Office. In 2003, SREL's funding was transferred to the Office of Science at the same time as the transfer of the EMSP. SREL conducts basic ecological research focused on understanding the effects on the local ecology of operations at the Savannah River Site (SRS). SREL carries out research into several areas that complement EMSP research efforts, such as bioremediation, phytoremediation, in situ immobilization of contaminants, and new technology for characterizing contaminated environments. The SREL offers significant opportunities for EMSP research, since the SRS is one of the major DOE cleanup sites.

Field Research Centers

The alignment of EMSP and NABIR also offers a particular opportunity for EMSP researchers to make use of the Field Research Center (FRC) in Oak Ridge, TN. The FRC was established in BER in 2000. It includes two thoroughly characterized areas totaling about 650 acres that have been set aside for long-term field research. Three research projects within the NABIR program are conducting studies at the FRC and some 30 other projects are using materials or data from the site. EMSP subsurface researchers are being encouraged to consider using the FRC as part of their projects, as it offers a stable platform for long-term field studies. This is especially important as laboratory research can only approximate field conditions. There are many complexities in the field that are not present in the laboratory, such as heterogeneity of the subsurface composition and variations in weather conditions. While discovery of new concepts often starts under controlled laboratory conditions, the applicability of these discoveries in the field is what will determine whether they can be used in DOE mission applications. Thus, the FRC serves as an intermediate testing ground between the laboratory and the actual cleanup sites.

The ERSD is planning to establish additional field research sites. The Biological and Environmental Research Advisory Committee (BERAC) adopted a report *Need for Additional Field Research Sites for Environmental*

Remediation Research in April 2004. The report contains these recommendations:

- Additional Field Research Centers be developed by ERSD;
- Additional Field Research Centers be focused on the conditions and environmental problems extant at DOE sites that differ from those at the ORNL site;
- New Field Research Centers have broad applicability to the research programs supported by the ERSD, and not simply the research focus of the NABIR program; and
- New Field Research Centers focus on scientific questions that arise from the need to remediate contamination due to radionuclides and mixed wastes (radionuclides with associated contaminants), contaminants that, within the United States, are predominantly the responsibility of DOE.

The third recommendation makes clear that EMSP research will be a factor in future sites.

The report also contains a recommendation "that 10–15% of available research funds be spent in the support of field research centers." It seems clear, therefore, that significant resources will be provided for supporting field sites suitable for EMSP research, enabling continued progress toward implementation in the DOE cleanup program of the concepts being studied in projects of the type described in this volume.

Conclusions

Remediation of subsurface contamination at DOE sites poses numerous challenges, because of the size of the problem and because of some unique contaminants and environments. The EMSP is responsible for basic research to help overcome these challenges. The projects described in this volume are addressing modeling, immobilization and extraction of contaminants, characterization, and monitoring. Thus, they represent a good cross section of the EMSP program in subsurface science. The progress reported in each chapter gives confidence that scientific research will contribute to the solution of many of the major problems facing the DOE cleanup program.

Further Reading

Much information is available on the Environmental Remediation Sciences Division web site (http://www.science.doe.gov/ober/ERSD_top.html). In addition, the EMSP web site (http://emsp.em.doe.gov/index.htm) offers detailed information on all projects funded in the program since it began in 1996, as well as texts of National Research Council reports and other reports on the program, past and current solicitations of research projects, and programs from conferences that included symposia devoted to the EMSP. Other pertinent reports include the Biological and Environmental Research Advisory Committee reports (http://www.science.doe.gov/ober/berac/Reports.html) and the report on *Need for Additional Field Research Sites for Environmental Remediation Research* (http://www.science.doe.gov/ober/berac/field_site_report.pdf).

Remedial Science
and Technology for Subsurface
Contamination

Chapter 2

Use of Sonication for In-Well Softening of Semivolatile Organic Compounds

Robert W. Peters[1], John L. Manning[2], Onder Ayyildiz[3], and Michael L. Wilkey[4]

[1]Department of Civil and Environmental Engineering, University of Alabama at Birmingham, Birmingham, AL 35294–4440
[2]Vanderbilt Medical Center, Vanderbilt University, Nashville, TN 37235
[3]Department of Chemical and Environmental Engineering, Illinois Institute of Technology, Chicago, IL 60616
[4]TechSavants, Inc., 211 East Illinois Street, Wheaton, IL 60187

This study examined an integrated sonication/vapor stripping system's ability to remove/destroy chlorinated organics from groundwater. Chlorinated solvents studied included carbon tetrachloride, trichloroethylene, trichlorethane, and tetrachloroethylene. Contaminant concentrations ranged from ~1 to ~100 mg/L. The sonicator had an ultrasonic frequency of 20-kHz; applied power intensities were 12.3, 25.3, and 35.8 W/cm^2. Batch reactions were operated for up to 10-min treatment time, with samples drawn for gas chromatography analysis every 2 min. Batch experimental results were obtained using sonication, vapor stripping, and combined sonication/vapor stripping. For the chlorinated solvents, the first-order rate constants were in the range of 0.02 to 0.06 min^{-1}, 0.23 to 0.53 min^{-1}, and 0.34 to 0.90 min^{-1} for sonication, vapor stripping, and combined sonication/vapor stripping. For the chlorinated organics (treatment time ~10 min), the fraction remaining after sonication and vapor stripping ranged from 62% to 82%, while < 3% remained from the combined sonication/vapor stripping system.

Introduction

Introduction/Description of Technology

The concept behind using sonication to "soften" halogenated organic compounds and polyaromatic hydrocarbons is relatively simple. By "softening," we mean partial degradation of volatile organic compounds (VOCs) and semivolatile organic compounds (SVOCs). Previous biodegradation studies have demonstrated relatively long periods of time are required to biologically remediate many aquifers contaminated with halogenated compounds. However, many of the corresponding nonhalogenated organic compounds are relatively easily microbially degraded to innocuous products such as carbon dioxide (CO_2). By using the sonication step to quickly remove the bound chloride ion (Cl$^-$) from a molecule, the by-product or remaining organic compound can be easily degraded under anaerobic or aerobic conditions.

This approach eliminates the costly proposition of using in-well sonication for a relatively long detention time to render the halogenated compounds innocuous to HCl and CO_2. Using sonication to enhance the relatively slow step in microbial degradation (i.e., dehalogenation) will allow remediation to occur at a much faster rate. This technology can be used at sites where natural attenuation is not a realistic option. At facilities where the contamination is moving "beyond the fence," a line of treatment wells could be set up to prevent halogenated compounds from moving offsite. The "by-products" can be degraded by natural attenuation.

The collapsing bubble interface results in the formation of hydroxyl and hydrogen radicals. These radicals destroy chlorinated organics and petroleum hydrocarbons very effectively.

The performance can be further enhanced through the addition of advanced oxidants (e.g., ozone [O_3], hydrogen peroxide [H_2O_2], etc.), and incorporation of recent advancements in the acoustic cavitation field. The destruction of organic pollutants can occur via several mechanisms. The organic pollutant inside the cavity and in the interfacial region can undergo pyrolysis reactions (or combustion reactions if oxygen is present) during the implosion. Free radicals (e.g., •OH, •H) formed due to thermolysis of water molecules may react with the organic in the interfacial region or in the solution near the interface. Three primary pathways have been identified for compound degradation, including: (1) hydroxyl radical oxidation, (2) direct pyrolytic degradation, and (3) supercritical water reactions. In aqueous solution, water vapor present in the microbubble is homolytically split during bubble collapse to yield •H and •OH radicals, while chemical substrates present either within or near the gas-liquid interface of the collapsing microbubble are subject to direct attack by •OH. Volatile compounds such as carbon tetrachloride (CCl_4), trichloroethylene (TCE), benzene, toluene, ethylbenzene, xylene, and MTBE readily partition into the vapor of the growing cavitation microbubbles and then undergo direct pyrolysis during transient collapse.

The in-well sonication process utilizes in situ ultrasonics/acoustics in a downhole well for destruction/conversion of organic contaminants from groundwater, described in an Invention Disclosure filed by Peters and Wu (1). Treatment is accomplished by circulating groundwater through the ultrasonic reactor and reinjecting the treated water into the unsaturated (vadose) zone rather than lifting the treated water to the ground surface (see Figure 1). Acoustic cavitation (commonly termed sonication) involves the application of sound waves being transmitted through a liquid as a wave of alternating cavitation cycles. Compression cycles exert a positive pressure on the liquid, pushing molecules together, while expansion cycles exert a negative pressure, pulling molecules away from each other. The chemical effects of sonication are a result of acoustic cavitation. During rarefaction, molecules are torn apart forming tiny microbubbles that grow to a critical size during the alternating cavitation cycles, and then implode, releasing a large amount of energy. Temperatures on the order of 5,000 K and pressures up to 500–1,000 atm have been observed in microbubble implosions, while the bulk solution stays near ambient. The ultrasonically treated water, partially or completely free of contaminant concentrations, infiltrates back to the water table. This process is continued until the contaminant concentrations are sufficiently reduced. Addition of oxidants/catalysts into the process may further enhance the process efficiency and reduce the required sonication period. Use of an in-well ultrasonic reactor, potentially coupled with megasonics, which is more one-directional in nature, may be required to treat more recalcitrant organic contaminants (1).

The in-well vapor stripping component of this combined sonication and stripping remedial technology is based on a special well design (see Figure 1) originally developed by Stanford University (2, 3) that removes VOCs from groundwater without bringing the water to the surface for treatment. The well itself is screened at depth below the water table and allows groundwater contaminated with VOCs to enter the well. An upper screened interval is located above the water table and allows water depleted in VOCs to be returned to the aquifer. An eductor pipe is installed inside the well casing, creating a "well-within-a-well." Inside the inner well, an air line is introduced into which air is injected. The air is released beneath the water table, creating bubbles that rise. A simple separator plate (or well packer) is located within the inner well at an elevation above the water table. As the water/bubble mixture hits this separator plate, the water is forced laterally into the outer well and exits into the vadose zone through the upper screened interval. From there, the water freely infiltrates back to the water table using a series of infiltration galleries/trenches.

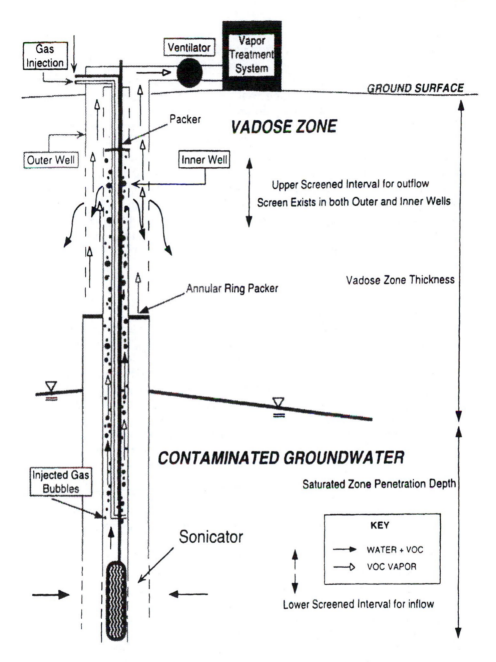

Figure 1. Schematic diagram of a single treatment well involving in-well sonication, vapor stripping, and in situ biodegradation.

The air bubbles, enriched in VOCs, are released into the outer well and are extracted to the ground surface under a vacuum.

In-well vapor stripping operates on two basic principles: groundwater recirculation and volatilization. Groundwater recirculation is accomplished using a dual-screened recirculation well. Air-lift pumping occurs when air is injected into the well. The density difference between the water outside the well and the water/bubble mixture within the inner well causes a lift to be created (*4*). Water and air rise within the inner well, forcing additional water to flow from the aquifer into the well through the lower screened interval. The water and bubble mixture flows upward in the annular space around the air line. Because the water enters the well at the lower screened interval and returned to the water table, a groundwater circulation cell is developed in the vicinity of the well. The second operating principle is that of volatilization (*5, 6*). When contaminated water enters the well at the lower screened interval, it encounters the injected air which has formed bubbles. The VOCs will volatilize and mass is transferred from the water to the gas phase. Given approximately 20 feet of contact distance between the contaminated water and the air bubbles, equilibrium partitioning occurs. The air within the well strips out the VOCs. During a demonstration conducted at Edwards Air Force Base demonstration, approximately 90% of the TCE was stripped from the water with each pass through the well. This air is separated from water using the separator plate located above the upper screened interval and the VOC-enriched vapor is vacuumed off and treated by sorption onto granular activated carbon. The water exiting the well at the upper screened interval has then been depleted of VOCs and is returned to the aquifer where microbial degradation can occur, depending upon the type of compound, and during sequential passes through the treatment well, additional VOCs are stripped and removed. In-well vapor stripping has efficacy at any level of contaminant concentration and can be particularly effective when VOC concentrations are high.

The combined remedial system will take contaminated water and remove a significant portion of the SVOCs and VOCs in the well (the treatment well). The system operates by employing groundwater recirculation as described previously. Contaminated water enters the well, and with each pass through the well contaminants are removed by both sonication and volatilization. A radial clean-up zone is created around the well. Previous work on recirculation wells (*5, 7–9*) has shown that a single well can achieve a zone of cleanup that extends radially 2 to 3 times the aquifer thickness. Groundwater entering this zone under regional flow conditions will be recycled and cleaned. It is expected that with each pass through the well, in-well vapor stripping will remove approximately 90 to 99% of the VOCs, and sonication will remove 90+%. Any residual VOCs may be biodegraded as the water enters the unsaturated zone. In addition, the VOC vapors that are produced by the in-well vapor stripping system may be treated at the ground surface or may be degraded in situ in the unsaturated zone through microbial activity.

The integrated technology involving in-well sonication, in-well vapor stripping, and in situ biodegradation has been described in an article published in Argonne National Laboratory's *Tech Transfer Highlights* (*10*). In that article, Argonne sought industrial collaborators with technical strengths in megasonics to help develop a prototype down-well reactor. More recently, the technology has been described in a publication of *HazTECH News* (*11*). Results are summarized for batch and continuous flow in-well sonication/in-well vapor stripping. Additionally, the results from this research were shared (on an annual basis) with other sonication researchers (funded through the U.S. Department of Energy [DOE] Environmental Management Science Program) in the annual sonication workshop that was held at Argonne each year of the project. The first annual workshop was held on Oct 25, 1996. These workshops brought together researchers to discuss results from their research projects, from the following organizations: University of Illinois, Purdue University, University of Washington, Syracuse University, and the University of California at Davis, in addition to Argonne.

Background

Boyd and Geiser (*12*) have pointed out the need to develop integrated system technologies; cleanup under the DOE Environmental Management Cleanup Program ties multiple technologies in an integrated solution to solve specific site/facility problems. Due to the complexity of environmental problems at these facilities, there are no "silver bullets" to solve these problems; it is essential that several technologies be combined together in a synergistic fashion. By coupling technologies, it is possible to collapse cleanup schedules drastically, enabling return on investments exceeding 10:1, and allow cleanup and closure of sites much more rapidly (*12*). The innovative technology in this research project couples in-well sonication, in-well vapor stripping, and biodegradation into an integrated process (see Figure 1). By partially destroying the SVOCs (e.g., opening up the benzene-ring structures), the ability to remove the resultant VOCs and biotreatment of the resultant organics is enhanced (over the case of biotreatment alone).

Background—Ultrasound

The introduction of high power ultrasound (i.e., sound energy with frequencies in the range of 15 kHz to ~1MHz) into liquid reaction mixtures is known to cause a variety of chemical transformations. In recent years, due to the growing need to eliminate undesirable chemical compounds, the utilization of high energy ultrasound for hazardous waste treatment has been explored with great interest (13–21).

Sonication of liquid media results in the formation of microbubbles that grow to a critical size and then implode. Temperatures at the collapsing bubble interface are about 5,000 K and pressures are near 500 atmospheres, but the bulk solution remains near ambient conditions. In these highly reactive conditions, hydroxyl radicals (•OH) and hydrogen ions (H•) are generated; they are very effective at degrading organic compounds. If organic compounds are present in the water, they are rapidly destroyed, either directly or by reacting with the free radicals. The intensity of cavity implosion, and hence the nature of the reactions involved, can be controlled by process parameters such as the ultrasonic frequency, ultrasonic intensity per unit volume of liquid medium, static pressure, choice of ambient gas, and addition of oxidants (e.g., H_2O_2, O_3).

The application of ultrasonic irradiation for treatment of hazardous chlorinated organic wastes started in the early 1980s. Argonne has been at the forefront of this research. To demonstrate the feasibility of the ultrasonic detoxification process and to obtain kinetic information about the process, two bench-scale batch-processing units, one pilot-scale batch/continuous-processing unit, and one continuous-flow unit were set up at Argonne. Research activities at Argonne have concentrated on the ultrasonic decomposition of hazardous organic compounds present in aqueous samples (including laboratory-prepared, laboratory-simulated, and real samples received from the field) from industrial waste streams and groundwaters. Results from these experiments indicated that ultrasonics or ultrasonics-enhanced advanced oxidation processes can convert the hazardous organic contaminants to nontoxic or less toxic, simpler organic compounds. For some simpler but toxic organic compounds (such as CCl_4, TCE, etc.), the introduced ultrasonic energy completely converts the compounds into nonhazardous compounds such as water, carbon dioxide, and hydrochloric acid. Of the process parameters investigated to date, the intensity of the ultrasonic-wave energy was found to have the largest effect on the destruction rate of the contaminant (which increases with the intensity). The results verified that the irradiation time required for a given degree of destruction can be further reduced by the addition of a small amount of chemical oxidant such as hydrogen peroxide. Some of these results have been published in the technical journals and conference proceedings.

Ultrasonic irradiation can easily be integrated with existing, conventional treatment systems, making it possible to simultaneously treat hazardous and nonhazardous waste streams. Successful development and deployment of this technology could completely change the treatment of wastes generated in the DOE complex.

Physical and Chemical Principles—Ultrasound

Ultrasonic irradiation of liquid reaction mixtures induces electrohydraulic cavitation, which is a process during which the radii of preexisting gas cavities in the liquid oscillate in a periodically changing pressure field created by the ultrasonic waves. These oscillations eventually become unstable, forcing the violent implosion of the gas bubbles. The rapid implosion of a gaseous cavity is accompanied by adiabatic heating of the vapor phase of the bubble, yielding localized and transient high temperatures and pressures (while the bulk solution remains near ambient conditions). Temperatures on the order of 4,200 K and pressures of 975 bar have been estimated (22). Experimental values of P = 313 atm and T = 3,360 K have been reported (23) for aqueous systems, while temperatures in excess of 5,000 K have been reported (24–26) for cavitation of organic and polymeric liquids. Recent experimental results on the phenomenon of sonoluminescence (27, 28) suggest that even more extreme temperatures and pressures are obtained during cavitational bubble collapse (29). Thus, the apparent chemical effects in liquid reaction media are either direct or indirect consequences of these extreme conditions.

Even though the basic physical and chemical consequences of cavitation are fairly-well understood, many fundamental questions about the cavitation site in aqueous solution remain unanswered. In particular, the dynamic temperature and pressure changes at the bubble interface and their effects on chemical reactions need further exploration. Since this region is likely to have transient temperatures and pressures in excess of 647 K and 221 bar for periods of microseconds to milliseconds, supercritical water (SCW) has been proposed that provides an additional phase for chemical reactions during ultrasonic irradiation in water (16). Supercritical water exists above the critical temperature, T_c, of 647 K and the critical pressure, P_c, of 221 bar and has physical characteristics intermediate between those of a gas and a liquid. The physicochemical properties of water such as viscosity, ion-activity product, density, and heat capacity change dramatically in the supercritical region. These changes favor substantial increases for rates of most chemical reactions.

Two distinct sites for chemical reactions exist during a single cavitation event (16, 18). They are the gas-phase in the center of a collapsing cavitation bubble and a thin shell of superheated liquid surrounding the vapor phase. The volume of the gaseous region is estimated to be larger than that of the thin liquid shell by a factor of $\sim 2 \times 10^4$.

During cavitation/bubble collapse, which occurs within 100 ns, water undergoes thermal dissociation within the vapor phase to give hydroxyl radical and hydrogen atoms as follows:

$$H_2O \xrightarrow{\Delta} H\bullet + \bullet OH$$

The concentration of •OH at a bubble interface in water has been estimated to be 4×10^{-3} M (30). Many of the chemical effects of ultrasonically induced cavitation have been attributed to the secondary effects of •OH and H• production.

For treatment of CCl_4-contaminated waters, Wu et al. (31, 32) speculated that the major reactions involved were bond-cleavage of water and of CCl_4 in the cavitation hole. Destructions greater than 99% were achieved. First-order kinetics were observed within the experimental concentration ranges studied (up to 130 mg/L of CCl_4). The residual CCl_4 concentration decreased for increasing pH in the range of 3 to 9 (for the same applied sonication period). Better sonication efficiency was observed at higher initial pH values; however, this improvement diminished for pH > 6. In the experimental conditions employed in their study, the researchers concluded that temperature and pH had little effect on the CCl_4 destruction rate; however, the destruction rate was significantly affected by the intensity of the ultrasonic energy. The destruction rate exhibited a linear relationship with the applied power intensity. Adding hydrogen peroxide as an oxidant had a negligible effect on the destruction rate (with or without ultrasonic irradiation), suggesting that the destruction of CCl_4 in water under ultrasonic irradiation was dominated by the high temperature dissociation reactions within the collapsing cavities (32).

Destruction of pesticides such as atrazine was shown to be effective using a Fenton's oxidation system, and was enhanced by the application of sonication (33). Removal efficiencies of atrazine exceeded 90%, even for treatment times as low as five minutes. Sonication improved the removal of atrazine by at least 10% for treatment times of 10 minutes or less. Higher dosages of Fe^{+2} and H_2O_2 resulted in higher removals of atrazine from solution. They investigated four separate systems for their ability to remove atrazine from solution: (1) Fenton's oxidation conducted without application of UV light; (2) Fenton's oxidation conducted in the presence of UV light; (3) Fenton's oxidation conducted using sonication in the absence of UV light; and (4) Fenton's oxidation conducted using sonication in the presence of UV light. Of these four systems, the sonication-enhanced Fenton's oxidation resulted in the best treatment of the Fenton's oxidation systems studied. That system promoted the formation of hydroxyl radicals (•OH) to more effectively destroy the organic contaminants (e.g., atrazine). Their work showed that addition of oxidants such as H_2O_2 or Fenton's Reagent could enhance organic destruction using sonication.

Sonochemical reactions in water are characterized by the simultaneous occurrence of supercritical water reactions, direct pyrolyses, and radical reactions especially at high solute concentrations. Volatile solutes such as CCl_4 (13) and hydrogen sulfide (20) undergo direct pyrolysis reactions within the gas phase on the collapsing bubbles or within the hot interfacial region as shown below:

$$CCl_4 \xrightarrow{\Delta} CCl_3\bullet + Cl\bullet$$

$$CCl_3\bullet \xrightarrow{\Delta} :CCl_2 + Cl\bullet$$

$$H_2S \xrightarrow{\Delta} HS\bullet + H\bullet$$

while low-volatility solutes such as thiophosphoric acid esters (*19*) and phenylate esters (*16, 19*) can react in transient supercritical phases generated within a collapsing bubble. In the case of ester hydrolysis, reaction rates are accelerated 102 to 104 times the corresponding rates under controlled kinetic conditions (i.e., same pH, ionic strength, and controlled overall temperature). This effect can best be illustrated by the catalytic effect of ultrasonic irradiation on the rate of hydrolysis of parathion in water at pH 7. The half-life for parathion hydrolysis at pH 7.4, in the absence of ultrasound at 25 °C is 108 days. However, in the presence of ultrasound, the half-life is reduced to 20 minutes (*19*).

Pyrolysis (i.e., combustion) and supercritical water reactions in the interfacial region are predominant at high solute concentrations, while at low solute concentrations, free radicals are likely to predominate. Depending on its physical properties, a molecule can simultaneously or sequentially react in both the gas and interfacial liquid regions.

In the specific case of hydrogen sulfide gas dissolved in water, both pyrolysis in the vapor phase of the collapsing bubbles and hydroxyl radical attack in the quasi liquid interfacial region occur simultaneously as follows:

$$H_2S \xrightarrow{\Delta} HS\bullet + H\bullet$$
$$H_2S + \bullet OH \rightarrow HS\bullet + H_2O$$

Hua and Hoffmann (*13*) investigated the rapid sonolytic degradation of aqueous CCl_4 at an ultrasonic frequency of 20 kHz and at an applied power of 130 W (108 W/cm^2). The rate of disappearance of CCl_4 was found to be first-order over a broad range of conditions, consistent with the results of Wu et al. (*31, 32*). The observed first-order degradation rate constant was 3.3×10^{-3} s^{-1} when $[CCl_4]_i = 195$ mM; k_{obs} was observed to increase slightly to 3.9×10^{-3} s^{-1} when $[CCl_4]_i$ was decreased by a factor of 10 (i.e., $[CCl_4]_i = 19.5$ mM) (*13*). Low concentrations of hexachloroethane, tetrachloroethylene, and hypochlorous acid (HOCl) in the range of 0.01 to 0.1 mM, were detected as transient intermediates, while chloride ion and CO_2 were found to be stable products.

The highly reactive intermediate, dichlorocarbene, was identified and quantified by means of trapping with 2,3-dimethylbutene. Evidence for involvement of the trichloromethyl radical was also obtained and was indirectly implied by the formation of hexachloroethane. The presence of ozone during sonolysis of CCl_4 did not affect the degradation of CCl_4 but was shown to inhibit the accumulation of hexachloroethane and tetrachloroethylene.

The following mechanism was proposed to account for the observed kinetics, reaction intermediates, and final products:

$$CCl_4 \xrightarrow{\Delta} \bullet CCl_3 + Cl\bullet$$

$$CCl_3 \xrightarrow{\Delta} :CCl_2 + Cl\bullet$$

Formation of dichlorocarbene, $:CCl_2$, is also thought to occur by the simultaneous elimination of two chlorine atoms:

$$CCl_4 \rightarrow :CCl_2 + Cl_2\bullet$$

A third mechanism for dichlorocarbene formation is disproportionation of the trichloromethyl radical that can be inferred from an analogous reaction between the trifluoromethyl radical and the hydrodifluoromethyl radical:

$$\bullet CCl_3 + \bullet CCl_3 \rightarrow CCl_4 + :CCl_2$$

All three pathways are possible at the hot center of the imploding microbubble. The trichloromethyl radical can also couple to form hexachloroethane:

$$\bullet CCl_3 + \bullet OH \rightarrow HOCCl_3$$

In the presence of oxidizing species, the trichloromethyl radical can act as a scavenger of hydroxyl radicals:

$$\bullet CCl_3 + \bullet OH \rightarrow HOCCl_3$$

or molecular oxygen:

$$\bullet CCl_3 + \bullet O\text{-}O\bullet \rightarrow O\text{-}OCCl_3$$

Based on analogous gas-phase mechanisms, the reactive intermediate $HOCCl_3$ appears to rapidly react to yield phosgene and other products as follows:

$$HOCCl_3 \rightarrow HCl + COCl_2$$
$$COCl_2 + H_2O \rightarrow CO_2 + 2\ HCl$$
$$\bullet O\text{-}OCCl_3 + H_2O \rightarrow HOCCl_3 + HO_2$$

Phosgene hydrolysis in water is rapid (*34*) under ambient conditions; and the rate constant is positively correlated with increasing temperature. Thus, the hydrolysis of this intermediate can be enhanced by the occurrence of supercritical water during cavitational microbubble collapse:

$$COCl_2 + H_2O \rightarrow 2\ HCl + CO_2$$

The dichlorocarbene can be coupled to form tetrachloroethylene:

$$2\ :CCl_2 \rightarrow C_2Cl_4$$

or hydrolyze to carbon monoxide and hydrochloric acid:

$$:CCl_2 + H_2O \rightarrow CO + 2\ HCl$$

Chlorine atoms can combine to form molecular chlorine, which hydrolyses to hypochlorous acid and chloride ion:

$$\overset{H_2O}{2\ \bullet Cl\ \rightarrow\ Cl_2\ \rightarrow\ HOCl + HCl}$$

The formation of the reactive intermediate, dichlorocarbene, is confirmed by the selective trapping of the carbene with 2,3-dimethyl-2-butene to form 1,1-dichloro-2,2,3,3-tetramethylcyclopropane. In a similar fashion, the trichloromethyl radical is trapped by 2,3-dimethyl-2-butene to yield 2-methyl-2-trichloromethyl-1-butene. These trapped intermediates can be identified and quantified by gas chromatography/mass spectroscopy (GC/MS) techniques.

In-Well Vapor Stripping

In-well vapor stripping has been demonstrated at Edwards Air Force Base (AFB) for the removal of dissolved TCE from groundwater. The well itself is screened at two horizontal intervals (*2, 3, 5, 35*). The air bubbles, enriched in VOCs, are released into the outer well and are extracted to the ground surface under a vacuum. As discussed previously, in-well vapor stripping operates on two basic principles. The first is that of groundwater recirculation and is accomplished using a dual-screened recirculation well. The second operating principle is that of volatilization (*5, 6*). During the Edwards AFB demonstration, approximately 90% of the TCE was stripped from the water with each pass through the well. It is expected that with each pass through the well, in-well vapor stripping will remove approximately 90 to 99% of the VOCs, and sonication will remove 90+%. Any residual VOCs may be biodegraded as the water enters the unsaturated zone. In addition, the VOC vapors that are produced by the in-well vapor stripping system may be treated at the ground surface or may be degraded in situ in the unsaturated zone through microbial activity.

In Situ Biotreatment

Chlorinated solvents have been biodegraded in situ by a variety of investigators (36–39). Unfortunately, the rate of degradation is very slow. Weathers and Parkin (40) have demonstrated that chlorinated solvent transformation was enhanced using methanogenic cell suspensions and iron filings (zero-valent metals). Although the exact mechanism of this activity is unknown, O'Hannesin (41) demonstrated the loss of tetrachloroethylene (PCE) and TCE (86% and 90%, respectively) and the near stoichiometric increase in dissolved chloride and detection of chlorination products. Matheson and Tratnyek (42) reported non-stoichiometric sequential reduction of CCl_4 to chloroform and dichloromethane by iron in laboratory settings. Under anaerobic conditions, however, Helland et al. (43) observed reduced CCl_4 transformation rates by iron metal.

Kriegman-King and Reinhard (44, 45) have reported on the transformation of CCl_4 in the presence of sulfide, pyrite, and certain clays. In some of their work, they demonstrated CCl_4 dechlorination to chloroform ($CHCl_3$) by pyrite. This transformation was blocked by sulfide ion.

The result of all this work is to encourage a physical/chemical method of removing the halogen groups, and the resultant "softened" organics can then be biodegraded at relatively enhanced rates. This approach minimizes the rate limiting step of dechlorination by microbial mechanisms and enhances the relatively easy nonhalogenated degradation of the remaining organic compounds.

The research work is consistent with the mission of providing the scientific understanding needed to develop methods for accelerating biodegradation processes for remediation of contaminated soils, sediments, and groundwater at DOE facilities. This project relates to in situ chemical transformation (e.g., softening of SVOCs), in-well vapor stripping, and biodegradation. The innovative technology couples in-well sonication, in-well vapor stripping, and biodegradation into an integrated process. By partially destroying the SVOCs (e.g., opening up the benzene-ring structures), the ability to remove the resultant

VOCs and biotreatment of the resultant organics is enhanced (over the case of biotreatment alone).

Project Objectives

The specific objective of the research project was to investigate the in situ degradation of SVOCs and to:

- Determine the system performance of the combined in-well sonication, vapor stripping, and biodegradation to destroy VOCs and change SVOCs into VOCs;
- Determine how the combined in-well vapor stripping, sonication, and in situ biodegradation remedial system functions together at the laboratory-scale to remove SVOCs and VOCs;
- Determine the chemical reaction mechanisms for destroying VOCs and changing the SVOCs into VOCs, and improve the overall system performance;
- Quantify the roles of the individual treatment components (sonication, in-well vapor stripping, and biodegradation) on the overall effectiveness of the remediation, and acceleration. It deals with novel bioremediation reactions, alternative electron acceptor conditions, and modeling the overall system;
- After water is treated in the well with sonication and VOCs are partially removed through in-well vapor stripping, determine the role of volatilization and microbial activity on water containing VOCs that is forced to infiltrate through the unsaturated zone;
- Determine the effect of sonication/megasonics on well corrosion, and
- Identify the appropriate system design for scale-up of the remedial system for demonstration in the field and deployment.

This project had as its goal the softening of the more recalcitrant organic compounds (e.g., SVOCs and nonvolatile organic compounds) in order to convert them into compounds that are more amenable to both vapor stripping and biological treatment. The SVOCs are not effectively removed from solution using air sparging techniques. Conversion of SVOCs to VOCs could allow effective removal of the organics with vacuum extraction techniques or in situ biotreatment. This project investigated the combined treatment using in-well sonication, in-well vapor stripping, and biodegradation. The research examined the use of sound-wave energies (e.g., ultrasonics) to transform the SVOCs to VOCs. Performing the softening in-well would permit the treated organics to be reinjected and percolated through the subsurface, thereby enhancing biodegradation rates by generating organics that are more easily biodegraded.

Successful implementation of such an approach would considerably reduce both the time and cost of in situ biotreatment. Pretreating groundwaters with sonication techniques would form VOCs that could be removed effectively by either bioremediation technologies or a dual vapor extraction technique (developed by Stanford University under the VOC-Arid Program, now part of the Plumes Focus Area). Sonication could also be coupled with technologies aimed at mobilizing dense nonaqueous phase liquids (DNAPLs) in the subsurface, such as surfactant flooding.

Research Tasks

Task 1. Quantify Representative VOC and SVOC Contaminants at DOE Facilities

In consultation with DOE and the Plumes Focus Area, representative target VOCs and SVOCs were determined. Potential VOCs and SVOCs present at various DOE facilities include TCE, trichloroethane (TCA), 1,2-ethylene dibromide (EDB), CCl_4, $CHCl_3$, vinyl chloride (VC), 1,1-dichloroethylene (DCE), 1,1,2,2-tetrachloroethylene (PCE), and polychlorinated biphenyls (PCBs). After determination of the target compounds, the subsequent tasks focused on the integrated in-well sonication/in-well vapor stripping, and biodegradation system.

Task 2. Conduct Batch Sonication Experiments to Determine Preliminary Optimal Conditions for Subsequent Continuous-Flow Experiments

Batch laboratory experiments were conducted to identify key process parameters, such as solution pH values, steady-state operating temperature, ultrasonic power intensities, and oxidant concentrations.

Task 3. Perform Batch Experiments for Measuring Hydroxyl/Hydrogen Radicals in Solution

Hydroxyl radicals can be trapped by direct reaction with terephthalic acid and were measured using a fluorimeter (the emission is measured at 425 nm). Measurement of the radical intermediates were conducted at California Institute of Technology (Caltech). The hydroxyl and hydrogen radicals concentrations were quantified for various operating conditions (e.g., solution pH values, steady-state operating temperature, ultrasonic power intensities, and oxidant concentrations).

Task 4. Conduct Continuous-Flow Experiments for Degradation of SVOCs and Identification or Quantification of By-Products

The degradation products were determined by GC (or GC/MS) analysis for various sonication operating conditions. This task was performed at Argonne and Caltech. The resulting degradation products were identified and characterized in terms of their volatility and biodegradability.

Task 5. Investigate the Chemical Reaction Mechanism to Improve System Performance

Model simulations of the chemical reactions were performed using the data obtained in Tasks 2 and 3. This model described in detail the chemical mechanisms being followed in the overall reaction. The model will include all of the identified organic by-products (from Argonne and Caltech) and radical intermediates (from Caltech) generated during the sonication process. Since the purpose of conducting model simulation is further understanding of the chemical mechanisms behind the reactions and to identify the optimal operating conditions resulting in an improvement of the decomposition efficiency of target contaminants, the model simulation was closely linked to the laboratory research.

Task 6. Perform Batch or Continuous-Flow Experiments to Determine the Effects of Oxidants (H_2O_2, Fenton's Reagent, etc.) on SVOC Degradation and the Biodegradability of the Resultant Product

Potential enhancements of the destruction of the parent SVOCs and VOCs by addition of oxidants (such as H_2O_2, Fenton's Reagent, etc.) were investigated in batch and/or continuous-flow sonication experiments. The influence of these oxidants was addressed in terms of the radical intermediates and organic degradation products. This information was used in conjunction with the reaction mechanism model developed.

Task 7. Identify and Quantify Corrosion Potential and Salt Formation

Metal coupons placed in the sonication reactors were analyzed for potential corrosion and scale formation (e.g., calcium carbonate, magnesium hydroxide, and metal salts). This information will be used in conjunction with the reaction mechanism model.

Task 8. Determine the Volatility and Biodegradability of the Treated Waters (and Compare to Untreated Waters)

The untreated water and the sonicated-treated water (containing degradation products, were compared in terms of the degradation products. The volatility and biodegradability of the resultant organic compounds were determined. The volatility of the target organic contaminants was determined by measuring the residual contaminant concentrations, using a laboratory-scale air stripping unit.

For the biodegradability of the resultant organic compounds, triplicate samples of untreated water and sonicated treated material were subjected to shake flask biodegradation studies. Initially, soil inoculum from sites contaminated with the parent compound(s) were needed to determine the biodegradability of the untreated and treated material. Comparisons were made by using GC/MS and HPLC (high-performance liquid chromatography) of material before and after biotreatment. These studies were conducted aerobically and anaerobically.

Task 9. Develop Computer Simulation Model to Describe Combined In-Well Sonication, In-Well Vapor Stripping, and Biodegradation

Technical Objectives

The technical objectives of the laboratory and modeling components of this project were to develop:

1. An understanding of the behavior of the combined sonication and in-well vapor stripping process, and
2. A predictive simulation model of the system that is tested ("validated") on the lab-scale system, used to design the field pilot system, and later used to interpret the operation of the field pilot system.

The physical model consisted of an instrumented sandbox approximately 3-m-long by 1-m-wide by 2-m-high. The physical model included a central recirculation well and would be instrumented with multi-level piezometers and sampling ports. The system can be run as a closed-flow system, which simulates the case of no (or low) regional groundwater flow, or as an open system, which can be used to simulate regional flow on the effects of nearby pumping. A central recirculation well can be run under conditions of in-well vapor stripping or conditions of combined in-well sonication and vapor stripping; in-well vapor stripping would employ a small blower. Air (or gas) injection rates can be monitored, as will temperature and pressure. VOC-enriched vapor can be vacuum extracted from the recirculation well and the vapor phase concentrations can be measured. The recirculation well water can be sampled for VOCs and other chemicals of interest at the lower inflow elevation and the upper outflow elevation. The tank was modeled to be filled with natural porous media to an elevation of 70%, leaving the upper portion (30%) of the water unsaturated. At the beginning of an experiment, a VOC or SVOC was simulated to be fed into the system by replacing the pore waters with the contaminant solution. Initial conditions would be carefully measured via the sampling ports. Additional sampling ports would be placed in the unsaturated zone to measure VOC vapor. Experiments would be run for approximately one to two weeks each, depending upon the removal rate of the contaminants. Once the system has been run under conditions representing sonication and in-well vapor stripping, additional experimental simulations could include biodegradation, which may involve the addition of nutrients and a carbon source to the lab system, all of which would need to be monitored.

The simulation model will be developed initially to represent the behavior of the physical model. The simulator will consist of a 3-D finite-difference code to simulate variably saturated flow and transport. The underlying model accounted for advection and dispersion, as well as equilibrium and kinetically controlled sorption. The model was developed as a series of modules with capabilities to represent the in-well SVOC and VOC removal processes, volatilization and diffusion of VOCs in the unsaturated zone, and biodegradation. The model built upon the simulator developed by Professor Gorelick at Stanford University and used to simulate laboratory and field in-well vapor stripping systems. Initial model development was aimed at design of the laboratory-scale system to help identify the best locations for sampling ports and to target optimal sampling intervals at each multilevel sampler. Finally, the simulator was used to scale up to a pilot-scale field system.

The laboratory integrated prototype system and modeling provided information that can be used directly in the simulation model. Given known boundary conditions, initial conditions, physical parameters, media properties, and system stresses (flow rates and VOC removal rates), the simulation model can be used to predict the behavior of the laboratory system. Measured parameters needed for incorporation into the simulation model are listed in Table I below. Concentrations and hydraulic heads are the dependent variables in the simulation model and are the variables that can be measured during each experiment. The physical and chemical processes in the model can be constructed so as to best represent the key observed changes observed in heads (flow) and concentration (transport). The data measured in the laboratory system can be compared with the simulation results. During different experiments, various physical processes can be "turned on" or "turned off" and the model results can be compared to measured values. For example, experiments with and without air injection, with and without in-well sonication, and with and without biodegradation (sterile or inoculated conditions) can be performed.

Table I. Measured Parameters Needed for Incorporation into Simulation Model

Porous-Media Properties	_Water Properties_	_Vapor Properties_
• Hydraulic conductivity • Porosity • Bulk density • Grain size distribution • Mineralogy (including organic carbon content) • Distribution coefficient • Unsaturated-media properties (conductivity vs saturation)	• Temperature • Chemical composition, including dissolved oxygen and pH	• Temperature • Chemical composition

Experimental Procedure

The compounds initially targeted for this study are CCl_4, TCE, PCE, 1,1,1-trichloroethane (TCA), and EDB. The effects of varying sonication time, ultrasonic power intensities, initial concentrations of target compounds, pH, and hydrogen peroxide were examined. The sonication experiments were performed using a Sonics and Materials Vibracell, operating at 20 kHz. The applied power could be varied to achieve power intensities ranging from ~12 to ~40 W/cm^2. The reactor was a glass vessel with an ultimate capacity of 1 L; normally the reactor is operated using a sample volume of ~500 mL. Batch experiments were performed separately on each of the chlorinated organic contaminants (CCl_4, TCE, TCA, and PCE). Initial contaminant concentrations ranged anywhere from ~1 to ~100 mg/L. Sonication/vapor stripping experiments were performed in a reactor used to treat the chlorinated organic contaminants in groundwater employing sonication alone, vapor stripping alone, and combined sonication/vapor stripping. The sonicator had an ultrasonic frequency of 20 kHz; the applied power intensity used in this study was 12.3, 25.3, and 35.8 W/cm^2. The batch reactions were operated normally for up to 10 minutes of treatment time, with samples drawn for GC analysis every 2 minutes. Air injection rates (for the vapor stripping) were nominally 0 (sonication alone), 500, 1000, and 1500 mL/min. The results of these tests aided in the determination of the operation conditions for a continuous-flowing ultrasonic irradiation system. In the continuous flow studies, the residence time in the reactor was set at 5, 8, and 10 minutes.

Procedural effects were also examined to better understand potential interferences. These included temperature increase of the sample due to sonication, and loss of the target compound due to volatilization. Considerable effort went into development of appropriate analytical procedures and methods associated with using volatile compounds. Analytical method development included selection and procurement of a Hewlett Packard 7694 Automatic Headspace Sampler for use with the Hewlett Packard 5890 Gas Chromatograph. Headspace analysis is well-suited for analysis of VOCs and SVOCs in water, and avoids the column degradation caused by liquid injection of water. The headspace sampling method includes 10 minutes equilibration of samples at 70 °C with shaking, programmed vial pressurization, venting, sample loop fill (1-mL loop volume), and 0.30-min injection time (*46, 47*). Equilibration time was selected by measuring area response for times ranging from 1 to 60 minutes. A 30-m megabore fused capillary DB-624 column was selected based on its sensitivity and selectivity in analysis of chlorinated organic compounds. The HP 5890 chromatograph is equipped with both flame ionization and electron capture detectors. In the analytical range for this project (0.1 to 100 ppm), flame ionization provided adequate resolution and reproducible detection. The electron capture detector was found to be too sensitive for detection of major components, but is useful in examining sonicated samples for minor breakdown products. The GC temperature program was set for 1 min at 90 °C, ramping 10 °/min to 140 °C, then 25 °/min to 200 °C, and held for 2 min at 200 °C. Standards were initially prepared in volumetric flasks. Due to analyte volatility, this was changed to injection by syringe through the vial septum of the chlorinated compounds into measured water mass. Planned sample size was 5 mL in 10-mL vials; however, this was reduced to 1 mL in 10-mL vials to avoid overloading the column with analyte. Response of CCl_4 standards held in sealed vials over a 0.25- to 48-hour time range was evaluated to determine how long samples could be stored prior to GC analysis. It was found that vials analyzed within 3 hours after sampling gave the most reproducible response. Vials showed a drop in response to approximately 80% at 15 h, and to 20-40% at 25 h (*46, 47*). Since the hydrolysis rate for CCl_4 in water is very low, there appears to be loss through the vial seal or septum. Therefore all subsequent samples and standards were analyzed as closely as possible to the actual sampling time, and all within three hours. Five analyte standards were prepared each day for the initial calibration curve, and check standards were performed late in the day, or when any questionable sample result was obtained.

Results and Discussion

Batch experiments were performed to help identify preliminary optimal conditions for subsequent continuous-flow experiments and to investigate the degradation of halogenated organic compounds by ultrasonic irradiation of a solution. The compounds initially targeted for this study were CCl_4, TCE, PCE, TCA, and EDB. For the sake of brevity of this paper, selected results from the batch sonication system testing only CCl_4 and TCA, are included below. Results of these two compounds are representative of the general trends observed. The effects of varying sonication time, ultrasonic power intensities, initial concentrations of target compounds, pH, and hydrogen peroxide were examined. Results from these studies have been presented at various conferences and are described in the technical literature (46–51). The results of these tests aided in the determination of the operation conditions for a continuous-flowing ultrasonic irradiation. Procedural effects were also examined to better understand potential interferences. These included temperature increase of the sample due to sonication, and loss of the target compound due to volatilization.

Temperature Increase Due to Sonication

Experiments performed to date were all started with an initial sample temperature of 20 °C. The sample was placed in a circulating constant temperature water bath to minimize temperature fluctuation during sonication. During the batch sonication experiments, sample temperatures increased from 6 to 8 °C (from ~20 °C to ~26–28 °C) over the 10-min period of sonication. No pattern was observed due to the presence of or concentration of the CCl_4. A higher power intensity used increased the sample temperature more rapidly than a lower power intensity. Slight variations in these results were likely due to the slight variation in the power output of the sonicator from run to run.

How this may effect the performance of sonication to break down the target compounds is questionable and probably insignificant. Higher temperatures could slightly decrease the effect of sonication on the compounds in the range of 20 to 30 °C. Also, if the temperature was near the boiling point of solvent (water in this case) the extreme condition of cavitation would disappear and the target compounds would not be dissociated because the water vapor would fill the cavitation bubbles and cushion the implosion phenomena.

Sonication of Target Compounds at Different Initial Concentrations

Once potential procedural interferences were established to be minimal (described above), samples spiked with the target compounds were sonicated to determine the effectiveness of ultrasonics on decreasing the concentration of the compound in the sample solution. Samples of 500-mL volume were sonicated for 10 minutes and sampled at certain intervals during sonication. Results indicated that the rate of reduction was greater with a lower initial concentration.

First-order plots of the natural logarithm of CCl_4 concentrations versus sonication time were generated for each experimental run. The experimental data fit the first-order reaction kinetic model. Degradation rate constants of CCl_4 for the first-order model were obtained from the slopes of the linear regression of the plots according to the equation:

$$- \frac{d[CCl_4]}{dt} = k \, [CCl_4]$$

Varying Power Intensities of the Ultrasonic Irradiation

The effect of varying the wattage introduced per probe tip surface area to the sample was also investigated. Three power intensities were investigated: 12.6, 25.3, and 35.8 W/cm^2. The samples were sonicated for 10 minutes. The results showed a clear effect of the power intensity on the rate of reduction. With an increase in the wattage introduced to the sample, there was an increase in the rate in which the CCl_4 concentration was decreased. The same effect was exhibited for initial CCl_4 concentrations of 8 and 1.6 ppm, respectively.

The Effect of Initial pH Values

The effect of the initial pH of the sample solution on the rate of contaminant reduction was investigated. Initial pH test values were between 3 and 9. Samples spiked at 8 ppm CCl_4 were sonicated for 10 minutes at 25.3 or 35.8 W/cm^2. Initial pH values between 5 and 9 did not appear to influence the rate of reduction. However, an initial pH value of 3 showed a significant decrease in the reduction rate. This phenomenon agrees with results of others' studies.

The Effect of Adding Hydrogen Peroxide

Using hydrogen peroxide (H_2O_2) [<0.1%] as an advanced oxidant to the sonication system resulted in negligible improvement over the ultrasonic system performance conducted in the absence of H_2O_2.

Duplication of Experimental Results

Most of the data generated in this project were based on one experiment per set of experimental conditions. However, a number of experiments were performed with three different sets of experimental conditions performed in replicate. The slopes representing the rate of CCl_4 decrease on the linear regression curves (of ln C/C_0 vs reaction time) for each of the experiments proved to be extremely reproducible. In each case, the slopes were within ~3% to 10% of the average slope. The data also indicate that with an increase in power intensity, the rate of decrease of the CCl_4 kinetics increases. In addition, with a lower initial CCl_4 concentration, the rate of decrease of CCl_4 was greater.

Degradation Products/Kinetic Modeling

Hua and Hoffmann (*13*) observed that the extent of CCl_4 degradation was greater than 99% after 90 minutes of sonication treatment. The initial concentration was held constant in their experiments at 0.20 ± 0.05 mM (~300 mg/L). Loss of CCl_4 due to volatilization was found to be less than 2% in separately run control experiments conducted in the absence of sonication (*52*). The pH after sonolytic degradation of CCl_4 was near 3.5; the principal products observed were OCl^-, Cl^-, C_2Cl_4, and C_2Cl_6 (*52*). The distribution of products and chemical intermediates after 90 minutes of ultrasonic treatment is summarized in Table II below.

Table II. Final Distribution of Chlorine Atoms in a Sonicated Solution of CCl_4 after 90 Minutes of Sonolysis (*13*)

	Concentration (μM)	
Species	*Initial*	*Final*
CCl_4	400	4.0
C_2Cl_4	Not Detected	3.1×10^{-3}
C_2Cl_6	Not Detected	8.4×10^{-2}
Cl^-	Not Detected	1100
$HOCl$	Not Detected	130

Source: Reproduced from reference 13. Copyright 1996 American Chemical Society.

In more recent work, Hung and Hoffmann (*52*) monitored the degradation products of CCl_4 and $CHCl_3$ degradation under sonolysis. Using a 0.2-mM solution of CCl_4, the highest concentrations of C_2Cl_4 and C_2Cl_6 observed were 80 nM and 25 nM, respectively, accounting for ~0.04% of CCl_4 appearing as C_2Cl_4 and <0.0125% of CCl_4 appearing as C_2Cl_6, respectively. The extent of degradation of CCl_4 exceeded 99% after 90 minutes of sonolysis (*52*). Hung and Hoffmann (*52*) noted that C_2Cl_6 and C_2Cl_4, produced as intermediates during the sonolytic degradation of CCl_4, are also readily degraded during aqueous-phase sonication.

In other studies, sonolysis of 10 mg/L of TCE produced Cl^-, H_2, and CO as major products, and small amounts of CO_2, CH_4, C_2H_6, and dichloroethylene (detected by GC-MS) (*53*). Sonication of high concentrations of TCE resulted in a higher number of intermediate compounds (*53*). During sonolysis of high TCE aqueous concentrations (~440 mg/L), Drijvers et al. (*54*) reported formation of chlorinated products such as C_2HCl, C_2Cl_6, C_4Cl_2, C_2Cl_4, C_4Cl_4, C_4HCl_5, and C_4Cl_6 in trace amounts. Sonication of aqueous chloroform with phenol present produced chlorophenols, and with benzene present produced phenol, chlorobenzene, and chlorophenols (*55*). Bhatnagar and Cheung (*56*) reported no chlorinated organic by-products being observed from sonolysis of C1 and C2 chlorinated compounds.

The sonication of CCl_4 follows simple pseudo-first order reaction kinetics (*13*). Slopes of standard linear regressions of the observed ln ($[CCl_4]_t/[CCl_4]_0$) vs time data correspond to the observed first-order rate constants. Hung and Hoffmann (*52*) conducted sonication experiments for degradation of CCl_4 using various ultrasonic frequencies (see Tables III and IV below); results from their studies are summarized below.

Hua and Hoffmann (*13*) have proposed the following mechanism for the sonolytic degradation of CCl_4 in water (Note:))) refers to sonolysis):

$$CCl_4 \xrightarrow{)))} \bullet Cl + \bullet CCl_3$$
$$CCl_4 \xrightarrow{)))} Cl_2 + :CCl_2$$
$$\bullet CCl_3 \rightarrow \bullet Cl + :CCl_2$$
$$\bullet CCl_3 + \bullet CCl_3 \rightarrow CCl_4 + :CCl_2$$
$$\bullet CCl_3 + \bullet CCl_3 \rightarrow C_2Cl_6$$

Table III. Results of Calorimetry Measurements (*52*)

Frequency (kHz)	Power Output (W)	Power Density (W/cm³)
200	62	0.65
50	48	0.10
205, 358, 618, 1078	35	0.06

Source: Reproduced from reference 52. Copyright 1999 American Chemical Society.

Table IV. Normalized Rate Constants for the Sonolytic Degradation of CCl$_4$ in Water at pH$_o$ ~ 7, pH$_\infty$ ~3.5, and T ~286 K with [CCl$_4$]$_0$ ~ 0.2 mM

Frequency (kHz)	K $_{[CCl4]}$ (min^{-1})
205	0.044
358	0.049
618	0.055
1078	0.039
20	0.025
500	0.070

Source: Reproduced from reference 52. Copyright 1999 American Chemical Society.

Dichlorocarbene formed in third equation above self-reacts to form tetrachloroethylene

$$:CCl_2 + :CCl_2 \rightarrow C_2Cl_4$$

or it can react with water to form carbon monoxide and hydrochloric acid

$$:CCl_2 + H_2O \rightarrow 2\ HCl + CO$$

C$_2$Cl$_6$ and C$_2$Cl$_4$ are produced as intermediates during the sonolytic degradation of CCl$_4$; these intermediates are also readily degraded during aqueous-phase sonication (*52*). Additional information related to the chemical mechanisms during sonolytic treatment of chlorinated organics has been described by Hoffmann and his researchers (*13, 52*).

Results from Batch Sonication and Batch Sonication/Vapor Stripping Experiments

Sonication/vapor stripping experiments were performed in a reactor used to treat the chlorinated organic contaminants in groundwater employing sonication alone, vapor stripping alone, or combined sonication/vapor stripping. The reactor is a glass vessel with an ultimate capacity of 1 L; normally the reactor is operated using a sample volume of ~500 mL. Batch experiments were performed separately on each of the chlorinated organic contaminants (CCl$_4$, TCE, TCA, and PCE). Initial contaminant concentrations ranged anywhere from ~1 to ~100 mg/L. The sonicator had an ultrasonic frequency of 20 kHz; the applied power intensity was 12.3, 25.3, and 35.8 W/cm². The batch reactions were operated normally for up to 10 minutes of treatment time, with samples withdrawn for GC analysis every 2 minutes. The batch experiments were operated using either sonication alone, vapor stripping alone, or using the combined sonication/vapor stripping technique. Air injection rates used in these experiments were nominally 0 (sonication alone), 500, 1000, and 1500 mL/min. Several hundred batch experiments were performed during this study; selected representative results are presented for the sake of brevity.

Figure 2 shows the results of CCl₄ removal using sonication alone and sonication+vapor stripping. The initial concentration of CCl₄ was nominally 50 mg/L. The applied power intensity was 12.3, 25.3, and 35.8 W/cm². The air injection rate shown in this figure for the combined sonication+vapor stripping system was 500 mL/min. Using sonication alone (for the entire range of power intensities), removal of CCl₄ after 10 minutes ranged from ~26% to ~47%, while removal of the combined sonication+vapor stripping system exceeded 99% for all three power intensities after 10 minutes of treatment time. After six minutes of treatment time, the combined system had CCl₄ removals ranging from ~88% to 98%, well above that by sonication alone. Using nearly identical operating conditions, Figure 3 presents analogous results for the TCE system. Using sonication alone, removal of TCE after 10 minutes ranged from 19% to ~38% for the power intensities studied, while removal of TCE in the combined sonication+vapor stripping system ranged from ~96% to ~99%. Even after six minutes of processing time, the combined system had removal efficiencies ranging from ~87% to ~93%, again considerably above that achieved by sonication alone. Figure 4 shows the first-order rate constants for removal of CCl₄ from groundwater via sonication alone, via vapor stripping alone (for an air injection rate of 500 mL/min), and via the combined sonication/vapor stripping system. The rate constants via sonication are on the order of 0.03 to 0.06 min⁻¹; this is lower than that by vapor stripping (k ~ 0.286 min⁻¹), while the rate constant for the combined system ranges from 0.35 to 0.65 min⁻¹. This data indicates that the combined sonication/vapor stripping system operates in a synergistic manner; the rate constant for the combined system is considerably greater than the sum of the individual rate constants for sonication and vapor stripping separately. The data for TCE is similar in nature to that shown for CCl₄. Figure 5 compares the removal efficiency of CCl₄ and TCE as a function of the processing time using the combined sonication/vapor stripping system. The data indicate that higher removals are achieved for CCl₄ compared to TCE for all processing times at comparable power intensities. This is not unexpected, as CCl₄ has a higher Henrys' law constant than does TCE, and hence is easier to strip or remove from solution than is TCE. Figure 6 shows a comparison of the individual first-order rate constants for CCl₄ and TCE over the range of power

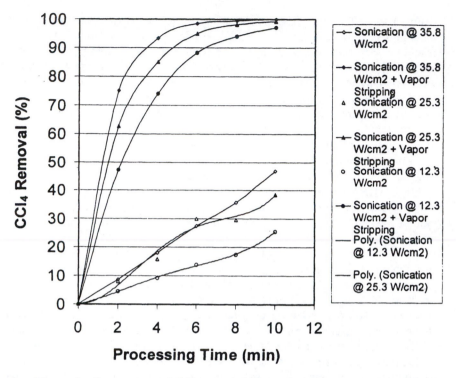

Figure 2. Comparison of CCl₄ removal by combined sonication + vapor stripping (air injection rate ~ 500 mL/min) to sonication alone (20 kHz).

intensities (12.3, 25.3, and 35.8 W/cm²). The rate constants for CCl₄ are consistently higher than that for TCE, for all the treatment systems (sonication alone, vapor stripping alone, and the combined sonication/vapor stripping system). The data again indicate that the combined sonication/vapor stripping system operates in a synergistic manner; the rate constant for the combined system is considerably greater than the sum of the individual rate constants for sonication and vapor stripping separately, both for CCl₄ and for TCE.

Figure 7 shows that the addition of air (vapor stripping) in the combined sonication/vapor stripping process greatly enhances the removal of TCE from solution; similar behavior was observed for the other chlorinated compounds studied (CCl₄, TCA, and PCE). For example, after six minutes of treatment time, removal of TCE was about 28% using sonication alone (without any vapor stripping); removal of TCE using sonication combined with vapor stripping was about 92% and 97% using air injection rates of 500 and 1,000 mL/min. This indicates that addition of a small amount of air greatly enhances the removal of the chlorinated compounds. However, doubling the air injection rates from 500 to 1,000-mL/min resulted in an improvement of only ~5% using the higher air injection rate.

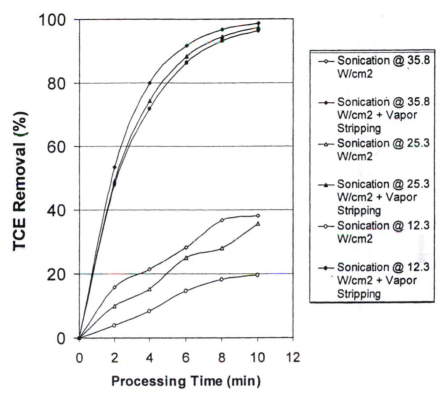

Figure 3. Comparison of TCE removal by sonication+vapor stripping (air injection rate ~500 mL/min) to sonication alone (20 kHz).

The next three sets of figures compare the removal of all of the various chlorinated organic contaminants (CCl$_4$, TCE, TCA, and PCE) at different power intensity levels. Figure 8 shows the normalized residual contaminant fraction remaining using sonication alone (20 kHz, 12.3 W/cm^2) and sonication+vapor stripping (using an air injection rate of 500 mL/min). For all the chlorinated organic species, the fraction remaining ranges from 74% to 84%, while the fraction of chlorinated organics remaining from the combined sonication/vapor stripping system is less than 4% in all cases after 10 minutes of processing time. This again indicates a highly synergistic system. The results using vapor stripping alone are only marginally better than that achieved by sonication alone. Figure 9 shows results comparable to those in the previous figure, except that the power intensity has been increased to 25.3 W/cm^2. For all the chlorinated organic species, the fraction remaining ranges from 62% to 82%, while the fraction of chlorinated organics remaining from the combined sonication/vapor stripping system is less than 3% in all cases after 10 minutes of processing time. Increasing the power intensity still further, to 35.8 W/cm^2 (see

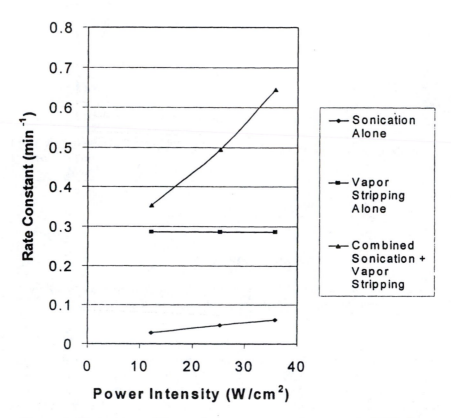

Figure 4. Comparison of first-order rate constants for removal of CCl_4 from groundwater using sonication, vapor stripping, and combined sonication/vapor stripping.

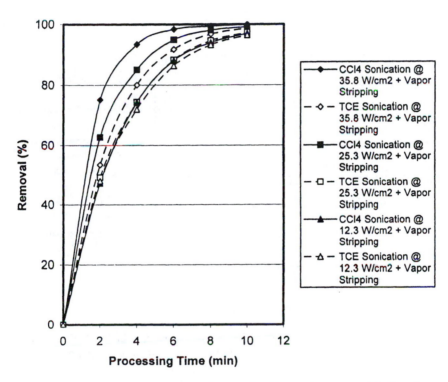

Figure 5. Removal efficiency of CCl₄ and TCE as a function of the batch treatment time using combined sonication/vapor stripping techniques.

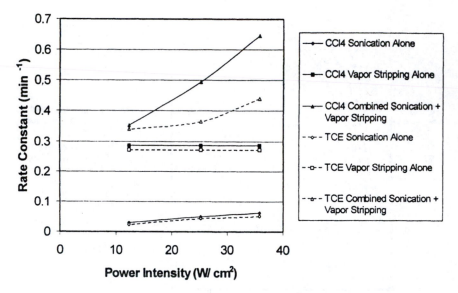

*Figure 6. Comparison of first-order rate constants for removal of CCl₄ and
TCE from groundwater using sonication, vapor stripping, and combined
sonication/vapor stripping.*

Effect of Air Sparging Rate on Removal of Trichloroethylene from Groundwater Using Sonication [20 kHz, 35.8 W/cm2; C~50 mg/L]

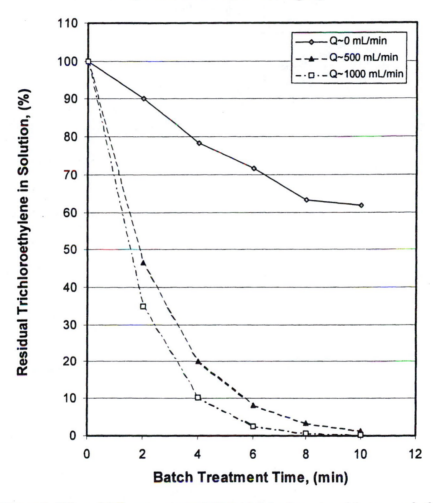

Figure 7. Effect of different vapor stripping air injection rates of the removal of TCE from solution using sonication techniques (20 kHz, 35.8 W/cm²).

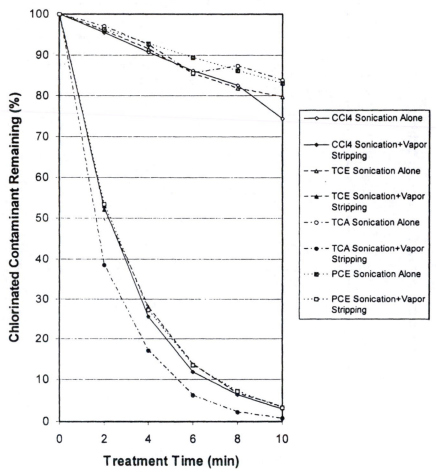

Figure 8. Fraction of chlorinated contaminant remaining in groundwater for various batch treatment times via sonication alone (20 kHz, 12.3 W/cm²) and via sonication + vapor stripping (air injection rate ~ 500 mL/min).

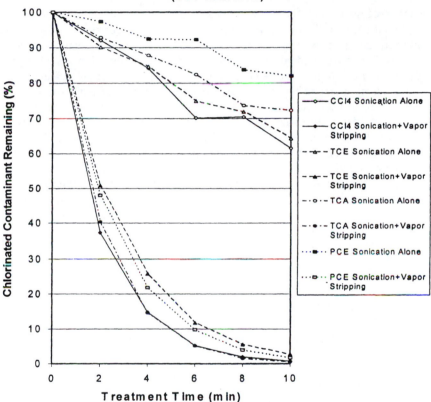

Figure 9. Fraction of chlorinated contaminant remaining in groundwater for various batch treatment times via sonication alone (20 kHz, 25.3 W/cm²) and via sonication + vapor stripping (air injection rate ~ 500 mL/min).

Figure 10), the fraction remaining ranges from 53% to 79%, while the fraction of chlorinated organics remaining from the combined sonication/vapor stripping system is less than 2% in all cases after 10 minutes of processing time. Even after six minutes of processing time, the removal of all of the chlorinated organic compounds in the combined sonication/vapor stripping system exceeds 90%. In Figures 8 to 10, removals of the chlorinated organic compounds follow the order involving their Henrys' law constant (CCl_4 > TCE > TCA > PCE).

Figure 11 shows a comparison of the various fractional residual concentrations of each of the chlorinated organic species as a function of the batch processing time for the various applied power intensities via sonication treatment only; as the power intensity is increased, the removal of each of the chlorinated organic species also increases. Figure 12 shows a comparison of the various fractional residual concentrations of each of the chlorinated organic species as a function of the batch processing time for the various applied power intensities in the combined sonication/vapor stripping system (using an air injection rate of 500 mL/min); as the power intensity is increased, the removal of each of the chlorinated organic species also increases. The decrease in concentration of the various chlorinated organic species tends to follow an exponential decay; hence the first-order kinetics are a reasonable description of the system kinetics. Figure 13 presents the first-order rate constants for removal of TCA from groundwater using sonication alone plotted as a function of the applied power intensity for various nominal initial contaminant concentrations. The first-order rate constants vary linearly with the applied power intensity. Likewise, the first-order rate constants are observed to vary slightly with the initial contaminant concentration; this phenomenon has also been observed by other researchers (*13, 31*).

Batch Reaction Rates

The results from these batch experiments are summarized in several tables that are discussed below. Table V compares the first-order rate constants by acoustic cavitation (sonication) for removal of CCl_4 and TCE from groundwater. The rate constants (min^{-1}) are listed for a variety of nominal initial concentrations and applied ultrasonic power intensities. There are several things worth noting. First, generally the first-order rate constant is higher for CCl_4 than for TCE. This is not unexpected, given that CCl_4 has a higher Henrys' law constant than does TCE. Secondly, as the power intensity is increased, the rate constant also increases in a linear fashion. Thirdly, as the initial concentration decreases, the rate constant increases (i.e., the removal efficiency by sonication increases as the contaminant concentration decreases). The first-order rate constants are observed to vary slightly with the initial contaminant

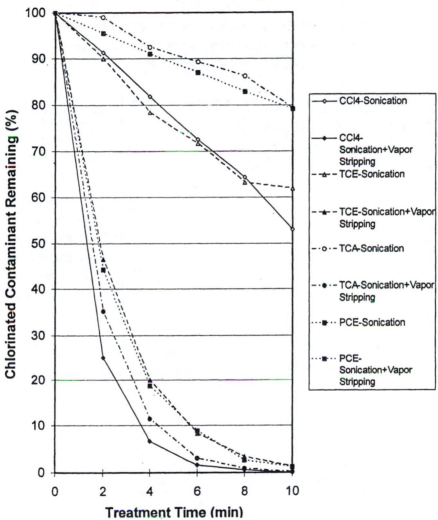

Figure 10. Fraction of chlorinated contaminant remaining in groundwater for various batch treatment times via sonication alone (20 kHz, 35.8 W/cm²) and via sonication + vapor stripping (air injection rate ~ 500 mL/min).

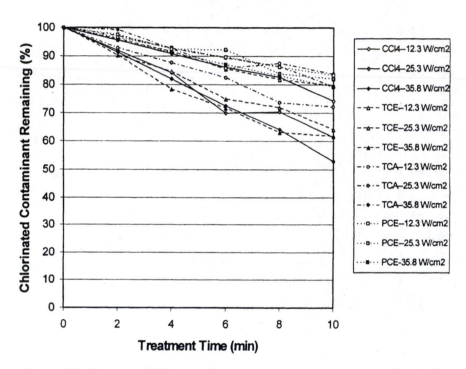

Figure 11. Fraction of chlorinated contaminant remaining in groundwater for various batch treatment times via sonication alone (20 kHz) for various applied power intensities (12.3, 25.3, and 35.8 W/cm²).

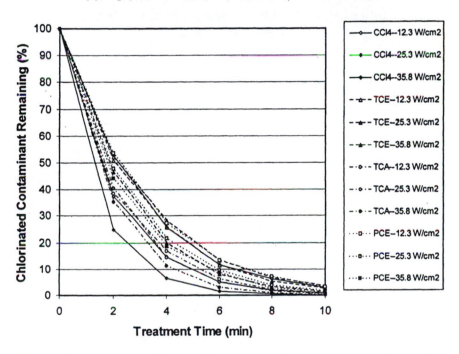

Figure 12. Fraction of chlorinated contaminant remaining in groundwater for various batch treatment times via combined sonication (20 kHz)/vapor stripping (air injection rate ~ 500 mL/min) for various applied power intensities (12.3, 25.3, and 35.8 W/cm²).

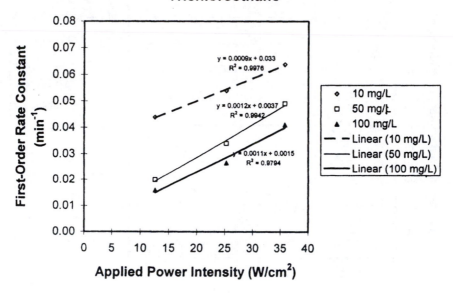

First-Order Rate Constant as a Function of Applied Power Intensity for Sonication Alone of Trichloroethane

Figure 13. First-order rate constants for removal of TCA from groundwater using sonication alone (20 kHz) for various applied power intensities and various initial TCA concentrations.

concentration; this phenomenon has also been observed by other researchers (*13, 31*).

Table V. Comparison of First-Order Rate Constants (min⁻¹) by Acoustic Cavitation for Removal of CCl₄ and TCE from Groundwater

Nominal Concentration (mg/L)	12.3 W/cm²		25.3 W/cm²		35.8 W/cm²	
	CCl₄	TCE	CCl₄	TCE	CCl₄	TCE
1.6	0.0345	0.0347	0.0574	0.0426	0.0788	0.0551
8.0	0.0308	0.0221	0.0489	0.0304	0.0673	0.0403
50	0.0282	0.0235	0.0488	0.0431	0.0619	0.0508
70	0.0193	0.0130	0.0376	0.0210	0.0577	0.0426

Table VI below compares the first-order rate constants obtained for removal of CCl₄ and TCE from groundwater using vapor stripping. Three different air injection flow rates were employed: 500, 1000, and 1500 mL/min. Table V cited above contains the first-order rate constants for an air flow rate of 0 mL/min (i.e., using sonication alone). The first-order rate constant by vapor stripping with 500 ml/min is significantly higher than the case with 0 mL/min (> 5x larger). For both CCl₄ and TCE, as the air injection flow rate is increased, the first-order rate constant also increases, but in a nonlinear fashion; for example, when the air flow rate is doubled from 500 to 1000 mL/min, the first-order rate constant does not double. As the air injection rate is further increased, the effect of the first-order rate constant is less pronounced. The first-order rate constants are higher for CCl₄ than for TCE, as expected due to the higher Henrys' law constant for CCl₄ as compared to TCE. Additionally, as the initial concentration is increased, the first-order rate constant generally increases.

Table VI. Comparison of First-Order Rate Constants for Removal of CCl₄ and TCE from Groundwater Using Vapor Stripping Techniques

Initial Concentration (mg/L)	CCl₄			TCE		
	Air Flow Rate (mL/min)			Air Flow Rate, (mL/min)		
	500	1000	1500	500	1000	1500
	k (min⁻¹)	k (min⁻¹)	k (min⁻¹)	k (min⁻¹)	k (min⁻¹)	k (min⁻¹)
10	0.2944	0.7126	0.5509	0.1538	0.5158	0.4501
50	0.2863	0.4110	0.6291	0.2426	0.4008	0.5458
100	0.3865	0.6583	0.8453	0.2044	0.4115	0.5559

Tables VII and VIII compare the first-order rate constants for removal of CCl$_4$ and TCE from groundwater, respectively, using sonication alone, vapor stripping alone (for an air injection flow rate of 500-mL/min), and by combined sonication/vapor stripping. These tables also list the first-order rate constants at three different applied power intensities. The first-order rate constant for the combined sonication/vapor stripping system is consistently the largest value. The effect is more pronounced as the applied power intensity is increased. It is also worth noting that the combination of sonication coupled with vapor stripping operates in a synergistic fashion; the rate constant of the combined system is much greater than the addition of individual treatments (sonication and vapor stripping). This is likely the result of much greater mixing being associated with sonication of the gas bubbles, resulting in much smaller (finer sized) bubbles being used in solution; this causes a more effective mass transfer of the organic components into the gas phase, and hence provides much greater removals (and hence larger first-order rate constants).

Table VII. Comparison of First-Order Rate Constants for Removal of Carbon Tetrachloride (CCl$_4$) from Groundwater

Applied Power Intensity for Sonication (W/cm^2)	$k_{sonication}$ (min^{-1})	$k_{vapor\ stripping}$ (min^{-1})	$k_{sonication+vapor\ stripping}$ (min^{-1})
12.3	0.0282	0.2863	0.3531
25.3	0.0488	0.2863	0.4947
35.8	0.0619	0.2863	0.6452

Note: Sonication frequency: 20 kHz; nominal initial CCl$_4$ concentration: 50 mg/L; and applied air flow rate: 500 mL/min.

Table VIII. Comparison of First-Order Rate Constants for Removal of Trichloroethylene (TCE) from Groundwater

Applied Power Intensity for Sonication, (W/cm^2)	$k_{sonication}$ (min^{-1})	$k_{vapor\ stripping}$ (min^{-1})	$k_{sonication+vapor\ stripping}$ (min^{-1})
12.3	0.0235	0.2724	0.3403
25.3	0.0431	0.2724	0.3662
35.8	0.0508	0.2724	0.4402

Note: Sonication frequency: 20 kHz; nominal initial TCE concentration: 50 mg/L; and applied air flow rate: 500 mL/min.

Table IX summarizes the first-order rate constants for the four major chlorinated organic contaminants studied (CCl$_4$, TCE, TCA, and PCE) for a variety of sonication conditions (applied power intensities of 12.3, 25.3, and

Table IX. First-Order Rate Constants for Removal/Destruction of Chlorinated Organic Compounds from Groundwater Using Sonication Alone, Vapor Stripping Alone, and Combined Sonication/Vapor Stripping in Continuous Flow Operations

Compound	Sonication Alone Power Intensity (W/cm^2)			Vapor Stripping Alone Air Injection Flow Rate (mL/min)		Combined Sonication/Vapor Stripping Air Injection Rate (mL/min) and Ultrasonic Power Intensity (W/cm^2)					
	12.3	25.3	35.8	500	1000	500/ 12.3	500/ 25.3	500/ 35.8	1000/ 12.3	1000/ 25.3	1000/ 35.8
CCl$_4$	0.028	0.049	0.062	0.286	0.411	0.353	0.487	0.645	0.729	0.870	0.866
TCE	0.019	0.022	0.024	0.243	0.401	0.356	0.356	0.440	0.516	0.583	0.633
TCA	0.020	0.034	0.049	0.228	0.531	0.341	0.549	0.640	0.691	0.864	0.904
PCE	0.019	0.021	0.022	0.282	0.454	0.483	0.411	0.447	0.492	0.412	0.494

35.8 W/cm^2) and air injection rates (500 and 1000 mL/min). First-order rate constants are listed for removal of the various contaminants using sonication alone, vapor stripping alone, and the combined sonication/vapor stripping system. The first-order rate constants are observed to be considerably larger for the combined system, as compared to using either sonication or vapor stripping alone. A plausible explanation for this behavior was described above. It is also worth noting that the same trends are observed for TCA and PCE as were reported previously for the CCl$_4$ and TCE systems.

It should be noted that while vapor stripping does a decent job in removing the chlorinated compounds from solution, it merely transfers the contaminant from the liquid phase to the gaseous phase; it does not destroy the organic contaminant. Using a vapor stripping system alone would require that a vapor treatment system (such as a granular activated carbon adsorber or a thermal oxidizer) be installed to treat the gaseous phase containing the transferred chlorinated solvent contaminant. The sonication and the combined sonication/vapor stripping system are much more effective in destroying the chlorinated solvent contaminants. As evidenced in the results presented for Task 5, the majority of the contaminants are destroyed (typically, greater than 99%), with only trace amounts of byproducts formed.

Results from Continuous Flow Experiments

Based on the promising results from the batch sonication/vapor stripping experiments, experiments were performed in a continuous flow mode. Results are presented for the cases involving CCl$_4$ and TCA as the chlorinated organic contaminant, although similar results were also obtained for the other chlorinated organic species (TCE and PCE). For these experiments, the ultrasonic frequency was 20 kHz, with an applied power intensity of 35.8 W/cm^2. The initial concentration of the chlorinated organic contaminant was nominally 50 mg/L. The residence time within the reactor was set at 5, 8, and 10 minutes. The reactor volume was 500 mL. Figure 14 shows the results of the residual concentration of TCA after a function of the number of residence times throughput through the reactor (well). Initially, the reactor was filled with artificial groundwater containing no contaminant; at the beginning of the run, the reactor was fed with two feed streams (one containing the uncontaminated groundwater; the other containing a saturated feed stream of the organic contaminant). The ratio of these two feed streams was determined by mass balance calculations to provide a nominal initial concentration of ~50 mg/L. In accordance with chemical reactor design, the reactor operated as a continuous stirred tank reactor. It takes about 3+ residence times to reach a "pseudo-steady-state" condition. This was observed in our experiments; after about 4 or 5 residence times throughput of both the artificial groundwater and the saturated contaminant feed stream, the concentration of the contaminant within the reactor had reached a steady-state condition. Afterwards, the sonicator was turned on, and the reactor effluent was analyzed for its residual TCA concentration. Similar experiments were performed in which both sonication and vapor stripping (at an air flow rate of 500 mL/min) were performed simultaneously. The results are shown in Figure 14. For the residence times ranging from 5 to 10 minutes, TCA removal by sonication alone ranged from 14.6% to 36.6%, while TCA removal by the combined sonication/vapor stripping system for the same set of residence times, ranged from 72.3% to 97.3%, showing a substantial enhancement of the combined system over that by sonication alone. Analogous results are presented in Figure 15 for the CCl_4 system, operating under nearly identical operating conditions. For the residence times ranging from 5 to 10 minutes, CCl_4 removal by sonication alone ranged from 17.1% to 17.9%, while CCl_4 removal by the combined sonication/vapor stripping system for the same set of residence times, ranged from 74.0% to 87.9%, again showing a substantial enhancement of the combined system over that by sonication alone.

Experiments were also performed using vapor stripping alone. The set up for these experiments was similar to that described above, in which a steady state concentration of the contaminant was achieved (~50 mg/L) prior to performing vapor stripping. Results from the vapor stripping experiments (with residence times ranging from 5–10 minutes) were obtained for the cases involving TCA and CCl_4. For the residence times ranging from 5–10 minutes, TCA removal by vapor stripping alone ranged from 55.6% to 75.4%, while TCA removal by the combined sonication/vapor stripping system for the same set of residence times, ranged from 72.3% to 97.3%, showing a significant enhancement of the combined system over that by vapor stripping alone. Similarly, for the residence times ranging from 5–10 minutes, CCl_4 removal by vapor stripping alone ranged from 68.4% to 80.3%, while CCl_4 removal by the combined sonication/vapor stripping system for the same set of residence times, ranged from 74.0% to 87.9%, again showing an enhancement of the combined system over that by vapor stripping alone. Although the vapor stripping alone results in improved residual concentrations of TCA and CCl_4 (comparing to Figures 14 and 15), the residual concentration of TCA and CCl_4 is significantly higher than that achieved with the combined system. In all cases, the combined system at any particular residence time resulted in improved performance over the individual cases involving sonication alone and vapor stripping alone. It should be noted that while vapor stripping does a decent job in removing the chlorinated compounds from solution, it merely transfers the contaminant from the liquid phase to the gaseous phase; it does not destroy the organic contaminant. The sonication and the combined sonication/vapor stripping system are much more effective in destroying the chlorinated solvent contaminants.

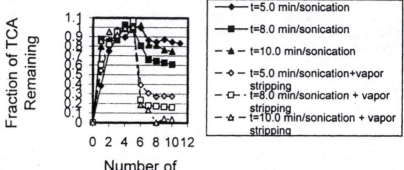

Figure 14. Comparison of fraction TCA remaining using sonication + vapor stripping (500 mL/min air injection rate) to sonication alone (20 kHz, 35.8 W/cm²).

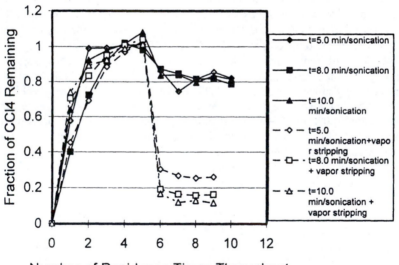

Figure 15. Comparison of fraction CCl₄ remaining using sonication + vapor stripping (500 mL/min air injection rate) to sonication alone (20 kHz, 35.8 W/cm²).

Summary and Conclusions

Batch and continuous flow experiments were performed using sonication alone, vapor stripping alone, and the combined sonication + vapor stripping system. The results show that the combined system is more effective than either of the separate sonication or vapor stripping systems. For example, while sonication is capable of removing ~30% of the TCA after 10 minutes of reaction time, the combined sonication/vapor stripping system can remove nearly 97% after 4 minutes of treatment time, and nearly 100% removal after 10 minutes. The rate constants for the combined system are nearly an order of magnitude higher than those for sonication alone (and for vapor stripping alone).

Based on these promising results, continuous-flow tests were conducted using a continuous flow, stirred tank reactor in which sonication and vapor stripping were performed simultaneously to treat the contaminated artificial groundwater. For residence times in the continuous flow system ranging from 5–10 minutes using sonication alone, removal of TCA ranged from 14% to 36%, while the combined system (using 500 mL/min air flow rate in combination with sonication) resulted in removals of TCA ranging from 72% to 97%, depending on the residence time used, once again showing a considerable enhancement in the removal of TCA from solution as a consequence of combining the sonication and vapor stripping technologies. These continuous flow tests showed that the system was very stable, resulting in steady-state operating conditions.

For continuous flow conditions, the residual TCA concentration following application of sonication alone and sonication+vapor stripping for reactor residence times ranging from 5–10 minutes, removals of CCl_4 and TCA were 74% to 88% and 72% to 97%, respectively, for a one-pass system. Sonication alone or vapor stripping alone results in removals ranging from ~20% to ~50%; however, combining sonication with vapor stripping enhances removal of CCl_4, TCE, TCA, and PCE as compared to removal by either system alone (sonication or vapor stripping) by nearly an order of magnitude, and results in synergistic behavior enhancing removal efficiency. The combination of sonication with vapor stripping results in much better mixing, creation of much finer bubbles, thereby enhancing mass transfer from the liquid phase to gaseous phase, and enhancing the destruction of the chlorinated organic compounds during the microbubble implosions.

In summary, the innovative technology couples in-well sonication, in-well vapor stripping, and in situ biodegradation into an integrated process. By partially destroying the SVOCs (e.g., opening up the benzene-ring structures), the ability to remove the resultant VOCs and biotreatment of the resultant organics is enhanced (over the case of biotreatment alone). Advantages of this technique include:

- Remediation is performed in situ (i.e., does not require handling or disposing of water at the ground surface);
- Treatment systems complement each other and their combination has the potential to drastically reduce or remove SVOCs and VOCs;
- System has the potential to add other innovative components (such as in situ chemical treatments or surfactants);
- Ability to convert hard-to-degrade organics such as chlorinated organics and heavy organics (e.g., polyaromatic hydrocarbons) into more volatile organic compounds;
- Ability to destroy chlorinated organic compounds;
- Ability to remove residual VOCs and softened SVOCs through the combined action of in-well vapor stripping and biodegradation;
- Improved in situ biotreatment of contaminated soils and groundwater; and
- Cost-effective and improved efficiency thereby shortening the time required to clean-up a contaminated site.

Disclaimer

The viewpoints expressed here are not necessarily those of Argonne National Laboratory, the University of Alabama at Birmingham, or their sponsors.

Acknowledgments

This research project was funded by the DOE Environmental Management Science Program (EMSP). This project was a collaborative effort involving Argonne National Laboratory's Energy Systems Division and Environmental Research Division, and the California Institute of Technology and Stanford University. Additionally, a subcontract was established with the Illinois Institute of Technology (IIT) enabling Mr. Po-Yao Kuo and Mr. Onder Ayyildiz, graduate students in the Department of Chemical and Environmental Engineering at IIT, to be active researchers on this project. Additionally, the lead author acknowledges the resources of the University of Alabama at Birmingham in further analyzing the experimental data and developing this manuscript.

References

1. Peters, R. W.; Wu, J.-M. *Invention Disclosure: In-Well Sonication for Destruction of Organic Contaminants;* ANL-IN-95-103; Argonne National Laboratory: Argonne, IL, 1995.
2. Gorelick, S. M.; Gvirtzman, H. In Situ Vapor Stripping for Removing Volatile Organic Compounds from Groundwater. U.S. Patent 5,180,503, 1993.
3. Gorelick, S. M.; Gvirtzman, H. In Situ Vapor Stripping for Removing Volatile Organic Compounds from Groundwater. U.S. Patent 5,389,267, 1995.
4. François, O.; Gilmore, T.; Pinto, M. J.; Gorelick, S. M. A Physically-Based Model for Air-Lift Pumping. *Water Resour. Res.* **1996**, *32* (8), 2383-2399.
5. Gvirtzman, H.; Gorelick, S. M. The Concept of In Situ Vapor Stripping for Removing VOCs from Groundwater. *Transport in Porous Media* **1992**, *8*, 71–92.
6. Pinto, M. J. Laboratory-Scale Analysis of Aquifer Remediation by In-Well Vapor Stripping: 2. Modeling Results. *Contaminant Hydrology* **1997**, *29* (1), 41–58.
7. Philip, R. D.; Walter, G. R. Prediction of Flow and Hydraulic Head Fields for Vertical Recirculation Wells. *Ground Water* **1992**, *30* (5), 765–773.
8. Kabala, Z. J. The Dipole Flow Test: A New Single-Borehole Test for Aquifer Characterization. *Water Resour. Res.* **1993**, *29* (1), 99–107.
9. Herrling, B. J.; Stamm, J.; Buermann, W. Hydraulic Circulation System for In Situ Bioreclamation and/or In Situ Remediation of Strippable Contamination. In *In Situ Bioreclamation*; Hinchee, R. E., Olfenbuttel, R. F., Eds.; Butterworth-Heinemann: Stoneham, MA, 1991; pp 173–195.
10. Argonne National Laboratory. New Technology Cleans Contaminated Groundwater: Seeking Collaborators for Further Development. *Tech Transfer Highlights* **1997**, *8* (1), 2.
11. HazTECH News. Sonication Combined with Vapor Stripping to Remove Organics from Ground Water. *HazTech News* **2000**, *15* (2), 11–12.
12. Boyd, G.; Geiser, D. Integrated Solutions – Meeting the Challenges of the Environmental Management Program. Plenary paper presented in the Topical Conference of "Environmental Remediation in the 21st Century: Integrated Systems Technologies," AIChE spring national meeting, Atlanta, GA, Mar 5–9, 2000.
13. Hua, I.; Hoffmann, M. R. Kinetics and Mechanism of the Sonolytic Degradation of CCl_4: Intermediates and Byproducts. *Environ. Sci. Technol.* **1996**, *30* (3), 864–871.

14. Hua, I.; Hoffmann, M. R. Optimization of Ultrasonic Irradiation as an Advanced Oxidation Technology. *Environ. Sci. Technol.* **1997**, *31* (8), 2237–2243.

15. Hua, I.; Höchemer, R.; Hoffmann, M. R. Sonochemical Degradation of p-Nitrophenol in a Parallel-Plate Near-Field Acoustical Processor. *Environ. Sci. Technol.* **1995**, *29* (11), 2790-2796.

16. Hua, I., Höchemer, R. H.; Hoffmann, M. R. Sonolytic Hydrolysis of p-Nitrophenyl Acetate: The Role of Supercritical Water. *J. Phys. Chem.* **1995**, *99* (8), 2335–2342.

17. Hoffmann, M. R.; Hua, I.; Höchemer, R.; Willberg, D.; Lang, P.; Kratel, A. Chemistry under Extreme Conditions in Water Induced by Electrohydraulic Cavitation and Pulsed-Plasma Discharges. In *Chemistry under Extreme or Non-Classical Conditions;* van Eldik, R., Hubbard, C. D., Eds.; Elsevier: Amsterdam, The Netherlands, 1996.

18. Kotronarou, A.; Mills, G.; Hoffmann, M. R. Ultrasonic Irradiation of p-Nitrophenol in Aqueous Solution. *J. Phys. Chem.* **1991**, *95* (9), 3630–3638.

19. Kotronarou, A.; Mills, G.; Hoffmann, M. R. Decomposition of Parathion in Aqueous Solution by Ultrasonic Irradiation. *Environ. Sci. Technol.* 1992, *26* (7), 1460–1462.

20. Kotronarou, A.; Mills, G.; Hoffmann, M. R. Oxidation of Hydrogen Sulfide in Aqueous Solution by Ultrasonic Irradiation. *Environ. Sci. Technol.* **1992**, *26* (12), 2420–2428.

21. Kotronarou, A.; Hoffmann, M. R. The Chemical Effects of Collapsing Cavitation Bubbles: Mathematical Modeling. In *Aquatic Chemistry: Interfacial and Interspecies Processes;* Huang, C. P., O'Melia, C. R., Morgan, J. J., Eds.; Advances in Chemistry Series 244; American Chemical Society: Washington, DC, 1995; pp 233–251.

22. Mason, T.; Lorimer, J. *Sonochemistry: Theory, Applications, and Uses of Ultrasound in Chemistry;* Ellis Norwood, Ltd.: New York, 1988.

23. Sehgal, C.; Steer, R. P.; Sutherland, R. G.; Verrall, R. E. Sonoluminescence of Aqueous Solutions. *J. Phys. Chem.* **1977**, *81* (26), 2618–2620.

24. Flint, E. B.; Suslick, K. S. Sonoluminescence from Nonaqueous Liquids: Emission from Small Molecules. *J. Am. Chem. Soc.* **1989**, *111* (18), 6987–6992.

25. Flint, E. B.; Suslick, K. S. The Temperature of Cavitation. *Science* **1991**, *253* (5026), 1397–1399.

26. Suslick, K. S.; Hammerton, D.; Cline, R. The Sonochemical Hot Spot. *J. Am. Chem. Soc.* **1986**, *108*, 5641.

27. Crum, L. A. *J. Accoust. Soc. Am.* **1994**, *95*, 559–562.

28. Crum, L. A.; Roy, R. A. Sonoluminescence. *Science* **1994**, *266* (5183), 233–234.

29. Putterman, S. Sonoluminescence: Sound into Light. *Sci. Am.* **1995**, *272* (2), 46–51.

30. Gutiérrez, M.; Henglein, A.; Ibañez, F. J. Radical Scavenging in the Sonolysis of Aqueous Solutions of I⁻, Br⁻, and NO_3^-. *J. Phys. Chem.* **1991**, *95* (15), 6044–6047.

31. Wu, J. M.; Huang, H. S.; Livengood, C. D. Ultrasonic Destruction of Chlorinated Compounds in Aqueous Solution. *Environ. Prog.* **1992**, *11* (3), 195–201.

32. Wu, J. M.; Huang, H. S.; Livengood, C. D. *Development of an Ultrasonic Process for Detoxifying Groundwater and Soil: Laboratory Research*; ANL/ESD/TM-32; Argonne National Laboratory: Argonne, IL, 1992.

33. Peters, R. W.; Wu, J.-M. Use of Enhanced Fenton-Like Reactions for Destruction of Atrazine. Paper presented at the AIChE spring national meeting, New Orleans, LA, 1996.

34. Manogue, W. H.; Pigford, R. L. The Kinetics of the Absorption of Phosgene into Water and Aqueous Solutions. *AIChE J.* **1960**, *6* (3), 494–500.

35. Gvirtzman, H.; Gorelick, S. M. Using Air-Lift Pumping as an In Situ Aquifer Remediation Technique. *Water Sci. Technol.* **1993**, *27* (708), 195–201.

36. Mohn, W. W.; Tiedje, J. M. Microbial Reductive Dehalogenation. *Microbial Review* **1992**, *56,* 482–507.

37. Semprini, L.; Hopkins, G. D.; McCarty, P. L.; Roberts, P. V. In Situ Transformation of Carbon Tetrachloride and Other Halogenated Compounds Resulting from Biostimulation Under Anoxic Conditions. *Environ. Sci. Technol.* **1992**, *26* (12), 2454–2461.

38. Criddle, C. S.; DeWitt, J. T.; Grbić-Galić, D.; McCarty, P. L. Transformation of Carbon Tetrachloride by *Pseudomonas sp* Strain ILC under Denitrification Conditions. *Appl. Environ. Microbiol.* **1990**, *56* (11), 3240–3246.

39. Tatara, G. M.; Dybos, M. J.; Criddle, C. S. Effects of Medium and Trace Metals on the Kinetics of Carbon Tetrachloride Transformation by *Pseudomonas sp* Strain ILC. *Appl. Environ. Microbiol.* **1993**, *59* (7), 2126–2131.

40. Weathers, L. J.; Parkin, G. F. Metallic Iron-Enhanced Biotransformation of Carbon Tetrachloride and Chloroform Under Methanogenic Conditions. In *Bioremediation of Chlorinated Solvents;* Hinchee, R. E., et al., Eds.; Battelle Press: Columbus, OH, 1995; pp 117–122.

41. O'Hannesin, S. F. A Field Demonstration of a Permeable Well for the In Situ Abiotic Degradation of Halogenated Aliphatic Organic Compounds. M.S. Thesis, University of Waterloo, Ontario, Canada, 1993.

42. Matheson, L. J.; Tratnyek, P. G. Reductive Dehalogenation of Chlorinated Methanes by Iron Metal. *Environ. Sci. Technol.* **1994**, *28,* 2045–2053.

43. Helland, B. R.; Alvarez, P. J. J.; Schnoor, J. L. Reductive Dechlorination of Carbon Tetrachloride with Elemental Iron. *J. Hazard. Mater.* **1995**, *41* (2-3), 205–216.

44. Kriegman-King, M. R.; Reinhard, M. Transformation of Carbon Tetrachloride in the Presence of Sulfide, Biotite, and Vermiculite. *Environ. Sci. Technol.* **1992**, *26* (11), 2198–2206.

45. Kriegman-King, M. R.; Reinhard, M. Transformation of Carbon Tetrachloride by Pyrite in Aqueous Solution. *Environ. Sci. Technol.* **1994**, *28* (4), 692–700.

46. Quinn, M. S.; Moehring, G. A.; Peters, R. W. Analyzing Headspace of Chlorinated Volatile Organic Compounds in Groundwater. Poster paper presented at the 4th Annual Governors State University Student Research Conference, University Park, IL, May 29, 1998.

47. Quinn, M. S.; Moehring, G. A.; Peters, R. W. Analyzing Headspace of Chlorinated Volatile Organic Compounds in Groundwater. Poster paper presented at the 9th Annual Student Research Conference of the Board of Governors Universities, Western Illinois University, Macomb, IL, April 3-4, 1998.

48. Peters, R. W.; Ayyildiz, O.; Quinn, M. S.; Wilkey, M. Removal of CCl_4 and TCE by Sonication, by Vapor Stripping, and by Combined Sonication/Vapor Stripping. In *Proceedings, Pan-American Workshop on Commercialization of Advanced Oxidation Technologies;* Science and Technology Network, Inc.: London, Ontario, Canada, 1998.

49. Peters, R. W.; Wilkey, M.; Ayyildiz, O.; Quinn, M.; Pierce, L.; Hoffmann, M.; Gorelick, S. Use of Sonication for In-Well Softening of Semivolatile Organic Compounds. Poster paper presented at the U.S. Department of Energy Environmental Management Science Program Workshop, Chicago, IL, July 27–30, 1998.

50. Peters, R. W.; Wilkey, M. L.; Furness, J. C.; Ayyildiz, O. Development of the Integrated In-Well Sonication/In-Well Vapor Stripping System to Treat Chlorinated Solvent-Contaminated Groundwater. Paper presented at the 12th International Conference on Bioremediation and Remediation, Orlando, Florida, Dec 10–12, 1999.

51. Peters, R. W.; Wilkey, M. L.; Furness, J. C.; Ayyildiz, O. Development of the Integrated In-Well Sonication/In-Well Vapor Stripping System to Treat Chlorinated Solvent-Contaminated Groundwater. Paper presented in the Topical Conference of "Environmental Remediation in the 21st Century: Integrated Systems Technologies," AIChE spring national meeting, Atlanta, GA, March 5–9, 2000.

52. Hung, H.-M.; Hoffmann, M. R. Kinetics and Mechanism of the Sonolytic Degradation of Chlorinated Hydrocarbons: Frequency Effects. *J. Phys. Chem. A* **1999**, *103* (15), 2734–2739.

53. Inazu, K.; Nagata, Y.; Maeda, Y.   Decomposition of Chlorinated Hydrocarbons in Aqueous Solutions by Ultrasonic Irradiation. *Chem. Lett.* **1993**, *1,* 57–60.
54. Drijvers, D.; Baets, R. D.; Visscher, A. D.; Langenhove, H. V. Sonolysis of Trichloroethylene in Aqueous Solution: Volatile Organic Intermediates. *Ultrason. Sonochem.* **1996**, *3,* S83–S90.
55. Kruus, P.; Beutel, L.; Aranda, R.; Penchuk, J.; Otsen, R.   Formation of Complex Organochlorine Species in Water Due to Cavitation. *Chemosphere* **1998**, *36* (8), 1811–1824.
56. Bhatnagar, A.; Cheung, H. M.   Sonochemical Destruction of Chlorinated C1 and C2 Volatile Organic Compounds in Dilute Aqueous Solution. *Environ. Sci. Technol.* **1994**, *28,* 1481–1486.

Chapter 3

Peroxidase-Catalyzed Oxidative Coupling of Phenols in the Presence of Geosorbents

Qingguo Huang[1] and Walter J. Weber, Jr.[2]

Departments of [1]Civil and Environmental Engineering and [2]Chemical Engineering, The University of Michigan, Ann Arbor, MI 48109–2125

The effects of geosorbents on the enzymatically catalyzed coupling reactions of phenol are quantified and interpreted on the basis of a conceptual model advanced earlier. The work reveals that the reactivity of different sorbents with respect to cross-coupling with phenol radicals correlates with the chemical characteristics of the sorbents studied. Cross-coupling reactivities of the sorbents were found to decrease in the order lignin > Chelsea soil > Lachine shale > Polymethylstyrene; i.e., from humic-type organic matter (highly amorphous) to kerogen-like materials (relatively highly condensed). The information provided and insights gained from this study have important implications for the behavior and efficiency of enzymatic phenol coupling reactions that occur in natural environments. As such, they have utility in guiding remediation practices involving soil/sediment materials of different characteristics and properties.

Introduction

Phenolic contaminants are of major environmental concern because of their widespread occurrence and multiple toxicity effects. Peroxidases comprise an important class of enzymes that can catalyze the oxidative coupling reactions of a broad spectrum of phenolic substrates in the presence of hydrogen peroxide (1–5). The catalysis dynamics of horseradish peroxidase (HRP) have been particularly well investigated (4–8). Phenolic substrates are converted into phenoxy radicals catalytically by peroxidase in the presence of hydrogen peroxide. The phenoxy radicals produced in the enzymatic reaction step can then, in post-enzymatic reactions, attack and inactivate the HRP enzyme itself, couple with each other (self-coupling), or bind to other reactive substances present in the system (cross-coupling). Self-coupling dominates in systems that lack appropriate substrates for cross-coupling reactions, leading to the formation of precipitated polymeric products that can be removed readily from water (9–16). Conversely, cross-coupling with other reactive substances tends to occur in aquatic systems that include soils and sediments, generally resulting in the incorporation of phenolic substrates in soil organic matter (SOM) (1, 17–22). Under appropriate conditions, such reactions can markedly reduce the environmental mobility and eco-toxicities of phenolic contaminants (1, 22). Enhancement of enzymatic coupling reactions has therefore been suggested as a promising means for subsurface immobilization of phenolic substances and soil remediation (1, 3, 22).

A conceptual model was developed earlier in our laboratory (17, 23, 24) to describe rate relationships for enzymatic phenol coupling in the presence of reactive sorbents, and a system of equations designed to articulate this conceptual model has been shown to successfully capture the rate behaviors observed experimentally for both enzyme inactivation and phenol reactions in systems containing reactive sorbent materials. This paper summarizes the essence and mechanistic basis of this model, and presents insights obtained by using the approach to examine recent experimental results with model sorbent materials. Two model sorbents, lignin and polymethylstyrene (PMS), having greatly different but well-defined characteristics are considered. In terms of their chemical compositions and characteristics, lignin and PMS respectively represent a diagenetically "young" humic substance that is more oxidized and reactive, containing more functionalities, and an "old" kerogen type material that is chemically reduced and relatively inert. The effect of each sorbent on the enzymatic phenol-coupling reaction was quantified and interpreted within the framework of the conceptual model we advanced. The sorbent effects observed were found to correlate with the chemical characteristics of the model sorbents studied. The information provided and insights obtained from this study have important implications for the behavior and efficiency of enzymatic phenol coupling reactions that occur in natural environments. As such, they have utility in guiding remediation practices involving soil/sediment materials of different properties.

Experimental Section

Materials

Lignin and PMS were investigated as model sorbents. The structures of the two organic polymers are presented in Figure 1. Poly(4-methylstyrene) purchased from Scientific Polymer Products Inc. (Ontario, NY) was washed thoroughly with methanol followed by milli-Q water, and dried prior to use. Organosolv lignin obtained from Aldrich Chemical Co. (Milwaukee, WI) was washed sequentially with 50% methanol and milli-Q water, and dried prior to use.

Extracellular horseradish peroxidase (type-I, $RZ = 1.3$), hydrogen peroxide (30.8%, ACS reagent), 2, 2'-azino-bis(3-ethylbenz-thiazoline-6-sulfonic acid) (ABTS) (98%, in diammonium salt form), and phenol-UL-^{14}C (51.4 mCi/mmol) were obtained from Sigma Chemical Co. (St. Louis, MO). Phenol (99+%, biochemical grade) was from Acros Organics (Belgium, NJ). ScintiSafe Plus 50% liquid scintillation cocktail and all other chemicals were obtained in the highest quality available from Fisher Scientific (Fairlawn, NJ).

Extent of Non-Extractable Product Formation

Non-extractable products (NEP) of phenol-coupling reactions are operationally defined here as products remaining in the solid phase after 50% methanol extraction and phase separation, as measured collectively by the radioactivity of ^{14}C-labeled species. Phenol-coupling reactions were performed at room temperature in 13×100-mm glass test tubes operated as Completely Mixed Batch Reactors (CMBRs). Each reactor contained 3 mL of 10-mM phosphate buffer (pH = 7.0) solution comprised by a 500-μM mix of ^{14}C-labeled and unlabeled phenol, 2-mM H_2O_2, and a predetermined dosage of HRP. Radioactivity contained in the reaction mixture was about 5 μci/L. The reactor also contained a model sorbent at one of a series of selected sorbent/water ratios. Triplicate experiments were conducted for each reaction condition. After a two-hour reaction period during which enzyme activity was completely depleted, three mL of methanol were added to each reactor and mixed with the contents for 30 minutes. After extraction, a 1.5-mL sample of solution was withdrawn from the reactor and centrifuged at 20,800g for 20 min to separate the liquid and solid phases. A 0.5-mL aliquot of the liquid phase was sampled and mixed with

3 mL of ScintiSafe Plus 50% liquid scintillation cocktail, and its radioactivity then measured using a Beckman LS6500 liquid scintillation counter (Beckman Instruments, Inc.). The radioactivity remaining in the solid phase was calculated by mass balance and the NEP concentration calculated and expressed in terms of molar equivalents of phenol in the original solution.

Constant initial phenol (500 μM) and H_2O_2 (2 mM) concentrations were used across this study. We have demonstrated that HRP can be saturated by both substrates at these initial concentrations regardless of their loss over the reaction when the HRP dosage is at or less than 1 unit/mL (24). We also found that the employed H_2O_2 concentration is still lower than the critical concentration that can cause significant side effects on enzyme efficiency in our experimental systems (24). It is known that H_2O_2 can cause several side effects leading to HRP inactivation and reduction of the enzymatic reaction efficiency via certain side-reaction mechanisms (25, 26). These H_2O_2-related side effects are suppressed in the presence of phenolic substrates due to competition effects (8, 26), and are insignificant when the H_2O_2 is below a certain critical concentration that depends upon the enzyme affinities of the phenolic substrates present in the system (8).

Blank control tests indicated that physical losses during the two-hour reaction time and subsequent extraction and analysis were negligible for the system employed.

Enzyme Activity Assessment

Enzyme activity was measured by the ABTS method described earlier (17, 27). In this method, 0.05 mL of sample are added to a cuvette containing a 3-mL volume of phosphate buffer solution (pH = 6.0), followed by addition of 0.3 mL of 20-mM ABTS and 0.3 mL of 10-mM hydrogen peroxide to initiate the assay. The absorbance change at 405 nm was monitored by a 6405 UV/Vis spectrophotometer (Jenway Inc., Princeton, NJ). One unit of peroxidase activity is defined as that amount catalyzing the oxidation of one μmol of ABTS per minute.

Results and Discussion

Conceptual Model

The conceptual model we use for characterizing rate relationships for peroxidase-mediated phenol coupling in the presence of reactive sorbents has been described in detail in earlier publications (17, 23, 24). As depicted schematically in Figure 2, the model presumes that phenoxy radicals (*AH•*) are generated in a reversible process during the enzymatic reaction step. These radicals then participate in several post-enzymatic reactions that include: (i) attack and inactivation of the enzyme; (ii) self-coupling of radicals; and/or (iii) cross-coupling of radicals with reactive sorbent materials. Self-coupling and cross-coupling processes can each lead to formation of NEP, and both are therefore of significance with respect to the treatment of phenolic wastes and the remediation of contaminated sites.

The reversible radical generation process mediated by HRP represented in Figure 2 involves radical-producing forward enzymatic reactions and backward "reverse electron transfer" processes in which phenoxy radicals are re-converted to their original substrate forms. The forward reactions produce phenoxy radicals (*AH•*) via single-electron transfers from phenol substrates (*AH₂*) to the enzyme (*5, 6*). The phenoxy radicals thus produced can form radical-pairs with the enzyme intermediates and abstract electrons from the enzyme via reverse electron transfer, essentially reversing the radical-producing reaction, as demonstrated by Taraban et al. (*28*).

The reversible radical generation reaction referenced above is actually mediated by HRP in a complex process. In this process several HRP intermediates are successively generated, producing a catalytic cycle referred to as a Chance-George mechanism (*8–10*). Based on the detailed analysis of HRP catalytic dynamics we presented earlier (*23*), the overall rate of radical generation, r_E , can be described in terms of a rate expression that is first-order with respect to total active enzyme concentration, [*E*], when both H_2O_2 and phenol substrates are present in excess, and the overall rate of reverse electron transfer, r_r, can be described by a rate equation that is first-order with respect to the phenoxy radical concentration, [*AH•*]. Assuming that the reverse electron transfer is the dominant mechanism for radical disappearance, these expressions for r_E and r_r can then be coupled to give an equation describing the pseudo-steady-state radical concentration in the reaction system at any instant in time; i.e.,

$$[AH\bullet] = \frac{k_E}{k_r}[E] \qquad (1)$$

Equation 1 indicates the proportionality between the radical concentration and the active enzyme concentration in the reaction system studied. As evident from Figure 2, this equation is a key to the rate analyses of all three post-enzymatic reactions involving the phenoxy radical; i.e., enzyme inactivation and NEP formation via self-coupling and/or cross-coupling reactions. Enzyme inactivation resulting from radical attack, for example, can be expressed as

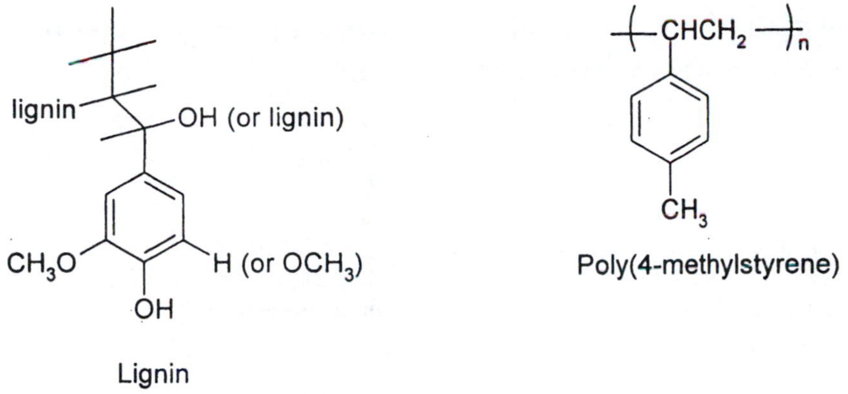

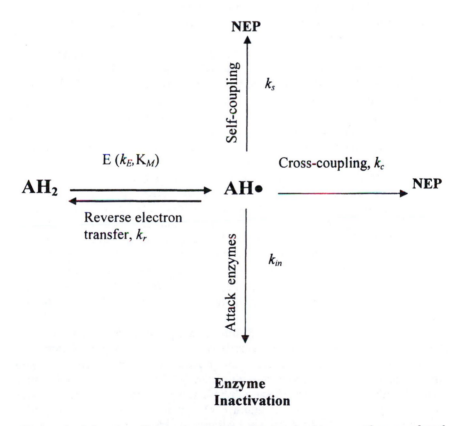

lignin—⌐—OH (or lignin)

CH₃O—⎯H (or OCH₃)

OH

Lignin

Poly(4-methylstyrene)

Figure 1. Structural characteristics of the organic polymers tested.

NEP

Self-coupling k_s

E (k_E, K_M) Cross-coupling, k_c

AH₂ ⟶ AH• ⟶ NEP

Reverse electron transfer, k_r

Attack enzymes k_{in}

Enzyme Inactivation

Figure 2. Schematic illustration of processes involved in peroxidase-catalyzed phenol coupling.

$-d[E]/dt = k_{in}[E][AH\bullet]$. Substitution of eq 1 into this expression for $[AH\bullet]$ and combining rate constants in k'_{in}, leads to

$$-\frac{d[E]}{dt} = k'_{in}[E]^2 \qquad (2)$$

Equation 2 was shown to provide a good description of the rate behaviors of enzyme inactivation observed in our earlier study, and will be applied in analyses of the rate information developed in the present study.

NEP Formation

Figure 3 presents concentrations of total NEP ($[NEP]_T$) formed in systems having varied initial HRP dosages and containing 0.1 g/L lignin, 25 g/L PMS and no sorbent, respectively. NEP formation in the absence of sorbents exhibits a rather evident linear dependence on the HRP dosage as shown in Figure 4. It is also evident in Figure 3 that NEP formation is significantly enhanced in the presence of both lignin and PMS at each HRP dosage tested. With the highest HRP dosage used in the experiment (1.0 unit/mL), the yields of NEP in the presence of 0.1 g/L lignin and 25 g/L PMS are respectively 356 and 313 μM phenol equivalents, while that in the absence of sorbents is 241 μM. Our studies also showed that such enhancement of NEP formation in the presence of lignin and PMS is dependent on the sorbent concentrations (data not shown); i.e., the higher the sorbent/water ratios, the greater the NEP formations.

Direct sorption of the parent phenol and the dissolved products of phenol self-coupling reactions on the model sorbent materials employed in this study was evaluated in our previous work (24). These species were found to be negligibly or very weakly sorbed at the sorbent/water ratios investigated in the current study. Preliminary tests on the 50% methanol extraction employed for NEP measurement in this study demonstrated essentially complete phenol and dissolved phenol-coupling product recoveries from the sorbent/solution systems investigated. Based on these results, it can be concluded that sorption processes per se were not significantly involved in the enhanced $[NEP]_T$ formation data shown in Figure 4.

The enhancement of NEP formation in the presence of sorbent materials illustrated in Figure 3 may be caused by one or both of two factors: i.e., mitigation of enzyme inactivation in the presence of sorbent materials, and/or cross-coupling of phenol to the sorbent materials. We previously demonstrated that the HRP inactivation effected by phenoxy radical attack is mitigated significantly in the presence of lignin (24), but no such enzyme inactivation mitigation occurs with the relatively hydrophobic PMS. Therefore, in the case

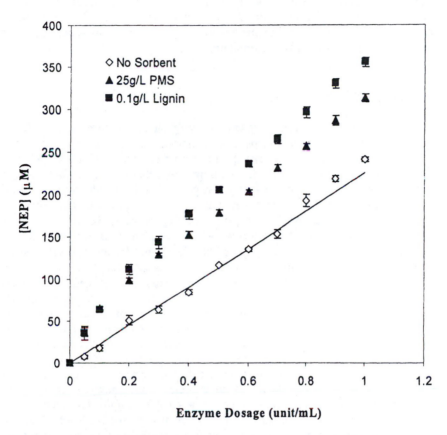

Figure 3. Extents of NEP formation in systems having varied HRP dosages and containing 0.1g/L lignin, 25 g/L PMS, and no sorbent. Initial phenol concentration = 500 μM, initial H_2O_2 concentration = 2 mM, reaction time = 2 h. Data points are the means of triplicate experiments with 1 SD error bars. The solid line represents a prediction using eq 6 and the rate coefficient values of k'_{in} = 1.6 mL/unit-min and k'_s = 361.4 μM-mL²/min-unit².

of PMS, cross-coupling between phenol and the sorbent material is believed to be responsible for the entire NEP enhancement observed. For the lignin-containing system, cross-coupling may also contribute to some extent to increased NEP production in addition to the contribution from HRP-inactivation mitigation. A further modeling analysis on the basis of the conceptual model we advanced and illustrate in Figure 2 provides a quantitative assessment of the relative contribution of enzyme inactivation mitigation and cross-coupling to the observed enhancement of NEP formation.

Modeling of NEP Formation

In HRP-mediated reaction systems lacking a reactive sorbent, NEP forms primarily via the self-coupling pathway illustrated in Figure 2; i.e., phenoxy radicals generated from the enzymatic reaction step couple with each other to form polyphenol type products. Coupling products that remain soluble undergo further coupling until larger polymers that precipitate from solution are formed (7, 9). Both phenol and dissolved phenol-coupling products thus serve as substrates from which NEP is formed. In this process, dissolved products precede formation of precipitate forms. As the concentrations of such dissolved product species increase, they compete with the phenol for enzyme active sites, thus reducing their own formation rates and increasing rates of their conversion to precipitates. When the formation and conversion rates of dissolved species approach each other, a pseudo-steady-state ratio of dissolved species to phenol is obtained. This ratio is unique to the intrinsic rate constants involved and to the associated enzyme affinities. As an approximation based on assuming that this condition is established quickly in the very early stages of the reaction, this pseudo-steady state can be applied to the whole reaction process. Because an essentially constant pattern of substrate distribution occurs at this pseudo-steady state, this approximation allows us in associated rate analyses to treat all dissolved substrate species collectively on a weight combination basis. This pseudo-steady-state approximation is fundamentally equivalent to one that has generally been adopted in phenol conversion rate analyses, whereby constant competition from the dissolved coupling-products with phenol for the enzymatic reactions is assumed and a fixed stochiometric ratio between phenol and hydrogen peroxide is employed for analysis of the entire phenol-coupling reaction process (7, 9).

Radical-radical coupling is a second-order process by nature, thus the rate expression for NEP formation via radical self-coupling reactions can be written

$$\frac{d[NEP]_S}{dt} = k_S [AH\bullet]^2 \qquad (3)$$

where k_S represents the self-coupling reaction rate constant, and $[NEP]_S$ the concentration of non-extractable products formed. Substitution of eq 1 into eq 3 leads to

$$\frac{d[NEP]_S}{dt} = k'_S[E]^2 \tag{4}$$

Equation 4 indicates that the NEP formation via self-coupling reaction pathways is second-order dependent on the active enzyme concentration present in the system. It is generally known that HRP inactivates during phenol-coupling processes via one or more of three possible mechanisms, i.e., (i) attack by phenoxy radicals, (ii) side-effects caused by excess H_2O_2, and (iii) sorption/occlusion of HRP by precipitated products formed in the coupling reactions. As explained earlier in the experimental section, our experimental conditions were selected at a region where H_2O_2–related side effects are insignificant. The third HRP-inactivation mechanism is also believed to be insignificant because the amounts of precipitated product formed under the reaction conditions employed were of the milligram magnitude, and negligible HRP sorption/occlusion by this relatively low quantity of precipitate was verified in our study (24). We therefore concluded that the HRP inactivation observed in this study was caused principally by radical attack, and can be described in terms of the second-order rate expression given in eq 2.

Division of eq 4 by eq 2 yields

$$\frac{d[NEP]_S}{d[E]} = -\frac{k'_S}{k'_{in}} \tag{5}$$

which defines the linear relationship between the rate of NEP formation by self-coupling and that of enzyme consumption, which are both second-order dependent on the active enzyme concentration. A linear type relationship between reaction outcome and enzyme consumption has in fact been widely observed, and accepted as an "empirical rule," in earlier studies of peroxidase-mediated reaction systems not involving cross-coupling processes (7, 9). Equation 5 now provides a mechanistic explanation for such observations. We noted in our studies that HRP inactivation occurred relatively quickly, so that enzyme inactivation was complete within the two-hour reaction period of our experiments. For this condition, integration of eq 5 over the reaction period gives

$$[NEP]_S = \frac{k'_S}{k'_{in}}[E]_0 \tag{6}$$

In eq 6, k'_{in} and k'_S are the two rate constants defined respectively by eqs 2 and 3, which values (k'_{in} = **1.6** mL/unit-min, k'_S = **361.4** μM-mL^2/min-unit^2) have been determined by appropriately designed rate experiments in our previous study (*17, 24*) employing systems similar to the current study. As demonstrated in Figure 3, eq 6 provides a good description of the NEP formation extents observed in this study with systems of varied HRP dosages. The linear equation $[NEP]_S$ = **225.9**$[E]_0$ for the line in Figure 3 does not represent a linear regression of that data, but rather an independent prediction of eq 6 using the values of k'_{in} = **1.6** mL/unit-min and k'_S = **361.4** μM-mL^2/min-unit^2 determined previously from separate rate experiments.

In HRP-mediated reaction systems containing a reactive sorbent, NEP is formed by phenoxy radicals via two pathways according to the process relationships schematicized in Figure 2: (i) self-coupling, and (ii) cross-coupling to sorbent materials. Total NEP formation under these conditions is given by

$$[NEP]_T = [NEP]_S + [NEP]_C \qquad (7)$$

where $[NEP]_T$ is the concentration of total NEP, and $[NEP]_S$ and $[NEP]_C$ the respective contributions of self-coupling and cross-coupling. NEP formation by self-coupling is a radical-radical combination process, thus second-order in nature. Conversely, cross-coupling is a process in which phenoxy radicals attack and subsequently attach to the reactive sorbent, manifesting a rate behavior that can appropriately be described as having first-order dependence on the radical concentration, i.e.,

$$\frac{d[NEP]_C}{dt} = k_c[AH\bullet] = k_c^*[S][AH\bullet] \qquad (8)$$

where k_c in the first right-hand term is a cross-coupling rate coefficient dependent on both sorbent type and sorbent concentration. By introducing the term $[S]$ for the sorbent concentration in the second right-hand term of the equation, we make the cross-coupling rate coefficient k_c^* specific only to the sorbent type. If rates of self- and cross-coupling reactions of phenoxy radicals can be expected to have different orders of dependence on radical concentration, it is logical to expect also that the two NEP formation pathways differ significantly in reaction behavior.

We demonstrated earlier that NEP formation by self-coupling is linearly correlated with initial enzyme concentration as long as sufficient substrates are supplied and the enzyme activity is depleted by the reaction (eq 6). The relationship between cross-coupling and enzyme consumption can be analyzed in a manner similar to that employed for self-coupling. We first substituted the radical concentration in eq 8 above with the proportionality relationship (eq 1) and divide by the rate equation for enzyme inactivation (eq 2). This exercise yields

$$\frac{d[NEP]_C}{d[E]} = -\frac{k_c^{**}[S]}{k_{in}'} \cdot \frac{1}{[E]} \tag{9}$$

where $k_c^{**} = \dfrac{k_c^* k_E}{k_r}$

is a composite rate coefficient relating to the cross-coupling reactivity of a sorbent. Integration of eq 9 leads to

$$[NEP]_C = \frac{k_c^{**}[S]}{k_{in}'} \cdot \ln\frac{[E]_0}{[E]_t} \tag{10}$$

in which $[E]_t$ is the residual enzyme activity at the end of the reaction. Integration of eq 2 in turn gives

$$\frac{[E]_0 - [E]_t}{[E]_t} = k_{in}' \cdot t \cdot E_0 \tag{11}$$

Assuming that enzyme activity is nearly completely depleted at the end of the two-hour reaction period (i.e., $[E]_t$ is negligible compared to $[E]_0$) and taking the natural logarithms of both sides of eq 11 yields an approximation for $\ln\dfrac{[E]_0}{[E]_t}$, allowing eq 10 to be rewritten in terms of $[E]_0$ alone as

$$[NEP]_C = \frac{k_c^{**}[S]}{k_{in}'} \cdot \ln[E]_0 + \frac{k_c^{**}[S]}{k_{in}'}\ln(k_{in}' \cdot t) \tag{12}$$

Equation 12 indicates that NEP formation via cross-coupling reactions, $[NEP]_C$, correlates with the logarithm of $[E]_0$.

Equations 7, 6 and 12 together provide a complete description of the quantitative relationships involved in HRP-mediated phenol coupling in systems containing reactive sorbents. As stated in eq 7, the total NEP formation

($[NEP]_T$) comprises the contributions of self-coupling and cross-coupling reactions, $[NEP]_S$ and $[NEP]_C$, respectively. Because the rates of self- and cross-coupling reactions have different orders of dependence on phenoxy-radical concentration, they differ significantly in their respective behaviors. Equation 6 indicates that $[NEP]_S$ correlates linearly with $[E]_0$; whereas eq 12 shows that $[NEP]_C$ is related to the logarithm of $[E]_0$. This difference in reaction behaviors provides a quantitative tool for differentiating between the NEP contributions of the two reaction mechanisms, and we employed this tool in the following data analyses and interpretations.

Figure 4 presents data for $[NEP]_T$ formed in systems containing 25-g/L PMS (A) and 0.1-g/L lignin (B) at varied enzyme dosages. The contribution of self-coupling, $[NEP]_S$, can be predicted with eq 6 as shown in Figure 4. For these predictions, the values of rate constants k'_S and k'_{in} that were determined in our earlier rate studies (24) were employed. The value of $k'_S = 361.4$ μM-mL^2/min-unit^2 was determined in a sorbent-free system (24), and this value was used in the predictions shown in Figure 4 for systems containing both lignin and PMS. Inherent to this practice is an assumption that the presence of lignin and PMS did not significantly interfere with the radical self-coupling reactions. The effects of this assumption on the prediction ability of our model will be discussed in depth later in this paper. The k'_{in} values of 1.54 and 1.37 mL/unit-min used in the predictions for the systems containing 25-g/L PMS and 0.1-g/L lignin, respectively, were, however, determined in our earlier rate studies employing systems containing these sorbents (24).

According to eq 7, the difference between $[NEP]_T$ and $[NEP]_S$ is the contribution of cross-coupling, i.e., $[NEP]_C$. As illustrated in Figure 4, the $[NEP]_C$ contribution, as estimated by $[NEP]_T - [NEP]_S$ values for the discrete $[E]_0$ corresponding to those used to collect the $[NEP]_T$ data, are simulated reasonably well by eq 12. The experimental values of $t = 120$ min, $[S] = 25$ and 0.1 g/L for PMS and lignin respectively, and k'_{in} values as specified above were used in the simulations presented in Figure 4. Thus, the cross-coupling rate coefficient, k_c^{**}, is the only parameter obtained by data fitting in this simulation. Corresponding k_c^{**} values for PMS and lignin are listed in Table I.

Predictions of $[NEP]_T$ formation in systems containing sorbents at any sorbent/water ratios can then be made using eqs 7, 6, and 12 and k_c^{**} values listed in Table I; i.e., $[NEP]_C$ is calculated from eq 12, and $[NEP]_S$ from eq 6, and $[NEP]_T$ is a linear combination of these two (eq 7). Predictions thus obtained using k'_S value referenced above are compared in Figure 5 with experimental results acquired for systems containing PMS and lignin at varying sorbent/water ratios and a constant enzyme dosage at 0.5 unit/mL. The k'_{in} values that were previously determined for lignin-containing systems (24) were employed in the prediction, but a value of $k'_{in} = 1.6$ mL/unit-min, determined for the sorbent-free system, was employed in these predictions for PMS to reflect the absence of any significant effect of PMS on HRP inactivation.

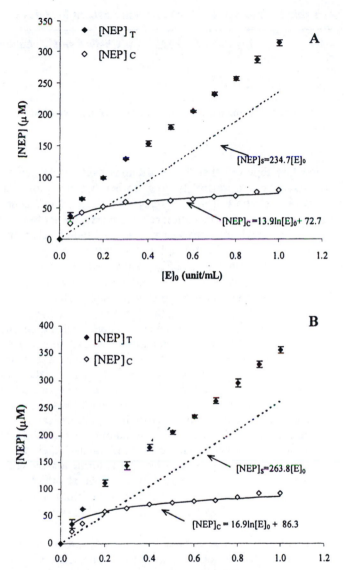

Figure 4. NEP formations at varied HRP dosages in a system containing 25-g/L PMS (A) and 0.1-g/L lignin (B). Initial phenol concentration = 500 µM, initial H_2O_2 concentration = 2 mM, reaction time = 2 h. Data points are the means of triplicate experiments with 1 SD error bars. Dashed lines represent predictions of eq 6, and solid lines simulations of eq 12.

Table I. Values of k_c^{**} for Different Sorbent Materials

k_c^{**}	Lignin	PMS	Chelsea soil	Lachine shale
(μmol·mL/unit·min·g)	231.8	0.86	11.8^a	5.0^a

a. k_c^{**} was calculated via normalizing k'_c to organic matter concentration contained in the geosorbent materials (17).

It is evident in Figure 5 that the predictions capture the behavior of the experimental data well at relatively low sorbent/water ratios, but deviate significantly at higher ratio values. This suggests that certain mechanisms not taken into account in the above modeling relationships may act to suppress NEP formation at relatively high sorbent/water ratios. One such mechanism might be the competitive suppression of self-coupling by cross-coupling. We calculated contributions of self-coupling $[NEP]_S$ using eq 6 and a value of $k'_S = 361.4$ μM-mL^2/min-unit^2 determined independently from a sorbent-free system. Thus an assumption inherent to this particular prediction is that self-coupling is not affected by cross-coupling reactions. In fact, self-coupling creates NEPs via a process involving repeated radical generation and radical-radical combination steps (7, 9). The radical may in fact be scavenged at any stage of such polymer-growth processes by a reactive sorbent, thus terminating further growth. Under such conditions, polymerized phenol products formed via self-coupling in the presence of reactive sorbents can be expected to generally have smaller sizes than those formed in sorbent-free systems. This is comparable to the kinetic-chain-length effect in a polymerization process via radical propagation (29, 30); i.e., with the addition of radical termination reagents sizes of polymer products tend to become smaller. Such a product size reduction effect would logically lead to less NEP formation by self-coupling. Perhaps this suppression effect is negligible or minor at lower sorbent/water ratios or lower cross-coupling reactivities, but becomes prominent when sorbent/water ratios or reactivities increase, giving the behavior shown in Figure 5.

Besides the aforementioned competitive suppression of self-coupling by cross-coupling, lignin may introduce additional suppression effects on NEP-formation by competitive inhibition of enzyme catalytic efficiency. Lignin contains phenolic functionalities in its monomeric structure, and these may serve as additional substrates for the HRP enzyme despite some steric hindrances. Such effects would lead to apparent reduction of enzyme catalytic efficiency, particularly at high lignin concentrations.

Our modeling analyses revealed that k_c^{**} values for lignin are two orders of magnitude greater than those for PMS (Table I), indicating a much greater cross-

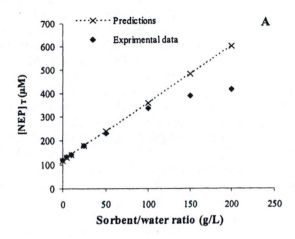

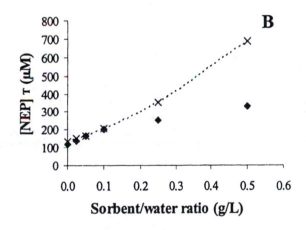

Figure 5. NEP formations at a constant HRP dosage (0.5 unit/mL) and varied sorbent/water ratios of PMS (A) and lignin (B). Initial phenol concentration = 500 μM, initial H₂O₂ concentration = 2 mM, reaction time = 2 h. Dashed lines represent predictions of [NEP]ₜ by eq 1, in which [NEP]𝒸 was calculated from eq 12 and [NEP]ₛ from eq 6.

coupling reactivity of lignin. Lignin and PMS, in terms of their chemical compositions and characteristics, represent a diagenetically "young" humic substance and a chemically reduced "old" kerogen type material, respectively. These two materials contain different types of substituent groups on their respective monomeric aromatic structures as shown in Figure 1, and it is known that aromatic substituents can significantly influence the reactivity of a chemical through electron induction or resonance effects. The several oxygen-containing functionalities on the monomeric aromatic structure of lignin may be responsible for enhancing the cross-coupling reactivity of the material.

The mechanistic interpretations of the effects of the chemical characteristics of sorbent materials on the enzymatic coupling of phenol observed in this study are consistent in terms of explaining the results we obtained earlier for different natural geosorbents (17). Those natural materials included Chelsea soil, a geologically young soil containing predominantly humic-type SOM rich in oxygen-containing functionalities, and Lachine shale, a diagenetically older material having predominantly kerogen-type SOM of relatively reduced chemical structure. The Chelsea soil was observed to have a greater cross-coupling reactivity than Lachine shale. In terms of the k_c^{**} values listed in Table 1 normalized to total organic matter content, the cross-coupling reactivity observed in the earlier studies and those reported here decreases in the order lignin > Chelsea soil > Lachine shale > PMS.

Acknowledgments

We thank Deborah Ann Ross, Carl W. Lenker and Tighe B. Herren for their diligent assistance in the experimental phases of this work. The research was supported by Grant No. DE-FG07-96ER14719 from the U.S. Department of Energy Environmental Management Science Program. The opinions, findings, conclusions, or recommendations expressed herein are those of the authors, and, do not necessarily reflect the views of DOE.

References

1. Bollag, J-M. *Environ. Sci. Technol.* **1992**, *26*, 1876–1881.
2. Bollag, J-M. *Met. Ions Biol. Syst.* **1992**, *28*, 205–217.
3. Nannipieri, P.; Bollag, J-M. *J. Environ. Qual.* **1991**, *20*, 510–517.
4. Job, D.; Dunford, H. B. *Eur. J. Biochem.* **1976**, *66*, 607–614.
5. Dawson, J. H. *Science*, **1988**, *240*, 433–439.
6. Dunford, H. B. In *Peroxidase in Chemistry and Biology*; Everse, J., Everse, K. E., Grisham, H. B., Eds.; CRC press: Ann Arbor, MI, 1990; Vol. II, pp 2–24.

7. Nicell, J. A. *J. Chem. Technol. Biotechnol.* **1994**, *60*, 203–215.

8. Choi, Y-J.; Chae, H. J.; Kim, E. Y. *J. Biosci. Bioeng.* **1999**, *88*, 368–373.

9. Wu, Y.; Taylor, K. E.; Biswas, N.; Bewtra, J. K. *J. Environ. Eng.* **1999**, *125*, 451–458.

10. Klibanov, A. M.; Tu, T. M.; Scott, K. P. *Science*, **1983**, *221*, 259–261.

11. Maloney, S. W.; Manem, J.; Mallevialle, J.; Fiessinger, F. *Environ. Sci. Technol.* **1986**, *20*, 249–253.

12. Nakamoto, S.; Machida, N. *Water Res.* **1992**, *26*, 49–54.

13. Aitken M. D.; Massey I. J.; Chen T. P.; Heck P. E. *Water Res.* **1994**, *28*, 1879–1889.

14. Nicell, J. A.; Bewtra, J. K.; Biswas, N.; Taylor, K. E. *Water Res.* **1993**, *27*, 1629–1639.

15. Cooper, V. A.; Nicell, J. A. *Water Res.* **1996**, *30*, 954–964.

16. Yu, J.; Taylor, K. E.; Zou, H.; Biswas, N.; Bewtra, J. K. *Environ. Sci. Technol.* **1994**, *28*, 2154–2160.

17. Huang, Q.; Selig, H.; Weber, W. J., Jr. *Environ. Sci. Technol.* **2002**, *36*, 596–602.

18. Hatcher, P. G.; Bortiatynski, J. M.; Minard, R. D.; Dec, J.; Bollag, J-M. *Environ. Sci. Technol.* **1993**, *27*, 2096–2103.

19. Roper, J. C.; Sarkar, J. M.; Dec, J.; and Bollag, J-M. *Water Res.* **1995**, *29*, 2720–2724.

20. Park, J-W.; Dec, J.; Kim, J-E.; Bollag, J-M. *Environ. Sci. Technol.* **1999**, *33*, 2028–2034.

21. Kim, J-E.; Wang, C. J.; Bollag, J-M. *Biodegradation* **1998**, *8*, 387–392.

22. Dec, J.; Bollag, J. M. *J. Environ. Qual.* **2000**, *29*, 665–676.

23. Huang, Q.; Selig, H.; Weber, W. J., Jr. *Environ. Sci. Technol.* **2002**, *36*, 4199–4200.

24. Huang, Q. *A Mechanistic Study of Peroxidase-Catalyzed Phenol Coupling in Water/Soil/Sediment Systems.* Ph.D. Dissertation, University of Michigan, Ann Arbor, 2002.

25. Buchanan, I. D.; Nicell, J. A. *Biotechnol. Bioeng.* **1997**, *54*, 251–261.

26. Arnao, M. B.; Acosta, M.; Del Rio, J. A.; Garcia-Canovas, F. *Biochim. Biophys. Acta* **1990**, *1038*, 85–89.

27. Putter, J.; Becker, R. In *Methods of Enzymatic Analysis,* 3rd ed.; Bergmeyer, H. U., Bergmeyer, J., Grassl, M., Eds.; Verlag Chemie: Weinheim, FL, 1983; Vol. 3, pp 286–293.

28. Taraban, M. B.; Leshina, T. V.; Anderson, M. A.; Grissom, C. B. *J. Am. Chem. Soc.* **1997**, *119*, 5768–5769.

29. Bevington, J. C. *Radical Polymerization*; Academic Press: New York, 1961.

30. Lazar, M.; Rychly, J.; Klimo, V.; Pelikan, P.; Valko, L. *Free Radicals in Chemistry and Biology*; CRC Press: FL, 1989.

Chapter 4

Using Phosphate to Control the Mn Oxide Precipitation During In Situ Chemical Oxidation of Chlorinated Ethylenes by Permanganate

X. David Li and Franklin W. Schwartz

Department of Geological Sciences, The Ohio State University, Columbus, OH 43210

In situ chemical oxidation (ISCO) with permanganate (MnO_4^-) has been developed recently as a technique for cleaning up aquifers contaminated by chlorinated organic solvents. As MnO_4^- oxidizes the contaminant in the subsurface, the reaction produces Mn oxide, a solid precipitate, which causes pore plugging and lowers the efficiency of the remediation scheme. This study utilized batch experiments to explore the feasibility of applying phosphate to slow the rate of formation of colloidal Mn oxide. Studies of the mineral structure and chemical composition of the Mn oxide determined that the mineral is a semi-amorphous potassium-rich birnessite. Its pzc (point of zero charge) was measured at 3.7 ± 0.4. UV-vis spectrophotometry and photon correlation spectroscopy were used to monitor the process of colloid formation as oxidation is occurring. The results show that phosphate can slow the formation of the colloids by tens of minutes. The process is thought to involve the formation of a phosphate-Mn(IV) complex in the aqueous phase before formation of the colloids. The results from this study provide an opportunity to control the distribution of Mn oxide precipitation and may help to achieve a complete site clean-up.

Introduction

Oxidation schemes using permanganate (MnO_4^-) for the remediation of subsurface contamination by chlorinated ethylenes have been developed and tested at various sites around the nation (*1-4*). These studies have shown that MnO_4^- is a very promising technique and capable of quickly destroying contaminants. MnO_4^- oxidizes common chlorinated organic solvents and produces nontoxic final products. In the case of trichloroethylene (TCE), the oxidation reaction can be written as

$$C_2HCl_3(TCE) + 2MnO_4^- \rightarrow 2MnO_{2(s)} + 2CO_2 + 3Cl^- + H^+ \qquad (1)$$

However, studies (*4-7*) have revealed that the solid product of eq 1, MnO_2, precipitates, particularly in zones with large dense nonaqueous phase liquid (DNAPL) saturations. Plugging of the pore space increases the heterogeneity of the medium. One result is a dramatic lowering of the flushing efficiency, and the DNAPL removal rate, especially later in the treatment period. The decrease in flushing efficiency due to the permeability changes is obviously a concern in the practical implementations of this technology.

To date, there have been few studies on the mineral structure and the environment fate of the Mn oxide precipitates that form within in situ chemical oxidation (ISCO) schemes with MnO_4^-. Identification of the structure of Mn oxide mineral, together with the determination of the chemical composition, is necessary to lay the foundation for further studies on how Mn oxide precipitation can be controlled. There are more than 30 different types of Mn oxide/hydroxide minerals with different structures and compositions.

Generally, there are two major structural forms for these minerals: chain or tunnel structures, and layer structures. All of these forms are comprised of MnO_6 octahedras. Water molecules and/or other cations (*8*) are often present at various sites in the structures. Mn oxides having a chain or tunnel structure include pyrolusite, ramsdellite, hollandite, romanechite, and todorokite. Typical structures for the chain or tunnel type Mn oxide mineral are presented in Figure 1. Lithiophorite, chalcophanite, and birnessite are examples of Mn oxide minerals having a layer structure. Typical structural maps are shown in Figure 2.

Identifying the particular Mn oxide mineral is not straightforward because the samples are usually not crystallized well enough for single-crystal diffraction studies, especially with synthetic oxides of the type formed here. In this study, we will apply powder x-ray diffraction and transmission electron microscopy (TEM), along with chemical analysis, to identify the type of structure and to determine the properties of the Mn oxide.

Remediation of DNAPL problems using MnO_4^- is facilitated when precipitation of Mn oxide occurs away from zones with large DNAPL

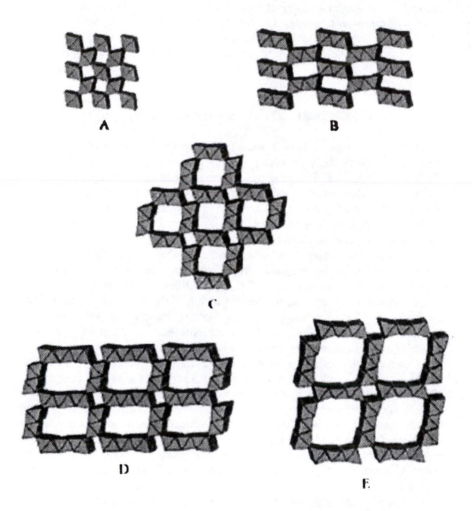

Figure 1. Polyhedral representations of crystal structures of Mn oxide minerals with chain or tunnel structures. (A) Pyrolusite. (B) Ramsdellite. (C) Hollandite. (D) Romanechite. (E) Todorokite. Adapted from (8).

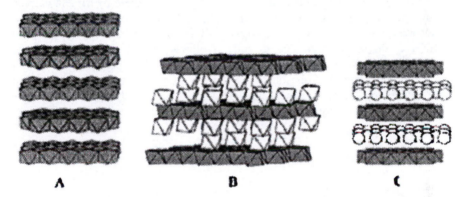

Figure 2. Polyhedral representations of crystal structures of Mn oxide minerals with layer structures. (A) Lithiophorite. (B) Chalcophanite. (C) Na-rich birnessite-like phase showing disordered H_2O /Na (light color ball) sandwiched between the Mn octahedral sheets. Adapted from (8).

saturations. One possible strategy for dealing with this plugging problem is to find ways of slowing the formation of the precipitates. The Mn oxide produced during the oxidation reaction forms colloids before more visible precipitates are evident. Coagulation of colloids leads to pore plugging. If the process of colloid coagulation can be controlled or delayed, Mn oxide colloids could be transported farther away from the treatment zone, to avoid local plugging in the DNAPL zone.

A colloid solution can be stabilized if the surface charge of the particles is increased. Changes to the surface charge of colloidal particles can be created by adding constituents with a large charge, which can sorb onto the colloids. Phosphate ions are ideal in this respect with a large charge and the tendency to bind to metal oxide colloids. Yao et al. (9) showed that manganese dioxide (MnO_2) is an important adsorbent for phosphate in natural waters. Mata-Perez and Perez-Benito (10) studied the reduction of permanganate by trimethylamine using UV-vis spectrophotometer analysis. They found that in aqueous phosphate buffers, the rate of conversion of the soluble Mn (IV) species into MnO_2 could be retarded due to surface effects related to the presence of phosphate. Stewart (11) also described the rather slow precipitation of Mn oxide in the presence of phosphate ion. In this study, we will try to understand the interaction between phosphate and Mn oxide colloids and to elucidate possible controls on colloid formation by phosphate ion.

Methods

A single batch of Mn oxide was prepared and used in all experiments. Preparation of the Mn oxide solid is documented in detail by Li (12). Briefly, saturated aqueous solution of TCE was mixed slowly with 30 g/L of $KMnO_4$ in a large flask. The solution was agitated with a magnetic stir bar throughout. Excess $KMnO_4$ was indicated by the purple color of the final solution. Precipitates formed in the solution were filtered and dialyzed to remove impurities. The dialysis bag (Spectral/Por Cellulose Ester Membrane, 3500 Dalton) containing the Mn oxide precipitates was placed in a glass column containing Milli-Q water, which was changed daily. The progress of the dialysis was monitored by measuring the specific conductance of the water in the glass column with a conductivity dip cell (K = 10.2) connected to a digital conductivity meter (1481-90 model, Cole-Parmer Instrumental Company). After dialysis, the solid sample was quick frozen in liquid nitrogen to remove the water (ice) from the sample.

The chemical composition of the Mn oxide was determined by thermal and inductively coupled plasma mass spectrometer (ICP-MS) analysis. A SEIKO SSC 5020 instrument (Haake Inst., Paramus, NJ) with a TG/DTA simultaneous analysis module, model 200 was used for thermal gravimetric (TG) and differentiated thermal analysis (DTA). An aliquot of Mn oxide powder sample was heated in the analyzer from 25 °C to 930.4 °C. A loss in weight of the sample at temperatures above 125 °C can be attributed to the loss of structural water. Dissolved Mn oxide was analyzed with ICP-MS to determine the concentration of manganese and potassium within the Mn oxide. A Philips CM-200 TEM (transmission electron microscope) was used to examine the fine structure of the mineral directly. X-ray diffraction powder analysis was performed with a Philips diffractometer (Philips Elec. Inst., Mahwah, NJ) using CuKα-radiation and vertical, wide-range goniometer equipped with a diffracted beam monochromator and a theta-compensating slit. The step scanning involved a counting time of 4 seconds with 2θ measurements from 5° to 70° in increments of 0.05°.

Specific surface area of the Mn oxide powder was measured using a liquid nitrogen adsorption (BET) method using Micromeritics FlowSorb II 2300 surface area analyzer. The point of zero charge (pzc) of the Mn oxide was determined using the method described by Sigg and Stumm (13), which is based on the following equation:

$$\sigma_0 = (C_A - C_B - [H^+] + [OH^-]) \frac{F}{a \bullet S} \qquad (2)$$

where F is the Faraday constant (C mol^{-1}), and S is the specific surface area (m^2 g^{-1}) of the solid. C_A and C_B are the concentrations of the strong acid and base (mol L^{-1}) added, respectively, and a is the concentration of solid in the suspension (g/L).

Experiments were conducted to examine the effect of the phosphate buffer on delaying colloid coagulation, by quantifying colloid growth as a function of time. A Varian Cary 1 UV-vis spectrophotometer was used for this purpose with continuous scanning between wavelengths of 350 nm and 700 nm. The experiments were run in the capped cuvette inside the spectrophotometer at 22 °C. The reactants were KMnO$_4$ and trichloroethylene (TCE), with the latter in excess. The concentrations of KMnO$_4$ and TCE were 0.158 mM and 3.8 mM for all the experiments. A buffer, comprised of equal quantities of KH$_2$PO$_4$ and K$_2$HPO$_4$, provided the necessary phosphate. The concentration of this buffer mixture was varied in each experiment. One experiment with no phosphate added provided a baseline for comparison. KCl was added as necessary to maintain all solutions at the same ionic strength. All chemicals were obtained from Fisher Scientific, and used as received. Details of the experiments are listed in Table I.

Table I. Experimental Conditions for Colloid Growth Involving the UV-Vis Spectrophotometer

Run	[KMnO₄] mM	[TCE] mM	Buffer (M) [KH₂PO₄] = [K₂HPO₄]	Ionic Strength (KCl) M	pH
1	0.158	3.8	0	0.23	varies
2	0.158	3.8	0.02	0.23	6.81
3	0.158	3.8	0.03	0.23	6.81
4	0.158	3.8	0.04	0.23	6.81
5	0.158	3.8	0.05	0.23	6.81

Colloid growth under different conditions was monitored using photon correlation spectroscopy, which measures the size of submicron particles. The size of colloid particles during the growth process was measured with 90 Plus Particle Sizer (Brookhaven Instrument Co., Holtsville, NY). The reactions were conducted inside the cuvettes supplied with the instrument. The experiments were run under conditions listed in Table II. Measurements were made automatically by the instruments at pre-set times. KCl was added to the reaction solution to provide the same ionic strength for all runs.

Table II. Experimental Conditions for Monitoring Colloid Particle Size Growth and Surface Charge with Photon Correlation Spectroscopy and Zeta-Plus

Runs	[KMnO₄] (mM)	[TCE] (mM)	[KH₂PO₄] (M)	[K₂HPO₄] (M)	[K₃PO₄] (M)	I (KCl) (M)
1	0.63	6.09	0	0	0	0
2	0.63	6.09	0	0	0	0.24
3	0.63	6.09	0.02	0.02	0	0.24
4	0.63	6.09	0	0	0.04	0.24

Results

TEM analysis found that the Mn oxide crystal is a cluster of small needles (Figure 3). The findings from TEM analysis indicated that the crystal structure

Figure 3. Results of the TEM analysis for the Mn oxide powder. The white fibrous material in the picture is a cluster of small needle-shaped crystals of the mineral. The image shows that crystals of the Mn oxide are not well developed.

of Mn oxide was poorly developed. TEM chemical composition analysis also identified K and Mn as the only major metal cations in the mineral.

The X-ray diffraction (XRD) spectrum (Figure 4) for the synthetic Mn oxide has no prominent peak. However, the pattern matched birnessite in the database with a characteristic small peak at 7.3 Å. Among all the Mn oxide minerals, the birnessite family has layer spacings of approximately 7 Å (*5, 14*) with Na, Ca, or K as the interlayer elements. Because the mineral is formed in a potassium rich environment, the mineral formed in our study is semi-amorphous potassium-rich birnessite. The basic structural unit for birnessite is a sheet of MnO_6 octahedra (Crystal structure C in Figure 2). The interlayer cations and water molecules generally are known to occupy different positions inside the mineral.

Thermal analysis showed that structural water in the mineral was 12.49% by weight of the total solid. The combination of ICP-MS analysis of the dissolved Mn oxide and a charge balance of the different elements provides a formula for the Mn oxide mineral as $K_{0.854}Mn_{1.786}O_4 \cdot 1.55H_2O$. This interpretation assumes that the oxidation state for all Mn inside the mineral is IV. As compared with other results (*14*), our interpretation seems reasonable. Four independent measurements using the BET method yielded a specific surface area for Mn oxide of 23.6 ± 0.82 m²/g. Potentiometric titrations of Mn oxide solid found the pzc for Mn oxide to be 3.7 ± 0.4, which is comparable to the range reported (2.0 to 4.5) for different forms of MnO_2 (*15*).

Spectrometric studies of the colloidal growth with and without phosphate show clear differences (compare Figure 5 and 6). In TCE oxidation by MnO_4^- without phosphate, colloids formed immediately, as indicated by an elevated absorbance across all the wavelengths scanned (Figure 5). After reaching a maximum, the spectrum declined to a near zero position. This pattern indicated that solid Mn oxide formed and precipitated out of solution. With the addition of phosphate, the absorbance spectrum shows a gradual decrease in the MnO_4^- peak until almost all the MnO_4^- was consumed. The spectrum also shows an increase in absorbance at all wavelengths, suggesting the formation of colloids, however at a later time and slower pace.

The results of colloid size measurements from photon correlation spectroscopy are shown in Figure 7. A distinctive feature of the data on this figure is that the colloid growth rate under conditions where no salts were added to the background solution is much smaller than that evident with the addition of either KCl or phosphate ions. This result indicates that ionic strength influences the colloid growth rate greatly. As expected, the start of colloid growth is delayed 35 minutes by the addition of phosphate. The results show that the presence of phosphate ions can slow the coagulation process, especially at early times. In addition, increasing of the ionic strength also increases the rate of colloid growth.

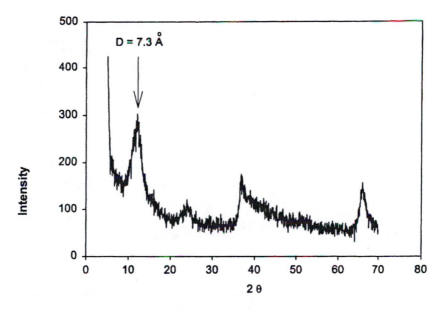

Figure 4. Spectrum obtained from the X-ray diffraction analysis of synthetic Mn oxide. Peak at 7.3 Å indicates the interlayer spacing of the mineral.

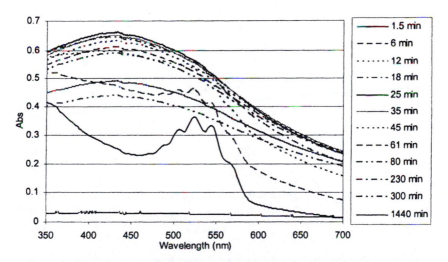

Figure 5. Sequential scanning of the reaction from 350 nm to 700 nm during the MnO_4^- reaction with TCE at conditions [$KMnO_4$] = 0.158 mM, [TCE] = 3.8 mM, I = 0.23 M(KCl), no phosphate added.

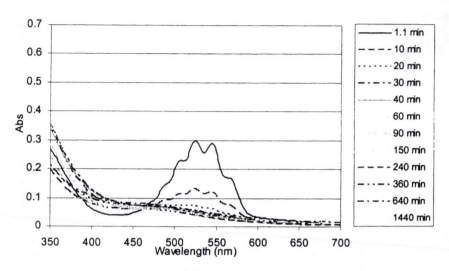

Figure 6. Sequential scanning of the reaction from 350 nm to 700 nm during the
MnO₄⁻ reaction with TCE at conditions [KMnO₄] = 0.158 mM, [TCE] =
3.8 mM, I = 0.23 M(KCl), [KH₂PO₄] = [K₂HPO₄] = 0.05 M.

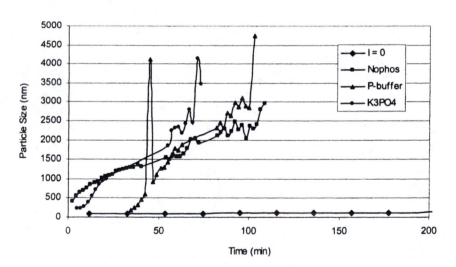

Figure 7. Result of the colloidal particle growth with time as monitored by
photon correlation spectroscopy.

Discussion

As mentioned earlier, the stability of a colloidal solution depends strongly on the surface charges of the particles. The pH of the solution in particular can influence the surface charge of the colloids. During the MnO_4^- treatment of TCE, the release of H^+ from the reaction (eq 1) reduces the pH of the solution from neutral to acidic in an unbuffered solution. Occasionally, the pH falls below 2, as observed in some experiments. Because the pzc of the Mn oxide is 3.7 ± 0.4, this change in pH enhances the coagulation of colloids. For pHs close to the pzc, the net charge on the surface of the particles approaches zero. The mutual repulsion of the colloids is reduced sufficiently to permit the inter-particle collisions, agglomeration, and subsequent sedimentation. Thus, the combination of H^+ production and a pzc for Mn oxide of 3.7 ± 0.4 enhances coagulation of the colloids and the precipitation of Mn oxide.

The presence of phosphate in solution can retard the onset of colloid precipitation. One possible explanation of this effect is the buffer capacity of phosphate. When phosphate is present in the oxidation reaction, it buffers the pH of solution away from the pzc of the Mn oxide. Therefore, surface charges of the colloid remain relatively high, which helps to slow the coagulation process.

Another explanation of the phosphate effect is possible phosphate-Mn(IV) interactions in aqueous phase. When chlorinated ethylenes are oxidized by MnO_4^-, soluble Mn(IV) forms before any colloids. The existence of the soluble Mn(IV) has been reported by many researchers (16, 17). Phosphate ion can react with soluble Mn(IV) species and reduce the formation of the colloid. The process is probably involved with the formation of a phosphate-Mn(IV) complex. As the complex forms, it keeps the Mn(IV) in the aqueous phase without forming colloids. Eventually, colloids and precipitates will be produced when the capacity of the phosphate effect has reached its limit. This mechanism is in agreement with our observation in the colloid growth experiments.

Conclusions

Experiments show that the Mn oxide formed in the in situ chemical oxidation with MnO_4^- is a disordered birnessite with the composition $K_{0.854}Mn_{1.786}O_4 \cdot 1.55H_2O$. Our measurement of the surface properties of this Mn oxide finds the pzc to be relatively low at 3.7 ± 0.4. This pzc may help to explain why precipitates form rapidly near the zone of high DNAPL saturation. Our studies indicate that phosphate ion can slow the formation of colloidal Mn oxide from TCE oxidation with MnO_4^-. The buffer capacity of phosphate likely contributes to this process. Another mechanism is that phosphate ion delays the formation of Mn oxide colloids by forming a soluble Mn(IV)-phosphate species.

Acknowledgments

This research is supported by the U.S. Department of Energy through the Environmental Management Science Program under contract No. FG07-00ER15115. The authors also want to thank Drs. Samuel J. Traina and Harold W. Walker for their kind help and useful suggestions.

References

1. Vella, P. A.; Veronda, B. In *Chemical Oxidation Technologies for the Nineties;* Echenfelder, W. W., Ed.; Technomic Publishing: Lancaster, PA, 1992.
2. West, O. R.; Cline, S. R.; Holden, W. L.; Gardner, F. G.; Schlosser, B. M.; Thate, J. E.; Pickering, D. A.; Houk, T. C. *A Full-Scale Demonstration of In Situ Chemical Oxidation through Recirculation at the X-701B Site*; ORNL/TM-13556; Oak Ridge National Laboratory: Oak Ridge, TN, 1997; p 101.
3. Yan, Y. E.; Schwartz, F. W. *J. Contam. Hydrol.* **1999**, *37*, 343–365.
4. Li, X. D.; Schwartz, F. W. In *Chemical Oxidation and Reactive Barriers: Remediation of Chlorinated and Recalcitrant Compounds;* Wickramanayake, G. B., Gavaskar, A. R., Chen, A. S. C., Eds.; Battelle Press: Columbus, OH, 2000; pp 41–47.
5. Li, X. D.; Schwartz, F. W. In *Innovative Strategies for the Remediation of Chlorinated Solvents and DNAPLs in the Subsurface;* Henry, S. M., Ed.; American Chemical Society Press: Washington DC, 2002, in press.
6. Reitsma, S., Marshall, M. In *Chemical Oxidation and Reactive Barriers: Remediation of Chlorinated and Recalcitrant Compounds;* Wickramanayake, G. B.; Gavaskar, A. R.; Chen, A. S. C., Eds.; Battelle Press: Columbus, OH, 2000; pp 25–33.
7. Thomson, N. R.; Hood, E. D.; MacKinnon, L. K. In *Chemical Oxidation and Reactive Barriers: Remediation of Chlorinated and Recalcitrant Compounds;* Wickramanayake, G. B.; Gavaskar, A. R.; Chen, A. S. C., Eds.; Battelle Press: Columbus, OH, 2000; pp 9–16.
8. Post, J. E., *Proc. Natl. Acad. Sci. U.S.A,* **1999**, *96*, 3447–3454.
9. Yao, W.; Millero, F. J. *Environ. Sci. Technol.* **1996**, *30*, 536–541.
10. Mata-Perez, F.; Perez-Benito, J. F. *Can. J. Chem.* **1985**, *63*, 988–992.

11. Stewart, R. Oxidation by Permanganate. *Oxidation in Organic Chemistry, Part A*; Wiberg, E. B., Ed.; Academic Press: New York and London, 1965; p 7.
12. Li, X. D. Ph.D. Dissertation, The Ohio State University, Columbus, OH, 2002.
13. Sigg, L.; Stumm, W. *Colloids Surf.* **1981**, *2*, 101–117.
14. Post, J. E.; Veblen, D. R. *Am. Mineral.* **1990**, *75*, 477–489.
15. Faure, G. *Principles and Applications of Geochemistry*, 2nd ed.; Prentice Hall Press: New York, 1998; p 219.
16. Simandi, L. I.; Jaky, M. *J. Am. Chem. Soc.* **1976**, *98*, 1995–1997.
17. Freeman, F.; Fuselier, C. O.; Armstead, C. R.; Daltion, C. E.; Davidson, P. A.; Karchefski, E. M.; Krochman, D. E.; Johnson, M. N.; Jones, N. K. *J. Am. Chem. Soc.* **1981**, *103*, 1154–1159.

Chapter 5

Strategies for the Engineered Phytoremediation of Mercury and Arsenic Pollution

Om Parkash Dhankher[1], Andrew C. P. Heaton[2], Yujing Li[2], and Richard B. Meagher[2]

[1]Department of Plant, Soil and Insect Sciences, University of Massachusetts, Amherst, MA 01003
[2]Department of Genetics, University of Georgia, Athens, GA 30602

We have developed genetics-based phytoremediation strategies for mercury and arsenic pollution. Plants engineered to use bacterial, animal, and plant genes take up and tolerate several times the levels of mercury and arsenic that would kill most plant species. Modified plants expressing the bacterial *merB* gene breakdown the most toxic and biomagnified methylmercury (MeHg) to ionic mercury (Hg(II)) and those expressing the *merA* gene detoxify ionic mercury to metallic mercury (Hg(0)). Using these and some other genes, mercury is stored below or above ground, or even volatilized as part of the transpiration process, keeping it out of the food chain. These remediation strategies also work in cultivated and wild plant species including canola, rice, yellow poplar, and cottonwood. For arsenic, we have developed a strategy in which the oxyanion arsenate is transported aboveground, electrochemically reduced to arsenite in leaves, and sequestered in thiol-rich peptide complexes. *Arabidopsis* plants expressing bacterial ArsC and γ-ECS, showed substantially greater resistance and accumulated more arsenic in shoot tissues than wild type plants. These arsenic and mercury phytoremediation strategies should be applicable to a wide variety of plant species.

Goals and Working Hypothesis

We are focused on developing environmentally friendly solutions to cleaning heavy metal- and metalloid-contaminated sediments and water. Our long-term goal is to develop and test highly productive, field-adapted plant species that have been engineered for the phytoremediation of mercury. Our current working hypothesis is that transgenic plants controlling the chemical species, electrochemical state, and aboveground binding of mercury and arsenic will (a) prevent elemental toxins from entering the food-chain, (b) remove mercury and arsenic from polluted sites, and (c) hyperaccumulate these toxic metals in aboveground tissues for later harvest. The strategies suggested by this hypothesis are outlined in Figure 1 as follows. (1) Uptake of mercury and arsenic will be enhanced by overexpression of transporters that recognize mercury and arsenate in roots. (2) Translocation of the elemental toxicant from root to shoot will be enhanced. In the case of arsenate, the endogenous reduction of arsenate to arsenite will be suppressed in roots, because arsenate (AsO_4^{-3}) will be more readily transported as a phosphate (PO_4^{-3}) analogue. In the case of mercury, Hg(II) would have to be converted to soluble nonreactive Hg(0), and Hg(0) should be most readily transported to the top of the plant. (3) The levels of toxic element sinks for Hg(II) and arsenite will be increased; for example, this could be accomplished by increasing the levels of γ-EC (γ-glutamylcysteine) through overexpression of γ-ECS (γ-glutamylcysteine synthetase). (4) Hg(0) and arsenate will be electrochemically reduced back to the reactive Hg(II) and arsenite (AsO_3^{-3}) species, respectively, in leaves for storage. (5) The transport of peptide metal/metalloid complexes into the vacuole for storage will be enhanced by overexpressing multiple drug resistance protein (MRP)-related glutathione conjugate pumps in leaves. Various parts of this hypothesis are being tested by examining different transgenes in model plants and comparing the results to control plants.

Introduction

Mercury and arsenic are extremely toxic heavy metal and metalloid pollutants that adversely affect the health of millions of people worldwide (1). These toxic pollutants have reached unacceptably high levels in the environment due to industrial, defense, agricultural, and municipal properties. The U.S. Department of Energy (DOE) and other government and industrial sites in the United States are heavily contaminated with mercury, arsenic, and other toxic metals such as cadmium, copper, lead, and zinc. Each of these elemental pollutants has common environmentally relevant electrochemical species that are thiol-reactive and thus relevant to the phytoremediation strategies outlined in Figure 1. Hundreds of Superfund sites in the United States are listed on the

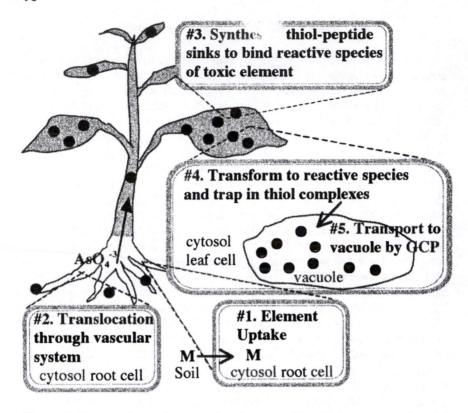

Figure 1. Strategy for engineering the phytoremediation of mercury and arsenic. Expanded diagrams of plant cells with critical activities are designated by large gray boxes. The element being taken up and concentrated is shown as black dots.

National Priority List (http://www.epa.gov/superfund/sites/npl/index.htm; accessed date July 9, 2004) as having unacceptably high levels of mercury and/or arsenic and are recommended for cleanup. In the majority of cases, these sites are not remediated because of the high costs associated with physical methods of cleanup. Physical remediation methods involving soil removal and reburial are expensive, impractical on the scale that is needed, and environmentally destructive. In contrast, phytoremediation, using plants to extract, detoxify, and sequester pollutants (*2, 3*), is an eco-friendly alternative or amendment to physical methods. Plants have several natural properties ideal for use in remediation (*2–4*). For example, their vast root systems are in intimate pervasive contact with the soil (*5*); they have an excess of reducing power from photosystem I; they can use solar energy to catalyze the remediation process; and they control more than 80% of the energy in most ecosystems (*2, 3*).

Phytoremediation of Elemental vs Organic Pollutants

Strategies for phytoremediation need to distinguish between the remediation of organic pollutants and the remediation of elemental pollutants (*3*). Organic pollutants include toxic chemicals such as benzene, benzo(a)pyrene, polychlorinated biphenyls, trichloroethylene (TCE), trinitrotoluene (TNT), and dichlorodiphenyltrichloroethane (DDT). Phytoremediation of organic pollutants is focused on their complete mineralization to harmless products and hence has few if any drawbacks. Native plants have some exceptional natural abilities to degrade organics, and there have been several reports of plants engineered for phytoremediation of organics dealing principally with the chlorinated solvent TCE (*6*) and the explosive TNT (*7, 8*). Elemental pollutants include heavy metals and metalloids (e.g., mercury, lead, cadmium, arsenic) and are immutable (i.e., they cannot be mineralized). Plants can be used to extract, sequester, detoxify, and transform elemental pollutants. Some exotic native plants have exceptional metal hyperaccumulation capabilities, but most are small and slow growing, and thus their potential for large-scale remediation of polluted sites is limited (*9*). Genetically engineered high biomass producing plants for the phytoremediation of elemental pollutants have been suggested for making phytoremediation a viable commercial technology (*3*).

Significance of Applying Phytoremediation to Mercury and Arsenic Pollution

Mercury Pollution

Mercury enters the environment primarily as liquid Hg(0) from industrial and defense related accidents or as mercury species (Hg(II)) bound to particulate matter from burning coal and trash and from volcanic activity, and as complex chemical derivatives released in industrial effluents (*10*). The geochemical mercury cycle and interconversion of various mercury species are discussed in our review (*4*). Although Hg(II) is relatively toxic, Hg(II) and Hg(0) have seldom been involved in serious incidents of human mercury poisoning without first being transformed into MeHg (*11*). The world first became aware of the extreme dangers of methylmercury (MeHg) after a large, tragic incident of human mercury poisoning in Japan in the 1950s at Minamata Bay (*12, 13*). In aquatic sediments, various mercury species are efficiently converted to MeHg by anaerobic bacteria (*14, 15*). Unfortunately, MeHg can be biomagnified by several orders of magnitude and has a greater toxicity than any other natural mercury compound (*16*). As a result, the fish-eating predatory animals and humans at the top of the food chain suffer the most severe MeHg poisoning.

There is currently a mercury-contaminated fish warning for children and pregnant women for the whole of the Southeastern U.S.A. Appropriately managed phytoremediation strategies may block Hg(II) from entering aquatic ecosystems or destroy MeHg in these systems, thus preventing MeHg from entering the food chain. We have strong preliminary evidence to support this claim.

Mercury is a particular problem for the DOE because it is found at hazardous levels at several large DOE sites and results in extensive contamination of associated water resources. One site in particular, at Oak Ridge National Laboratories, still contains millions of pounds of mercury spread over miles of a river valley even after extensive and costly physical remediation efforts of excavation and reburial off site.

Arsenic Pollution

Arsenic is an extremely toxic metalloid pollutant that adversely affects the health of millions of people worldwide (1). Inorganic arsenic is classified as a "group A" human carcinogen causing skin lesions; lung, kidney, and liver cancers; and damage to the nervous system (17-19). Arsenic is probably carcinogenic due to arsenate's properties as a phosphate analog and arsenate's negative effects on DNA repair systems (20-23). However, arsenite has greater toxicity and a lower LD_{50} in most organisms. Normal levels of arsenic in the earth's crust are about 1-2 ppm (mg/kg) (24), and soil levels several-fold above this are considered unacceptable depending upon bioavailability primarily because arsenic leaches into drinking water. For the last 20 years the maximum level of arsenic in drinking water in the USA was set at 0.05 µg/liter (50 ppb), but a lower standard of 0.01 µg/liter (10 ppb) was established by EPA in 2001 (*http://www.epa.gov/safewater/ars/arsenic_finalrule.html*; accessed date Aug 16, 2004). The EPA has recently released new data suggesting that a large area covering four states in the Southwestern United States may expose millions of Americans to unacceptably high levels of arsenic (> 10 ppb) in drinking water. Previous studies have shown that moderate levels of arsenic in drinking water in New Mexico led to a higher incidence of cancer (25). In many parts of India and Bangladesh, the contaminated-water arsenic concentrations reach 2 ppm (2 mg/L) (26). The use of such arsenic contaminated water for irrigation has increased the soil level of arsenic up to 80 mg/kg (80 ppm), which is 40- to 80-fold higher than the normal levels of arsenic in soil. Such high levels of arsenic in drinking water and soil will lead to 200,000–300,000 cancer related deaths in Bangladesh alone. In India and Bangladesh, more than 449 million people are at risk of arsenic poisoning due to high levels of arsenic in drinking water (27).

At many Superfund sites arsenic is a primary contaminant, and a few examples will be given. An agricultural site in southeastern North Dakota contains ~330,000 pounds of arsenic pesticide and covers 700 square miles, many parts of the site exceeding 50 ppm arsenic. Monsanto's industrial dump site for pesticides in Augusta and Richmond counties in Georgia contains 1,500 pounds of arsenic trisulfide spread over 75 acres with hot spots exceeding 100 ppm. The Rhone-Poulenc, Inc. / Zoecon Corp. manufacturing site in East Palo Alto, CA, covers 13 acres and has arsenic levels as high as 5,000 ppm. Groundwater arsenic leaching from these sites usually reaches unacceptable levels, but the efficiency of leaching is strongly dependent upon arsenic speciation in soils of different compositions (*1*). Several DOE sites such as Oak Ridge and Hanford report significant arsenic contamination of soil and water (*28*). The vast majority of arsenic contaminated sites have not been cleaned because the cost in both dollars and environmental damage from physical remediation methods is too high.

Environmental Speciation of Arsenic

Most arsenic in surface soil and water exists primarily in the oxidized form, the oxyanion arsenate (AsO_4^{-III}), where arsenic has a valence of V. Arsenate is an analogue of phosphate and potentially can be taken up and translocated up the plant vascular system in place of phosphate (*29, 30*). In anaerobic soils (i.e., several inches below the surface of many soils) the reduced oxyanion arsenite (AsO_3^{-III}), which has a valence of III, predominates. Arsenite is a highly thiol-reactive reagent that has the potential to be trapped aboveground in leaf and stem tissues as peptide thiol-complexes. However, we believe that plants trap arsenite below ground to prevent access to aboveground reproductive tissues to avoid possible mutagenic consequences. Both species of arsenic are also bound in a variety of chemically insoluble matrixes that may be available to plant roots mining nutrients like phosphate.

Results and Discussions

Mercury Detoxification and Remediation

Our laboratory made use of two genes from the well-characterized bacterial *mer* operon, *merA* and *merB*, in order to engineer a mercury transformation and remediation system in plants (*31, 32*). The bacterial *merB* gene encodes an organomercury lyase that is responsible for the protonolysis of compounds like MeHg into reduced organic molecules (in this case methane) and ionic mercury Hg(II) as shown in Reaction #1. The bacterial *merA* gene encodes an NADPH-dependent mercuric ion reductase that converts ionic mercury (Hg(II)) to elemental, metallic mercury (Hg(0) (Reaction #2)). Hg(0) is nearly two orders of magnitude less toxic to eukaryotes than ionic mercury and is readily eliminated by plants due to its volatility.

MerB catalyzed $CH_3Hg^+ + H+ \rightarrow CH_4 + Hg(II)$ (1)

MerA catalyzed $Hg(II) + NADPH + OH^- \rightarrow Hg(0) + NADP^+ + H_2O$ (2)

Arabidopsis Plants Expressing MerB Confer Strong Resistance to Methylmercury

We successfully engineered model plant *Arabidopsis thaliana* by using a slightly modified bacterial *merB* gene, methylmercury lyase, to convert methylmercury to much less toxic ionic mercury. These plants germinate, grow, and set seed at normal growth rates on levels of methylmercury (2 µM) that are lethal to normal plants, as shown in Figure 2 (*32*). Control plants lacking the expression of *merB* gene were severely inhibited and died quickly after seed germination at this concentration of methylmercury. The 100-fold range in concentration (0.01-2 ppm) over which merB plants are more resistant than wild type controls far exceeds concentrations commonly found in the environment (0.01–0.2 ppm). These results showed that MerB plants efficiently protonolyzed organic mercury, and thus produce a more tolerable mercury species (HgII) in the plants.

MerA Plants Confer Strong Resistance to Ionic Mercury

Arabidopsis thaliana and tobacco plants were engineered to express a highly modified bacterial mercuric ion reductase gene, *merA*, to further detoxify ionic mercury (Hg(II)) to metallic mercury (Hg(0)). These plants germinate, grow vigorously, and set seed at normal growth rates on media containing 20–50 ppm ionic mercury ($HgCl_2$), levels of Hg(II) that are toxic to normal seeds and plants (*2, 31*). These Hg(II) levels include and exceed what is present at 90% of most contaminated sites (1–500 ppm). For example, Figure 3 shows the result of transplanting wild-type (left of each pair of pots) and *merA* expressing (right of each pair) tobacco plantlets into heavily mercury-contaminated soil. The *merA* plants perform exceedingly well. In five replications of this experiment, 100% of the *merA* plants achieved full size and seed set, while all the controls died shortly after planting. Our MerA-expressing *Arabidopsis* and tobacco plants thrive on heavily mercury contaminated media that kills wild-type plants.

No methylmercury   2 µM methylmercury

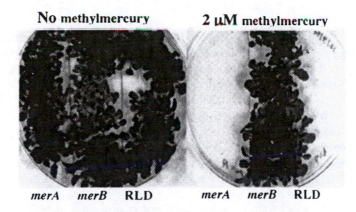

merA   merB   RLD   merA   merB   RLD

Figure 2. MerB expressing Arabidopsis resistant to methylmercury. Seeds from wild-type control (RLD), merA, and merB expressing lines were germinated on 1/2 MS (Murashige and Skoog) medium without (left) and with 2 µM methylmercury and grown for three weeks. Only merB expression allows normal growth on methylmercury.

Hg(II) Concentrations

0 ppm   100 ppm   500 ppm

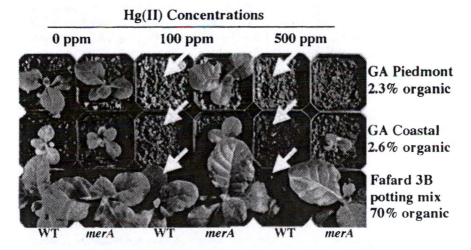

GA Piedmont
2.3% organic

GA Coastal
2.6% organic

Fafard 3B
potting mix
70% organic

WT   merA   WT   merA   WT   merA

Figure 3. MerA confers strong resistance to ionic mercury in soil. Three-week-old wild-type (WT) tobacco plantlets and transgenic plantlets expressing merA were transplanted into two different native Georgia soils (GA) and Fafard poting mix soil without (0 ppm) or with mercury at the concentrations indicated. This photograph was taken another three weeks later. The WT controls all severely inhibited or died on mercury and are indicated by arrows.

Coupling MerA and MerB Expression

We focused our recent efforts on studying the efficiency of methylmercury processing in plants. Plants expressing both MerA and MerB proteins can detoxify methylmercury in two steps to the least toxic metallic form, Hg(0) (*33*). These plants are even more resistant to methylmercury (> 10 ppm) than plants expressing MerB alone. Combining *merA* and *merB* genes in *Arabidopsis* resulted in a five-fold increased resistance to MeHg$^+$ and more importantly a 100- to 1000-fold increase in the rate of conversion of methylmercury to Hg(0) as compared to wild-type controls, as shown in Figure 4 (*33*). This gene combination provided the most complete detoxification of methylmercury possible by a single plant, converting it to the least toxic form of mercury, Hg(0).

A detailed biochemical and physiological examination of 60 independently transformed *merA/merB Arabidopsis* plants demonstrated that MerB enzyme levels were rate limiting in methylmercury processing as shown in Figure 5. MerA levels were in excess even when MerA was expressed at only 1% to 5% of the highest levels. This indicated that we should focus our efforts on improvement in merB expression. Toward this end, we demonstrated that targeting the MerB enzyme to either the endoplasmic reticulum (ER) or both the ER and cell wall resulted in even more efficient MeHg processing (*34*). It lowered the required levels of MerB protein 20-fold below that required to achieve the same levels of MeHg processing with standard cytoplasmic MerB expression. Our results suggested that combing *merA* and *merB* genes together in field species could prevent the flow of highly toxic methylmercury into the food chain.

Need for High Biomass Fast Growing Conservation Plants

Arabidopsis and tobacco are excellent plants for laboratory testing, but they lack the size and longevity necessary for on-site phytoremediation. For this reason, once gene constructs have been tested and proved in *Arabidopsis* and/or tobacco, these genes are moved into higher biomass conservation plants (e.g., yellow poplar, cottonwood, rice, canola, etc.) for field research and cleanup. An example of effective Hg(II) resistance that the *merA* expression confers on rice is shown in Figure 6. Many mercury contaminated sites are wetlands where rice would grow effectively. We have further extended our results in tree species such as cottonwood, and these results have been published elsewhere (*35*).

Creating Sinks for Mercury

Hyperaccumulation of mercury depends upon making plants that are both highly tolerant to mercury and have the capacity to store enormous amounts of mercury aboveground. Thiol-rich chemical sinks for Hg(II) are being tested in

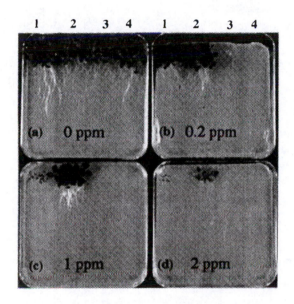

Figure 4. Coupling MerA and MerB expression in Arabidopsis provides complete detoxification of methylmercury. Seeds from lines expressing merB alone (lane 1), merA/merB (lane 2), merA alone (lane 3), and wild-type control (RLD) (lane 4) were germinated on 1/2 MS medium without (a) and with 1 μM /0.2 ppm (b), 5 μM / 1 ppm (c), and 10 μM / 2 ppm (d) methylmercury and grown for three weeks.

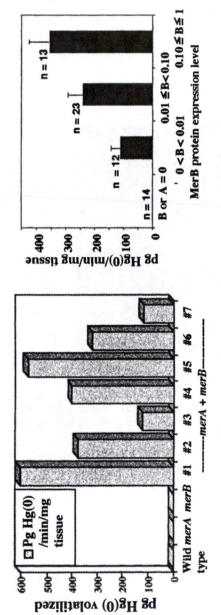

Figure 5. The concerted expression of merA and merB genes in plants is required for most efficient processing of methylmercury. (A) The two-step conversion of methylmercury to Hg(0) was assayed on several transgenic plants expressing merA and merB and controls. Neither plants with merA or merB alone, nor control WT plants can convert significant amounts of methylmercury to metallic mercury. (B) Methylmercury lyase levels (MerB) limits the two-step conversion of methylmercury to least toxic elemental mercury in merB/merA plants. Using antibodies to MerB and MerA, protein levels were assayed in over 60 independent merA/merB transgenic plants and normalized to phosphoenol pyruvate (PEP) carboxylase and total protein levels. The conversion of methylmercury to Hg(0) was a direct function of increasing MerB levels.

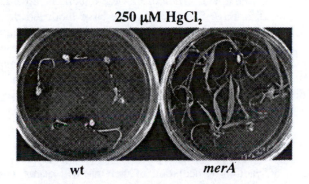

Figure 6. Resistance of merA transgenic rice to high levels of ionic mercury in their medium. WT and merA transgenic rice seeds were germinated on 250 μM Hg(II) in their medium and grown for two weeks.

Three step enzymatic synthesis of phytochelatins	
γ–ECS (γ-glutamylcysteine synthetase) catalyzed	
Glu + Cys → γ-Glu-Cys (EC)	**Reaction #3**
GS (Glutathione synthetase) catalyzed	
EC + gly → Gly-γGlu-Cys (GSH)	**Reaction #4**
PCS (phytochelatin synthetase) catalyzed	
GSH + EC → Gly(γGlu-Cys)$_n$ (PC)$_n$	**Reaction #5**

Figure 7. Three step enzymatic synthesis of phytochelatins.

transgenic plants as modeled in Figure 1, and each has the potential to confer mercury tolerance. Phytochelatins (PCs) are synthesized enzymatically from amino acids in a three-step enzymatic pathway shown in Figure 7. We have cloned the γ-glutamylcysteine synthetase (γ–ECS) and glutathione synthetase (GS) genes by polymerase chain reaction (PCR) from E. coli DNA starting with sequence from the public DNA database. Phytochelatin synthetase (PCS) genes from fission yeast, wheat, and Arabidopsis were obtained from Julian Schroeder (University of California – San Diego). We have modified the sequences of these five genes by PCR to have the appropriate sites for cloning and testing in E. coli and plants. We made monoclonal antibodies specific to each of the three enzymes γECS, GS, and PCS (fission yeast), to monitor their expression (36). We have cloned all five enzymes individually in Arabidopsis under control of a variety of different promoters (Li and Meagher, manuscripts in preparation).

Constitutive and Aboveground Mercury Sinks

The most efficacious remediation of mercury will occur when plants can extract mercury from soil, transport it to aboveground organs, and hyperaccumulate it aboveground in leaves and stems as shown in Figure 1. We recently developed a vector for expressing metal-binding peptides and other proteins in aboveground tissues. The expression of the soybean gene encoding the light-regulated small subunit of ribulose bisphosphate carboxylase (rubisco), *SRS1*, was thoroughly characterized in our laboratory nearly a decade ago (37). RNA steady-state levels, transcription run-off assays, and promoter-reporter fusion assays, all demonstrated that this gene was strongly expressed in aboveground green tissues in the light (38, 39). For strong aboveground expression of mercury and other metal ion sinks, we have recently constructed a new vector, SRS1pt, driving light-induced expression from both the SRS1 promoter and homologous terminator sequences, separated by a multicloning site region (40).

In order to demonstrate independently that the new SRS1pt vector had the desired aboveground specificity and no significant root activity, the vector was used to drive the expression of a GUS (β-glucuronidase) reporter gene and expressed in transgenic *Arabidopsis*. The SRS1pt/GUS fusion vector confers strong expression in leaves and stems, but not in roots as shown in Figure 8 (41). Quantitative measurements suggest the GUS reporter protein is at least 100 times more strongly expressed aboveground. This promoter will now be used to express and test those genes with potential for creating mercury sinks aboveground.

Arsenic Detoxification and Hyperaccumulation

The most efficacious remediation of arsenic requires that plants extract arsenic from soil or water and hyperaccumulate it aboveground in leaves and stems. To accomplish this we must control the electrochemical species of arsenic in different parts of the plant. The bacterial *arsC* gene encodes a 141 amino acid arsenate reductase, arsC (Accession #J02591) and uses glutathione (GSH) as hydrogen/electron donor in the electrochemical reduction of the oxyanion arsenate (AsO_4^{-3}), to the oxyanion, arsenite (AsO_3^{-3}) (*42*). While arsenate is a phosphate analogue, arsenite is chemically very different, being a highly reactive species with strong affinity toward thiol-groups such as those in γ-EC, GSH, and PCs (Figure 9).

We overexpressed two bacterial genes, *arsC* and *γ-ECS*, under control of a light-regulated, leaf-specific rubisco small subunit promoter (*SRS1p*), and a constitutively expressed actin promoter/terminator cassette (*ACT2pt*), respectively (*41*). The bacterial *arsC* gene product directs leaf- and stem-specific reduction of arsenate to arsenite. This confers to plants the potential to trap more arsenic in arsenite-thiol complexes aboveground (Figure 1). The *γ-ECS* gene product directs the synthesis of the dipeptide γ-EC, the first and proposed limiting enzyme in the phytochelatin synthetic pathway. Transgenic plants expressing bacterial ArsC alone were hypersensitive to arsenate (Figure 10, column 1). This is most likely due to the reduction of arsenate to more toxic arsenite in leaves. Plants expressing *γ*-ECS alone were moderately resistant to arsenate (Figure 10, column 2) compared to wild-type controls (column 4), whereas doubly transformed plants expressing ArsC and γ-ECS together were significantly more resistant to arsenate (Figure 10, columns 3a and 3b). After three weeks of growth on 200 μM arsenate, the hybrid plants (ArsC+*γ*-ECS) attained three-fold more biomass than plants with *γ*-ECS alone, six-fold more than wild-type, and ten-fold more than ArsC alone. Additionally, when the plants were allowed to grow for up to four weeks, the hybrid plants accumulated four-fold more biomass than γ-ECS alone and approximately 17-fold more than ArsC alone (*41*).

For determining the total arsenic concentrations in shoots, two hybrid lines expressing ArsC and γ-ECS together and lines expressing SRS1p/ArsC9 or ACT2p/ECS1 alone, and wild-type controls were grown on 125 μM arsenate for three weeks. Both hybrid lines expressing ArsC+γ-ECS showed three-fold higher concentrations of arsenic in their shoots than wild-type, ACT2p/ECS1, and SRS1p/ArsC9 plants alone (Table I). These results can be explained as follows. The coexpressed *SRS1p/ArsC* and *ACT2p/ECS* transgenes complemented each other's activities resulting in increased arsenic resistance. The *γ-ECS* gene product directs the synthesis of the dipeptide *γ*-EC, the first and proposed limiting enzyme in the PC synthetic pathway (*43*). All downstream thiol-peptide compounds in the phytochelatin pathway (e.g., *γ*-EC, GSH, and PCs) may be increased, and all three can bind arsenite and potentially contribute to arsenic tolerance and accumulation.

ACT2pt driven constitutive expression of γ-ECS enzyme should have resulted in increased thiol-sink peptides in all major vegetative organs (e.g., roots, stem, leaves, petals, sepals) (*44*) and resulted in plants that can grow on

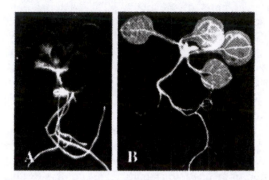

Figure 8. Strong light induced GUS-reporter expression is driven from a new SRS1pt soybean rubisco small subunit vector. (A) The SRS1pt/GUS fusion construct was transformed into Arabidopsis and one of several stable transgenic lines is shown after staining for GUS expression. The indigo blue product of GUS activity was developed by incubating the plant in the colorless x-gluc substrate, after 24 hours of fixation and bleaching of the tissue in ethanol. The GUS reporter was fused between the SRS1 promoter and terminator to test specificity of the new SRS1pt vector. (B) A wild-type Arabidopsis plant is similarly stained for GUS activity, but shows no color.

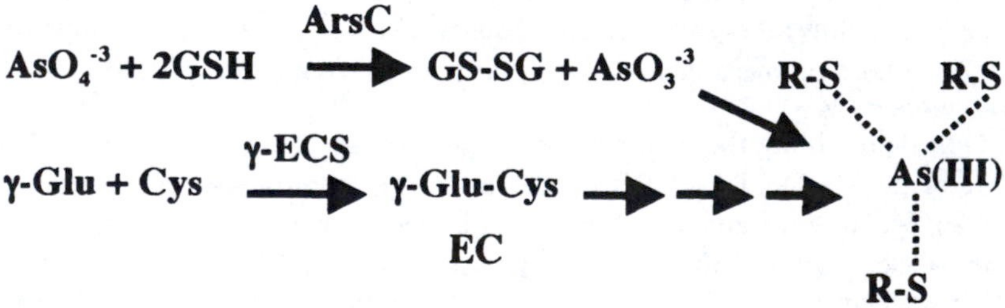

$$AsO_4^{-3} + 2GSH \xrightarrow{\text{ArsC}} GS\text{-}SG + AsO_3^{-3}$$

$$\gamma\text{-Glu} + Cys \xrightarrow{\gamma\text{-ECS}} \gamma\text{-Glu-Cys} \rightarrow \rightarrow \rightarrow As(III)$$

EC

R-S   R-S

R-S

Figure 9. ArsC and γ-ECS catalyzed reactions. The bacterial arsenate reductase (ArsC) catalyzes the electrochemical reduction of arsenate to arsenite. The bacterial γ-glutamylcysteine synthetase (γ-ECS) catalyzes the formation of γ-glutamylcysteine (γ-EC) from the amino acids glutamate and cysteine and is the committed step in the synthesis of glutathione (GSH) and phytochelatins, PCs (indicated by three arrows). Reduced arsenite can bind organic thiols (RS) such as those in γ-EC, GSH, and PCs through the replacement of oxygen by organic sulfur species.

200 µM Arsenate

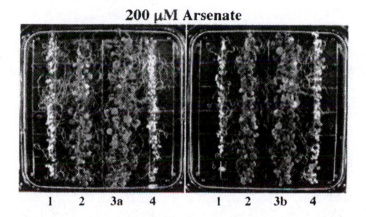

1   2   3a   4   1   2   3b   4

Figure 10. Arsenic resistance of plants expressing ArsC9 and γ-ECS. Each arsenate containing plate has a column of seeds from plants that express (1) light regulated SRS1p/ArsC/Nost; (2) constitutively expressed ACT2p/γECS/ACT2t; (3a, 3b) doubly transformed lines expressing both transgenes, γECS and ArsC; and (4) Wild-type (WT), grown on 200 µM sodium arsenate for four weeks. Wild-type (column 4) showed intermediate resistance between the two lines with single transgenes (column 1 and 2). Seeds were germinated and grown for four weeks on 200 µM arsenate containing ½ strength MS media.

arsenic contaminated media. γ-ECS overexpression complemented the hypersensitivity of *SRS1p/ArsC* plants. In addition, the overexpression of both ArsC and γ-ECS proteins together in hybrid plants further enhanced arsenic tolerance and hyperaccumulation of arsenic in the aboveground parts, far beyond what γ-ECS expression could achieve alone. We suggest this is due to more electrochemical reduction of arsenate to arsenite by the activity of ArsC and binding of arsenite to the immediate or downstream products of γ-ECS (e.g., γ-EC, GSH, PCs). The resistance of these engineered plants to 200 μM (~15,000 ppb) of soluble arsenate suggests that similarly engineered fast

Table I. Total Arsenic Accumulation in Shoots of Wild-Type (WT), Act2p/ECS1, SRS1p/ArsC9, and two hybrid lines (ArsC9+ECS1 and ArsC9+ECS10) *Arabidopsis* seedlings grown on 125 μM (~10 ppm) Sodium Arsenate

Plant Lines	Total Shoot Arsenic (μg/g dry wt)
WT	214± 16
Act2pt/ECS1	265± 22
SRS1p/ArsC9	308± 10
ArsC9+ECS1	663± 27
ArsC9+ECS10	705± 19

NOTE: Values shown are the average and standard deviation (±) of three replicates of 50 plants for each lines.

growing, high biomass plant field species would grow on soil and water contaminated with these environmentally relevant concentrations of arsenic. These plants could be used to hyperaccumulate arsenic aboveground. Currently we are in the process of extending this arsenic strategy to other higher plants including tobacco and cottonwood. However, we know that another two or three genes will be needed (Figure 1) to make this system commercially relevant.

Conclusions

Several strategies for the phytoremediation of arsenic and mercury pollution, using bacterial, animal, and plant genes, have been tested in model plants including *Arabidopsis* and tobacco. We have demonstrated that plants can be engineered to take up and tolerate several times the levels of mercury and arsenic that would kill most plant species. We have engineered plants that detoxify, volatilize, or sequester these elements (*31, 32, 41*). Using these modified plants, methylmercury and/or ionic mercury is detoxified, stored below or above ground, or even volatilized as part of the transpiration process, thus keeping it out of the food chain. In the case of volatilization, the approximate Hg(0) atmospheric residence time of 0.5 to 2 years (*11*) is likely to facilitate the dilution of volatilized Hg(0) to negligible concentrations. The volatilization of Hg(0) may prove useful particularly in wetland areas where biomagnification of the more toxic methylmercury is a serious problem. Transgenic mercury remediation strategies also worked efficiently in cultivated and wild plant species like canola, rice, yellow poplar, and cottonwood. Initial efforts with arsenate demonstrate that it can be taken up, transported aboveground, electrochemically reduced to arsenite in leaves, and sequestered in thiol-rich peptide complexes (*41*). Considering our success so far, we are hopeful that by combing several genes together as described in Figure 1, we will be able to demonstrate dramatic increases in mercury and arsenic resistance and hyperaccumulation aboveground. We will test various logical combinations of genes that optimize resistance and hyperaccumulation. Genomic surveys suggest that hundreds of other bacterial, animal, and plant genes encoding enzymes that transform, transport, and sequester elemental toxic pollutants may be available to help us enhance these strategies in the near future (*45*). We are examining several other potential genes that might further advance toxic metal tolerance, uptake, and hyperaccumulation. These transgenic technologies and the understanding of the physiology of mercury and arsenic behavior in plants should lead to the development of fast-growing, high-biomass mercury and arsenic hyperaccumulator plants for field use.

Though the potential of the engineered plants for toxic metal hyperaccumulation has raised new hopes for reducing heavy metals and metalloids toxicity of contaminated water and soil, much remains to be done. Attention should be given to engineer aquatic plants and plants that are acclimatized to the agro-climatic conditions of the contaminated sites on one hand and develop eco-friendly technology to remove the toxic metals from the green biomass on the other hand. Natural plant hyperaccumulators have much to contribute to both initial clean up efforts and to our understanding of important physiological mechanisms of toxic element processing. Also, there is a strong need to address public concerns about the processing and handling of metal-laden biomass and to educate the public about the benefits of genetically modified organisms (GMOs) in environmental restoration programs.

Acknowledgments

This work was supported by a DOE Environmental Management Science Program (EMSP) grant DE-FG07-96ER20257 and DOE grant DE-FC09-93R18262 to R.B.M.

References

1. Nriagu, J. O. *Arsenic in the Environment*, Part 1: Cycling and Characterization; Wiley: New York, 1994; Vol. 26.
2. Meagher, R. B.; Rugh, C. L. In *OECD Biotechnology for Water Use and Conservation Workshop;* Organization for Economic Co-Operation and Development: Cocoyoc, Mexico, 1996; pp 305–332.
3. Meagher, R. B. *Curr. Opin. Plant Biol.* **2000**, *3*, 153–162.
4. Meagher, R. B.; Rugh, C. L.; Kandasamy, M. K.; Gragson, G.; Wang, N. J. In *Phytoremediation of Contaminated Soil and Water;* Terry, N., Banuelos, G., Eds.; Lewis Publishers: Boca Raton, FL, 2000; pp 203–221.
5. Dittmer, H. J. *Am. J. Bot.* **1937**, *24*, 417–420.
6. Doty, S. L.; Shang, T. Q.; Wilson, A. M.; Tangen, J.; Westergreen, A. D.; Newman, L. A.; Strand, S. E.; Gordon, M. P. *Proc. Natl. Acad. Sci.* **2000**, *97*, 6287–6291.
7. French, C. E.; Rosser, S. J.; Davies, G. J.; Nicklin, S.; Bruce, N. C. *Nat. Biotechnol.* **1999**, *17*, 491–494.
8. Hannink, N.; Rosser, S. J.; French, C. E.; Basran, A.; Murray, J. A.; Nicklin, S.; Bruce, N. C. *Nat. Biotechnol.* **2001**, *19*, 1168–1172.
9. Ebbs, S. D.; Lasat, M. M.; Brady, D. J.; Cornish, J.; Gordon, R.; Kochian, L. V. *J. Environ. Qual.* **1997**, *26*, 1424–1430.
10. Adriano, D. C. *Trace Elements in the Terrestrial Environment*; Springer Verlag: New York, 1986.
11. Keating, M. H., Mahaffey, K. R.; Schoeny, R.; Rice, G. E.; Bullock, O. R. Jr.; Ambrose, R. B.; Swartout, J.; Nichols, J. W. *Mercury Study Report to Congress. Volume I: Executive Summary*; EPA-452/R-97-003; U.S. Environmental Protection Agency: Washington, DC, December 1997.
12. Harada, M. *Crit. Rev. Toxicol.* **1995**, *25*, 1–24.
13. Allchin, D. *The American Biology Teacher* **1999**, *61*, 413–419.
14. Begley, T. P.; Walts, A. E.; Walsh, C. T. *Biochemistry* **1986**, *25*, 7192–7200.
15. Begley, T. P., Walts, A. E., and Walsh, C. T. *Biochemistry* **1986**, *25*, 7186–7192.
16. Watras, C. J.; Bloom, N. S. *Limnol. Oceanogr.* **1992**, *37*, 1313–1318.

17. Brown, K. G., and Ross, G. L. Arsenic, Drinking Water, and Health: A Position Paper of the American Council on Science and Health. *Regul. Toxicol. Pharmacol.* **2002**, *36*, 162–174.

18. Kaiser, J. Second Look at Arsenic Finds Higher Risk. *Science* **2001**, *293*, 2189.

19. Lewis, D. R.; Southwick, J. W.; Ouellet-Hellstrom, R.; Rench, J.; Calderon, R. L. Drinking Water Arsenic in Utah: A Cohort Mortality Study. *Environ. Health Perspect.* **1999**, *107*, 359–365.

20. Okui, T.; Fujiwara, Y. *Mutat. Res.* **1986**, *172*, 69–76.

21. Okada, S.; Yamanaka, K.; Ohba, H.; Kawazoe, Y. *J. Pharmacobiodyn.* **1983**, *6*, 496–504.

22. Leonard, A.; Lauwerys, R. R. *Mutat. Res.* **1980**, *75*, 49–62.

23. Hughes, M. F. *Toxicol. Lett.* **2002**, *133*, 1–16.

24. Emsley, J. *The Elements;* Oxford University Press: New York, 1991.

25. Siegel, M.; Frost, F.; Tollestrup, K. *Ann. Epidemiol.* **2002**, *12*, 512.

26. WHO. *Guideline for Drinking Water Quality, Recommendation*, 2nd ed.; World Health Organization: Geneva, Switzerland, 1992; Vol. 1.

27. Chakraborti, D.; Mukherjee, S. C.; Pati, S.; Sengupta, M. K.; Rahman, M. M.; Chowdhury, U. K.; Lodh, D.; Chanda, C. R.; Chakraborti, A. K.; Basu, G. K. *Environ. Health Perspect.* **2003**, *111* (9), 1194–1201 (Accessed online, February 2003; available via http://dx.doi.org/ using the reference doi:10.1289/ehp.5966).

28. Riley, R. G.; Zachara, J. M.; Wobber F. J. *Chemical Contaminants on DOE Lands and Selection of Contaminant Mixtures for Subsurface Research*; DOE/ER-0547T; U.S. Department of Energy, Washington, DC, 1992; pp 21–26.

29. Meharg, A. A.; Macnair, M. R. *Heredity* **1992**, *69*, 336–341.

30. Meharg, A. A.; Macnair, M. R. *New Phytol.* **1990**, *116*, 29–35.

31. Rugh, C. L.; Wilde, D.; Stack, N. M.; Thompson, D. M.; Summers, A. O.; Meagher, R. B. *Proc. Natl. Acad. Sci.* **1996**, *93*, 3182–3187.

32. Bizily, S.; Rugh, C. L.; Summers, A. O.; Meagher, R. B. *Proc. Natl. Acad. Sci.* **1999**, *96*, 6808-6813.

33. Bizily, S.; Rugh, C. L; Meagher, R. B. *Nat. Biotechnol.* **2000**, *18*, 213–217.

34. Bizily, S.; Kim, T.; Kandasamy, M. K.; Meagher, R. B. *Plant Physiol.* **2003**, *131*, 463–471.

35. Che, D. S.; Meagher, R. B.; Heaton, A. C. P.; Lima, A.; Merkle, S. A. *Plant Biotech.* **2003**, *1*, 311–319.

36. Li, Y.; Kandasamy, M. K.; Meagher, R. B. *Plant Physiol.* **2001**, *127*, 711–719.

37. Berry-Lowe, S.; Meagher, R. B. *Mol. Cell. Biol.* **1985**, *5*, 1910–1917.

38. Shirley, B. W.; Berry-Lowe, S. L.; Rogers, S. G.; Flick, J. S.; Horsch, R.; Fraley, R. T.; Meagher, R. B. *Nucleic Acids Res.* **1987**, *15*, 6501–6514.

39. Shirley, B. W.; Meagher, R. B. *Nucleic Acids Res.* **1990**, *18*, 3377–3385.

40. Dhankher, O. P.; Rosen, B. P.; Fuhrman, M.; Meagher, R. B. *New Phytol.* **2003**, *159*, 431–441.
41. Dhankher, O. P.; Li, Y.; Rosen, B. P.; Shi, J.; Salt, D.; Senecoff, J. F.; Sashti, N. A.; Meagher, R. B. *Nat. Biotechnol.* **2002**, *20*, 1140–1145.
42. Rosen, B. P. *Trends Microbiol.* **1999**, *7*, 207–212.
43. Hell, R.; Bergmann, L. *Planta* **1990**, *180*, 603–612.
44. An, Y.-Q.; McDowell, J. M.; Huang, S.; McKinney, E. C.; Chambliss, S.; Meagher, R. B. *Plant J.* **1996**, *10*, 107–121.
45. Cobbett, C.; Meagher, R. In *The Arabidopsis Book;* Meyerowitz, E., Somerville, C., Eds.; Cold Spring Harbor Laboratory Press: Cold Spring Harbor, NY, 2002, pp 1–22. (also available via http://www.aspb.org/publications/arabidopsis/toc.cfm).

Chapter 6

Microbially Mediated Subsurface Calcite Precipitation for Removal of Hazardous Divalent Cations: Microbial Activity, Molecular Biology, and Modeling

Frederick S. Colwell[1], Robert W. Smith[2], F. Grant Ferris[3],
Anna-Louise Reysenbach[4], Yoshiko Fujita[1], Tina L. Tyler[1,5],
Joanna L. Taylor[1,5], A. Banta[4], Mark E. Delwiche[1],
Travis L. McLing[1], and Mary E. Watwood[5]

[1]Idaho National Engineering and Environmental Laboratory,
Idaho Falls, ID 83415–2203
[2]University of Idaho, Idaho Falls, ID 83402
[3]University of Toronto, Toronto, Ontario M5S 3B1, Canada
[4]Portland State University, Portland, OR 97207–0751
[5]Idaho State University, Pocatello, ID 83209

Current approaches for remediating hazardous divalent cations in aquifers are costly, can generate large volumes of waste, and focus on the small amounts of contaminants in the water rather than the larger reservoir of contamination sorbed to the aquifer matrix. An alternative to waste removal and repackaging is to encourage in situ biogeochemical processes to permanently bind the contaminants in the mineral matrix of an aquifer. Our research involves one such approach in which we accelerate calcite precipitation (an on-going geochemical process in arid western aquifers) and the assisted co-precipitation of cationic contaminants like strontium-90 using biologically driven urea hydrolysis to increase aquifer pH and alkalinity. This paper describes progress related to stimulating and measuring indigenous urease activities in aquifer microbes and how these activities can be modeled for application in an aquifer of concern to the U.S. Department of Energy. Experiments using ^{14}C-labeled urea indicated that microbial communities from the Snake River Plain aquifer (SRPA) of eastern Idaho hydrolyzed urea at rates higher than those measured for a model urea hydrolyzing bacterium (Bacillus pasteurii) under similar conditions, if they were provided a source of organic carbon along with the urea. By using a phylogenetic approach for analyzing urease gene sequences we developed polymerase chain reaction primer pairs that detected the ureC gene in urease positive microbial isolates. In a field test where molasses and urea were added to the SRPA, the ca. 400 base pair ureC fragment was amplified from DNA extracted from aquifer cells. Amplification and sequencing of bacterial 16S rDNA gene fragments from the aquifer before and after the molasses and urea additions indicated measurable changes in the communities as a result of the treatment. Rate constants derived from urease activity experiments were used to simulate the calcite precipitation process in the SRPA. The model predicts that field application would result in three distinct geochemical reaction phases: a condition where urea hydrolysis rates exceed calcite precipitation rates, a condition where calcite precipitation rates exceed urea hydrolysis rates, and finally a condition where the two rates are equal. The model also indicates that most of the metals that are precipitated as carbonates will come from the aquifer matrix, not the groundwater. These two modeling observations suggest that when the rates of calcite precipitation and urea hydrolysis are equal, the entire process can be described by a pseudo-first order kinetic model. In this model the calcite precipitation rate is controlled by the urea hydrolysis rate and is independent of the concentration of calcium in the groundwater. The use of these techniques for determining the response of microbial communities to urea additions, as well as the predictive capabilities of the model, will allow better control and evaluation of pending field experiments to test calcite precipitation as an approach for contaminant removal from aquifers.

Introduction

Radionuclide and metal contaminants are present in the vadose zone and groundwater at many of the U.S. Department of Energy (DOE) sites (1). Although the protection of groundwater is of primary concern, any remedial treatment must account for the large amount of contaminant that is sorbed to the aquifer matrix and serves as a continuing source of groundwater contamination. In situ containment and stabilization of these contaminants in vadose zones or groundwater is a cost-effective treatment strategy (2). However, implementing in situ containment and stabilization requires definition of the mechanism that controls contaminant sequestration. One mechanism for sequestration of metals and radionuclides is co-precipitation in authigenic calcite and calcite overgrowths (3). Calcite, a common mineral in many aquifers and vadose zones in the arid western United States, can incorporate divalent metals such as strontium, cadmium, lead, and cobalt into its crystal structures by the formation of solid solutions. For strontium-90, a radioisotope of concern to DOE, the calcite precipitation approach can be particularly effective. Because strontium-90 has a radioactive half-life of about 30 years, a period of only 300 years is required to eliminate > 99.9% of its radiologic hazard. Thus, removal of this radioisotope from the groundwater by securing it in a calcite mineral phase followed by control of the groundwater geochemistry to prevent calcite dissolution can be an effective way to treat this contaminant.

Our approach, shown schematically in Figure 1, for encouraging calcite precipitation and co-precipitation of divalent cations relies upon addition of a carbon source (e.g., molasses) to an aquifer followed by the addition of urea. While calcite precipitation occurs naturally in these aquifers urea hydrolysis by the in situ microbial community causes an acceleration of calcite precipitation (and trace metal co-precipitation) by increasing groundwater pH and alkalinity. This process has been demonstrated under laboratory conditions with *Bacillus pasteurii*, a microorganism that constitutively hydrolyzes urea (4). Urea hydrolysis and calcite precipitation have also been demonstrated, albeit more slowly, using microbial isolates acquired from the Snake River Plain Aquifer (SRPA), a groundwater system of significance to the DOE in the context of contamination (3). Because many western aquifers are saturated with respect to calcite and the precipitation processes are irreversible for these aquifers, the co-precipitated metals and radionuclides will be effectively removed from groundwater. The newly formed calcite should be stable as the aquifer returns to pre-treatment conditions and as long as the pH in the aquifer remains at normal values (> 7). The urea hydrolysis approach has the added advantage that the ammonium ions produced by the reaction can potentially exchange with radionuclides sorbed to subsurface minerals, thereby enhancing the susceptibility

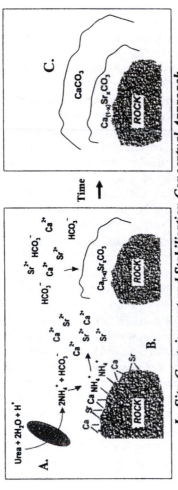

In Situ Containment and Stabilization Conceptual Approach

Net Reaction

$Solid - \left(^{90}Sr^{2+}\right) + H_2N(CO)NH_2 + 2H_2O \xrightarrow{urease} {}^{90}SrCO_{3, (calcite)} + Solid - (NH_4^+)_2$

A. Enzymatically catalyzed hydrolysis of introduced urea produces HCO_3^-, NH_4^+, and raises pH

$$H_2N(CO)NH_2 + H^+ + 2H_2O \xrightarrow{urease} 2NH_4^+ + HCO_3^-$$

B. HCO_3^- promotes irreversible precipitation of calcite and co-precipitation of ^{90}Sr NH_4^+ promotes desorption of ^{90}Sr from aquifer minerals

$$\chi\,{}^{90}Sr^{2+} + (1-\chi)\,Ca^{2+} + 2HCO_3^- \Leftrightarrow Ca_{(1-\chi)}\,{}^{90}Sr_{\chi}CO_3 + H_2O + CO_2$$

$$Solid - \left(^{90}Sr^{2+}\right) + 2NH_4^+ \Leftrightarrow Solid - (NH_4^+)_2 + {}^{90}Sr^{2+}$$

C. Continued precipitation of clean calcite isolates ^{90}Sr from contact with groundwater

Figure 1. Conceptual model for in situ containment and stabilization of divalent cations in calcite minerals using microbial-based urea hydrolysis.

of the radionuclides for re-capture in a more stable solid phase (co-precipitation rather than adsorption).

Although the coupling of calcite precipitation and trace metal partitioning is theoretically a useful means to control divalent contaminants, a fundamental understanding of how urea hydrolysis-based calcite precipitation occurs in aquifers and vadose zone environments is lacking. In this study we had two specific research objectives. First we attempted to determine the urease activities of groundwater consortia from the aquifer and model these activities to estimate how the process may work in situ, prior to the actual field test. To achieve the first objective we relied upon groundwater samples collected from research well (UP-1) with characteristics representative of the uncontaminated SRPA. Second, we wanted to determine the usefulness of urease-specific polymerase chain reaction (PCR) primers and 16S rDNA-specific primers for detecting molecular level changes in the aquifer communities. To achieve the second objective we characterized groundwater communities before and after a field experiment that was designed to maximize community shift and conducted in a well (Second Owsley) with characteristics representative of a contaminant plume (e.g., low dissolved O_2) in the SRPA.

Materials and Methods

Field Sites and Aquifer Experiment

The SRPA is a large, freshwater, semi-confined aquifer that exists beneath the semi-arid high desert in southeastern Idaho. The Quaternary volcanic-sedimentary stratigraphic sequence in the subsurface consists of multiple monogenetic tholeiitic basalt flows interbedded with thin sedimentary zones (5). Typically, the basalts are porphyritic containing phenocrysts or olivine and plagioclase in a fine-grained matrix that consists of interlocking plagioclase, augite, Fe-Ti bearing oxide minerals and glass (see 6).

Preliminary field experiments were conducted at Second Owsley (43° 48' 19.495" N latitude, 112° 38' 09.014" W longitude), a 94.4 m deep well that accesses the SRPA where the depth to groundwater is 68 m below land surface (mbls). Prior to the investigation the water in Second Owsley had a pH of 7.8 to 8.0, a temperature of 13.9 °C, a dissolved oxygen concentration of 0.5 mg/L, and a conductivity of 328 μS/cm^3 as determined by a DataSonde 4/minisonde (Hydrolab Inc., Austin, TX). Control samples were obtained from Second Owsley on two occasions prior to the start of the field experiment so that background microbial community and urease characterizations could be compared to those obtained after the field experiment.

For the actual experiment, a straddle packer pump was placed in the open borehole to isolate the interval between 82.6 and 94.4 mbls where the experiment would be confined. In order to stimulate the indigenous microbial communities, molasses (Grandma's Molasses, Riverton, NJ) and urea were added to the aquifer in 1000 L of groundwater that had previously been pumped from the Second Owsley well. Final concentrations of the molasses and urea were 0.1% and 30 mM, respectively. Potassium bromide was also added as a conservative tracer to determine the amount of the injected water that was retrieved. The added Br⁻ ion equaled 5.3 mM compared to background concentrations of Br⁻ in the SRPA of approximately 0.0084 mM. Bromide concentrations were measured continuously using a combination sure flow bromide electrode (Model 9635BN, Thermo Orion, Beverly, MA) and an ion-specific meter (Model 290A, Orion, Beverly, MA).

Following the injection, the well was allowed to rest for 13 days. Subsequently, the well was pumped and sampled intermittently during the next 22 days until the concentration of bromide reached background levels. On any given sampling day the amount of water removed from the well was limited to 200 L to minimize the amount of urea and molasses solution that was recovered. Removing too much water on a single day would limit the downhole incubation period of the reactants. A total of about 3000 L of water was pumped from the aquifer during 15 sampling dates.

Urease Activity of *Bacillus pasteurii* and SRPA Consortia

The rates of urea hydrolysis by *B. pasteurii* (a constitutive urease producer and model microbe for calcite precipitation), by *Escherichia coli* (a microbe that is typically unable to hydrolyze urea), and by unenriched SRPA consortia were determined using the tracer method with ^{14}C-labeled urea (7). Briefly, *B. pasteurii* was grown at 26 °C in Brain Heart Infusion medium (Difco, Inc.) with 2% urea added for 48 h. The grown cells were collected by centrifugation and washed three times using sterile synthetic groundwater. Using synthetic groundwater, the washed cell suspension was adjusted to an absorbance reading of 0.01 at 600 nm, a turbidity equivalent to approximately 5×10^6 cells/mL (L. Petzke, personal communication). *E. coli* was treated in the same manner, except that it was originally grown in Trypticase Soy Broth (Difco, Inc.). SRPA consortia were obtained from UP-1, a research well located on the University Place, Idaho Falls campus of the University of Idaho and distant from Second Owsley. Total cell numbers in this consortium were approximately $1.3 - 1.7 \times 10^4$ cells/mL of untreated groundwater (T. Tyler, unpublished data).

A 256 µCi/mL stock solution of ^{14}C-urea (7.7 mCi/mmol, Sigma) was prepared in sterile nanopure water. Nine milliliters of groundwater or washed *B. pasteurii* cell suspension were added to each 25-mL flask and then amended with 5.9 nmol ^{14}C-urea (equivalent to 100,000 dpm). Prior experiments suggested the need to add a carbon source along with the urea in order to stimulate urea hydrolysis in indigenous microbial communities. For the activity experiments described herein, we added 0.01%, 0.001%, and 0.00075% molasses to the flasks containing groundwater consortia and 0.001% molasses to the flasks containing *B. pasteurii*. The 0.00075% molasses addition results in a final dissolved organic carbon (DOC) concentration that is just higher than the 1mg/L DOC typically found in the SRPA. Negative controls, all with ^{14}C-urea added, included groundwater with no added molasses, filtered (0.2 µm pore size) groundwater with 0.001% molasses, and *E. coli* with 0.001% molasses. All experimental treatments were performed in triplicate series. Flasks were incubated at room temperature. For each series, ^{14}C-CO_2 was collected at 0, 24, or 48 h in order to collect rate data. Selection of unique populations within the groundwater samples during the incubation periods may have occurred but this was not measured.

Activity in the flasks was terminated by addition of 1 mL 2N H_2SO_4. The ^{14}C-labeled CO_2 produced from urea hydrolysis was collected on filter paper suspended in the flasks as previously described (*8*) except that 2N NaOH was used to collect the CO_2. The amount of ^{14}C-labeled urea that was converted to ^{14}C-labeled CO_2 was quantified using scintillation fluid (ScintiSafe Plus 50%, Fisher Scientific) in a liquid scintillation counter (1220 Quantulus Wallac; Turku, Finland).

Urease and Community Molecular Biology Studies of Snake River Plain Aquifer Microbes

For molecular studies, microbial cells were filtered from Second Owsley groundwater through the course of the field experiment. At each sampling time, three separate 0.2 µm capsule filters (Pall Gelman Laboratory, Ann Arbor, MI) were obtained, each containing cells filtered from 100 L of groundwater. The Second Owsley groundwater contains approximately 10^5 cells/mL (D. Cummings, personal communication); thus, approximately 10^{10} cells were adhered to each filter and available for the subsequent DNA extractions. Filters were stored frozen at -80 °C prior to DNA extraction.

DNA from the capsule filters was extracted for whole community 16S rDNA and urease gene characterization using a modified sucrose lysis filter extraction method (*9, 10*). Briefly, 40 mL of lysis buffer (20 mM EDTA, 400 M NaCl, 50 mM Tris, and 0.75 M sucrose, pH 9), 4 mL 10% SDS, and

4 mg Proteinase K were added to the capsule filter (previously thawed at 4 °C for 2 h), incubated at 37 °C for 2 h and then 55 °C for 1 h, rotating every 20 min. The supernatant was extracted with chloroform, phenol:chloroform, and chloroform again. DNA was precipitated with sodium acetate and ethanol, and then resuspended in 10 mM Tris-HCL pH 8.0.

In order to detect microbes in the aquifer that possess the ability to hydrolyze urea, we developed primers designed to amplify *ureC*, a gene that codes for a conserved subunit of the urease enzyme. Several sets of PCR oligonucleotide primers were designed based on an alignment and phylogenetic analysis of known *ureC* genes from 44 bacterial species (*11*). A phylogenetic analysis was used to divide the sequences into three subgroups. All possible combinations of oligonucleotide primer pairs from each group were evaluated for their accuracy and specificity in PCR using DNA derived from several bacterial species belonging to each of the three subgroups identified. PCR products derived using the L2F/L2R primer pair having the predicted ca. 400 base pair length (Figure 2) were sequenced and found to have high similarity to known *ureC* sequences. Subsequent to these primer design studies, the L2F/L2R primer pair was used to determine the presence of urease genes in complex microbial communities collected from the SRPA before, during, and after the groundwater had been stimulated using molasses and urea in Second Owsley.

To determine changes in the aquifer microbial communities as a result of the molasses and urea amendments we used denaturing gradient gel electrophoresis (DGGE) of 16S rDNA gene fragments. Approximately 350 nt fragments of the 16S rRNA gene were amplified by PCR genes using the bacterial-specific primers 338FGC (CGCCCGCCGCGCCCCGCGCCCGTCCCG CCGCCCCCGCCCTCCTACGGGAGGCAGCAG) and 690R (TCTACGCATTTCACC). PCR conditions were as follows: Promega PCR buffer-1X final, 2 mM MgCl$_2$, 0.4% BSA, 0.2 mM dntps, 20 pmol each primer, and 1 U Promega Taq DNA polymerase. Thermal cycling conditions were as follows: 94 °C, 5 min, 35 cycles of 94 °C, 30 s, 50 °C, 30 s, and 72° C, 60 s, and a final incubation of 72 °C, 7 min. Duplicate DNA extractions from each of the SRPA sampling time points (normalized to filtrate amount) were used as template. PCR products were separated on a 30-70% urea/formamide-6% acrylamide gel and visualized with SYBR green (*12*). Bands with intensity changes between time points were selected for further analysis. DNA was eluted from the bands into 10 mM Tris pH 8 buffer. 16S rRNA gene fragments were amplified from this buffer using the primers 358F (CTACGGGRGGCAGCAGTG) and 690R. PCR products were purified using the Ultraclean PCR cleanup kit (MoBio) and sequenced with the same primers using the Big Dye Terminator V.2 cycle sequencing kit (Applied Biosystems) according to the manufacturer's protocol. Sequences of complementary strands

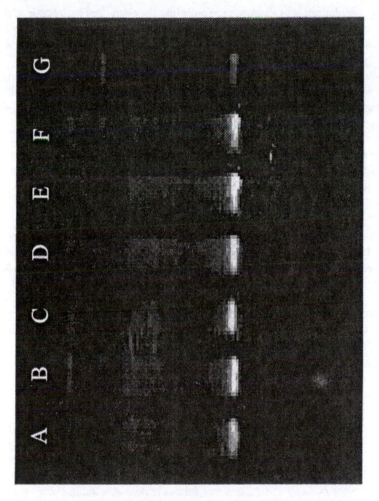

Figure 2. Urease genes amplified from seven unique, pure bacterial cultures isolated from Snake River Plain Aquifer groundwater. Lanes A through G show the characteristic PCR product (ca. 400 base pair length) amplified from these isolates.

were assembled and edited using Autoassembler (Applied Biosystems). Sequences were compared to GenBank using BLAST.

Modeling

Scoping calculations that illustrate the coupling between urea hydrolysis and calcite precipitation for a groundwater system similar to the SRPA were conducted using the Geochemist Workbench (13), a commercially available geochemical reaction path computer code.

The aquifer conceptual model is a mixed equilibrium-kinetic system. Dissolution reactions involving aquifer host rock were assumed to be negligible relative to the time frames of the simulation and were ignored. Cation exchange reactions involving major groundwater cations and NH_4^+ were assumed to be at equilibrium. Reactions other than the hydrolysis of urea that occurred within the groundwater phase were also assumed to be in equilibrium. Urea hydrolysis and calcite precipitation were both treated as kinetic reactions using rate laws as described below. First-order kinetics were used to describe urea hydrolysis. The rate constant was estimated from a measured rate of 0.6 mmol/L/day for groundwater amended with 50 mM urea (unpublished data). Calcite precipitation was described using a second-order chemical affinity based rate law (14). The rate constant was estimated from the result of an aquifer-scale hydrochemistry evaluation of the SRPA (15). Using inverse modeling and a steady state assumption, McLing (15) estimated that 0.3 mmol/L of calcite precipitates over the approximately 50 years that groundwater traverses the SRPA beneath the Idaho National Engineering and Environmental Laboratory. To estimate the rate constant, a constant (saturation index of 0.06) but slight degree of supersaturation was assumed.

Results and Discussion

Experiments using ^{14}C-labeled urea indicated that the microbes that make up the pristine SRPA communities possess considerable potential for urease activity under the correct conditions (Figure 3). Less than 6% of the ^{14}C-labeled urea was metabolized in each of the treatments shown in Figure 3 indicating that the urea available to the cells for hydrolysis was probably not depleted in such a way that the rates would be altered. The highest rate of urea hydrolysis (> 800 fmol urea hydrolyzed/mL/h) was noted for SRPA consortia that were supplemented with 0.01% molasses. The same consortia provided with less molasses (0.001% and 0.00075%) still showed marked urea hydrolysis (ca. 400 fmol urea hydrolyzed/mL/h). These values are higher than those measured for B. pasteurii

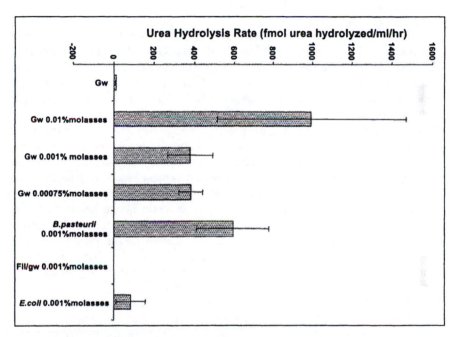

Figure 3. Mean urea hydrolysis rates for Snake River Plain Aquifer groundwater consortia provided with 0 (GW), 0.01%, 0.001%, and 0.00075% molasses as a carbon source supplement along with 5.9 nmol ¹⁴C-urea. For comparison, mean urea hydrolysis rates are also shown for B. pasteurii, E. coli, and filtered groundwater, all with 0.001% molasses. Each treatment was performed in triplicate, and error bars depict the standard deviation.

provided the same amount of molasses. The concentration of cells in the *B. pasteurii* suspension is more than 300 times the concentration of cells in the SRPA consortia. This suggests that the specific activity of the aquifer microbes is higher than that of *B. pasteurii* under similar conditions. The control flasks with SRPA communities and no added molasses showed only trace levels of urea hydrolysis. This suggests the need to amend the indigenous SRPA microbial communities with a source of carbon in addition to the urea if urea hydrolysis is to be detected at any reasonable level. Cultures of *E. coli* showed some evidence of urea hydrolysis even though these cells are not supposed to be able to enzymatically cleave urea.

The *ureC*-specific PCR primer set L2F/L2R was tested with bacterial community DNA collected and extracted from Second Owsley groundwater samples. Samples for DNA analysis were taken throughout the course of a four week study during which molasses and urea were added to the subsurface in order to stimulate urease positive communities. The predicted ca. 400 base pair fragment of *ureC* was amplified from all of the samples regardless of whether they were obtained from the well before, during, and after the molasses/urea addition (Figure 4). Although the microbial communities obtained from the well prior to the experiment showed evidence of *ureC*, the bands associated with this gene appeared to be more intense following the molasses and urea treatment. No attempt was made to quantify *ureC* copy number in the samples and so we cannot say whether the treatment caused an increase in the number of these genes in the groundwater. Our results indicate that the primers designed to specifically amplify the *ureC* gene work on known urease positive isolates, including *B. pasteurii*, and also on microbial communities that are extracted from groundwater before and after aquifer treatments that are designed to stimulate urea hydrolysis by indigenous communities.

The results of the DGGE molecular analysis of the microbial community in Second Owsley indicate that there was a shift in community members as a result of the molasses and urea treatment (Figure 5). One microorganism (represented by Band A in Figure 5) was present throughout the sampling periods although the DNA bands corresponding to this microbe were most intense in samples acquired prior to the addition of molasses and urea. When the 16S rDNA sequence from this microbe was compared by BLAST to 16S rDNA sequences in GenBank, it was 94% similar to sequences from known microbes in the Nitrospira group, specifically an uncultured clone "GOUTA19" (16). The Nitrospira group contains microbes that are chemolithotrophic nitrite-oxidizing bacteria (17) and some appear to be well adapted to low nitrite and oxygen concentrations (18). If ammonia resulting from urea hydrolysis was oxidized to nitrite then these bacteria could play a role in the conversion of nitrite to nitrate in the groundwater. One of the other prominent bands in the DGGE analysis (represented by Band B in Figure 5) had a 16S rDNA sequence that was most

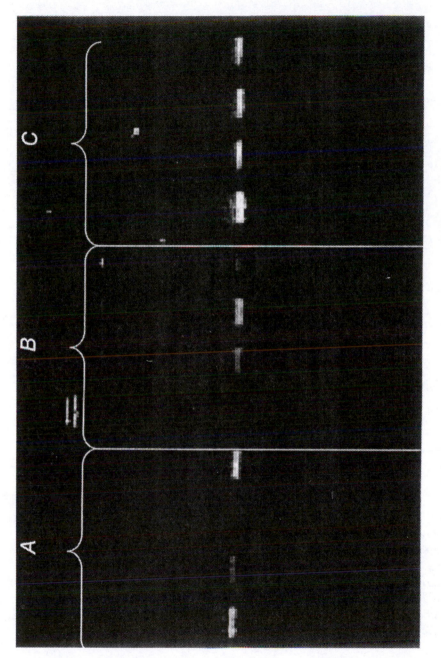

Figure 4. Urease gene products amplified from complex microbial communities filtered from Second Owsley groundwater before (A), during (B), and after (C) the field experiment. Different lanes within one of the sampling times represent replicate filters obtained from the well.

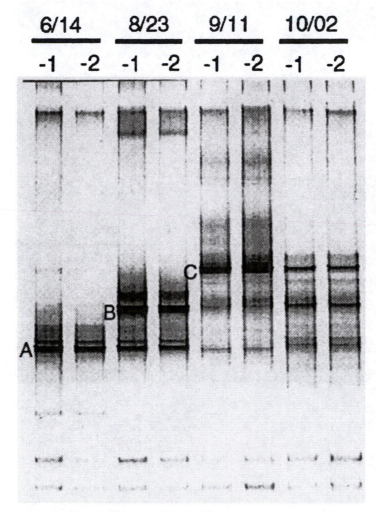

Figure 5. DGGE gel depicting 16S rDNA sequences amplified from water samples taken before (6/14 and 6/23), during (9/11), and after (10/02) the Second Owsley experiment. Based on sequence similarity to GenBank entries, the microbes from which the three prominent bands A, B, and C originated are most closely related to an uncultured bacterium, Pseudomonas putida, and Clostridium botulinum, respectively. "-1" and "-2" designations above the lanes indicate, respectively, undiluted and 2x diluted DNA extract used in the experiment.

closely related to that from *Pseudomonas putida* (99% similarity). This microbe was not an evident member of the community during the first sampling time (6/14) but appeared subsequently in the Second Owsley groundwater (on 8/23) still prior to the amendments. This suggests that there is some variability in the pretreatment groundwater communities that may be controlled by factors of which we are unaware. Band C in Figure 5 contains a sequence that appears to be related to that of several microbes in the genus *Clostridium*, particularly *Clostridium botulinum* (99% similarity). The sequence from this microbe was evident in samples taken on 6/14 but not in samples taken on 8/23. Following the addition of molasses and urea this microbe appeared to become a prominent member of the community before becoming less prominent on the 10/02 sampling point. Clostridia are all strict anaerobes, and the presence of Clostridia in the post treatment phase of the field experiment strongly suggests that our treatment caused the well to become even more anoxic than at the beginning of the study.

The evaluation of microbial responses to the molasses and urea additions to the aquifer at Second Owsley are an important aspect of our effort to track changes in the aquifer. These molecular tools are consistent with those recommended as being essential for determining whether bioremediation is occurring (*19, 20*). Ultimately, we hope to determine that *ureC* genes are actually being expressed (presence of mRNA that is specific to urease) as a way to confirm the activity of urease positive cells (*21*).

The results of a one-year kinetic simulation conducted using the Geochemist Workbench are presented in Figure 6. In this simulation, 10-mM urea was added to a representative groundwater. The groundwater pH was adjusted to an initial value of 7.3 so that the water was undersaturated with respect to calcite. The simulated aquifer had a porosity of 12% and a cation exchange capacity of 0.5 meq/100 g, both of which are consistent with observations for the SRPA. All simulation results are for a closed system and based on 1 kg of water (~1 L of solutions) and 20 kg of aquifer matrix.

The simulation shows that the pH increased rapidly to a value of approximately 8.7 due to urea hydrolysis (Figure 6a). The pH then decreased when calcite precipitation began. Figure 6b shows the behavior of some of the solution components. Urea decreased throughout the simulation in accordance with first-order kinetics. Initially, total NH_4^+, Ca^{2+}, and dissolved inorganic carbon (DIC expressed as HCO_3^-) increased. Ammonium and DIC were released by urea hydrolysis. Exchange of NH_4^+ ions for Ca^{2+} on cation exchange sites (Figure 6c) accounted for the increase in Ca^{2+} concentration. At longer times, calcium and DIC concentrations decreased and approached near constant values as calcite precipitated via the overall reaction:

$$CO(NH_2)_2 + 2H_2O + >X_2:Ca^{2+} \rightarrow CaCO_3 + 2 >X:NH_4^+$$

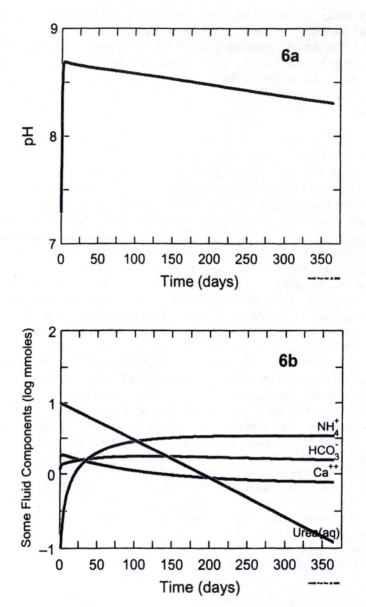

Figure 6. Results of a one-year kinetic simulation of urea hydrolysis for the Snake River Plain Aquifer. (a) pH. (b) Dissolved urea, dissolved inorganic carbon (as HCO₃), calcium, and ammonium. (c) Amount of calcium and ammonium on aquifer matrix exchange sites. (d) Amount of calcite precipitated.

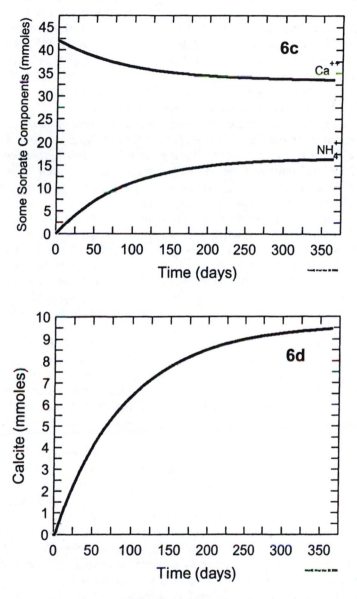

Figure 6. *Continued*.

where $>X_2:Ca^{2+}$ and $>X:NH_4^+$ are cations on ion exchange sites. The stoichiometry of the above reactions suggests that as long as there is sufficient urea and $>X_2:Ca^{2+}$ is available in excess, the urea hydrolysis rate and the calcite precipitation rate should be the same. In fact for most of this simulation (days 2 through 285), the absolute value of the two rates was within 10% of each other. At longer time the urea was depleted causing decoupling of the precipitation of calcite and the hydrolysis of urea. In addition, the rates of both reactions decreased significantly (less than 5% of the maximum rate) as background aquifer conditions and processes were approached.

During the course of the simulation, almost 10 mmol of calcite precipitated (Figure 6d) corresponding to the amount of urea initially present. It is important to note that only 1.75 mmol of Ca^{2+} was present in the initial groundwater and at the end of the simulation there was 0.77 mmol present in solution. This indicates that approximately 90% of the calcium in the precipitated calcite originated from calcium that was sorbed onto the aquifer matrix.

Although the simulation results presented here are not based on the results of field experiments, important generalizations regarding urea hydrolysis and calcite precipitation can be made using the simulation. First, because the rate constants used in this study are based on (as much as possible) pristine aquifer conditions, any engineered system that includes electron donor amendments at a contaminated location will likely be faster. Second, any application in the field would probably exhibit the three phases observed in this simulation, namely (1) urea hydrolysis rate greater than calcite precipitation rate, (2) calcite precipitation rate greater than urea hydrolysis rate, and (3) the two rates being equal. Third, the majority of the metals precipitated as carbonates will come from the aquifer matrix, not the groundwater. Fourth, during the period in which significant urea was available for hydrolysis, the entire hydrolysis-precipitation process can be described by a simple pseudo-first order kinetic model with the precipitation rate controlled by the urea hydrolysis rate and independent of the concentration of calcium in the groundwater.

In this paper we have presented results of investigations prerequisite to the use of calcite precipitation as a means to remove hazardous divalent cations (e.g., ^{90}Sr) from groundwater. Our proposed method relies on the acceleration of calcite precipitation by encouraging aquifer microbes to hydrolyze urea and thereby increase the pH and alkalinity of the system. We have demonstrated that aquifer microbial communities possess the ability to hydrolyze urea at rates that are consistent with a known urea hydrolyzing microbe (*B. pasteurii*) and possibly even at higher rates under similar conditions. Community urea hydrolysis was most rapid when the cells were simultaneously provided with a source of organic carbon (molasses) as opposed to when no exogenous organic carbon was supplied. This suggests that field experiments may require an organic carbon amendment if reasonable rates of urea hydrolysis are to be expected. In order to trace the effect of accelerated calcite precipitation in the field, we designed *ureC* gene primer pairs in order to use PCR to detect the gene in environmental samples. We confirmed the efficacy of these primers with isolates from the Snake River Plain aquifer, one of the candidate aquifers for calcite precipitation-mediated remediation. Our molecular ecology experiments at an aquifer field site indicated that the complex microbial communities in this aquifer possess the *ureC* gene and that we can detect the gene in DNA extracts from free-living cells collected from the aquifer. Furthermore, using 16S rDNA studies we measured changes in the aquifer microbial community as a result of introducing molasses and urea in a preliminary test of our method. Because two distinct wells were used in these experiments the microbial communities in the two settings may have been different. Nevertheless, both of the wells access the same aquifer defined by fractured basalt geology and based on prior studies in the same aquifer (*3*), we believe that cells capable of urea hydrolysis are generally abundant in the SRPA.

A kinetic simulation of the microbial-facilitated calcite precipitation was conducted using geochemical conditions consistent with the SRPA, realistic concentrations of urea, and rates of microbial urea hydrolysis that are consistent with those measured on complex microbial communities. In the simulation, urea hydrolysis and then calcite precipitation proceeded at maximal rates. We expect that in a real field test of our process in which an electron donor (e.g., molasses) is added to the system to stimulate urea hydrolysis, the process would proceed even more rapidly. The modeling suggested that over the long periods that the process would occur in an aquifer, the precipitation rate would be controlled by the urea hydrolysis rates and would be independent of the concentration of calcium in the aquifer.

In future field experiments in which we will attempt to accelerate calcite precipitation, we will closely track the water chemistry of the aquifer during the experiment to determine whether calcite is precipitated. Our newly designed urease oligonucleotide primer sets will be used to quantify *ureC* gene copy number in groundwater samples using MPN-PCR or real-time PCR. Quantification of *ureC* gene copy number before, during, and after in situ urea amendment to groundwater will provide supporting evidence for the efficacy of this overall remediation approach by showing that we can increase the number of urea hydrolyzing bacteria. Use of 16S rDNA community profiling over the course of a field experiment and beyond will provide data that can indicate how aquifer microbes change during the experiment and indicate the longevity of the changes in the community. These techniques will help to evaluate subsurface remediation of divalent cations (e.g., ^{90}Sr) using this biogeochemical manipulation as a promising approach for arid western DOE sites such as Hanford and the Idaho National Engineering and Environmental Laboratory, where calcite precipitation is already an ongoing geochemical process.

136

Acknowledgments

The DOE Environmental Management Science Program provided funding for this work. Research at the Idaho National Engineering and Environmental Laboratory is performed under the DOE Idaho Operations Office Contract DE-AC07-99ID13727.

References

1. Riley, R. G.; Zachara, J. M.; Wobber, F. J. *Chemical Contaminants on DOE Lands and Selection of Contaminant Mixtures for Subsurface Science Research*; DOE/ER-0547T; U.S. Department of Energy, Office of Energy Research, Subsurface Science Program: Washington, DC, 1992.
2. National Research Council. *In Situ Bioremediation. When Does it Work?* National Academy Press: Washington, DC, 1993.
3. Fujita, Y.; Ferris, E. G.; Lawson, R. D.; Colwell, F. S.; Smith, R. W. *Geomicrobiol. J.* **2000**, *17*, 305–318.
4. Warren, L. A.; Maurice, P. A.; Parmar, N.; Ferris, F. G. *Geomicrobiol. J.* **2001**, *18*, 93–115.
5. Knutson, C. F.; *Communities*; Kluwer Academic Publishers: The Netherlands, 1996; Vol. 3.4.4, pp 1–23.
13. Bethke, C. M. *The Geochemist's Workbench: A User's Guide to Rxn, Act2, Tact, React, and Gtplot*; Release 4.0 ed.; Hydrogeology Program, University of Illinois: Urbana, IL, 2002.
14. Teng, H. H.; P.M., Dove; De Yoreo, J. J. *Geochim. Cosmochim. Acta* **2000**, *64*, 2255–2266.
15. McLing, T. L. *The Pre-Anthropogenic Groundwater Evolution at the Idaho National Engineering Laboratory, Idaho*; Idaho State University: Pocatello, ID, 1994; pp 62.
16. Alfreider, A.; Vogt, C.; Babel, W. *Appl. Microbiol.* **2002**, *25*, 232-240.
17. Teske, A.; Alm, E.; Regan, J. M.; Toze, S.; Rittmann, B. E.; Stahl, D. A. *J. Bacteriol.* **1994**, *176*, 6623–6630.
18. Loy, A.; Daims, H.; Wagner, M. *Activated Sludge - Molecular Techniques for Determining Community Composition*; Bitton, G., Ed.; Wiley: New York, 2002; Vol. 1, pp 26–43.
19. Madsen, E. L. *Environ. Sci. Technol.* **1991**, *25*, 1663–1673.
20. Madsen, E. L. *Environ. Sci. Technol.* **1998**, *32*, 429–439.
21. Wilson, M. S.; Bakermans, C.; Madsen, E. L. *Appl. Environ. Microbiol.* **1999**, *65*, 80–87.McCormick, K.A.; Smith, R.P.; Hackett, W.P.; O'Brien, J. P.; Crocker, J. C. *FY89 Report Radioactive Waste Management Complex*

Vadose Zone Basalt Characterization; EG&G Idaho Report EGG-WM-8949; EG&G: Idaho Falls, ID, 1990.

6. Morse, L. H.; McCurry, M. *Genesis of Alteration of Quaternary Basalts within a Portion of the Eastern Snake River Plain Aquifer*. In *Geology, Hydrology, and Environmental Remediation*; Link, P. K., and Mink, L. L., Eds.; Idaho National Engineering and Environmental Laboratory: Eastern Snake River Plain, Idaho, 2002; Special Paper 353; p 213–224.

7. Wright, R. T.; Hobbie, J. E. *Ecology* **1966**, *47*, 447–464.

8. Peters, G. T.; Colwell, F. S. *Hydrobiologia* **1989**, *174*, 79–87.

9. Britschgi, T. B.; Giovannoni, S. J. *Appl. Environ. Microbiol.* **1991**, *57*, 1313–1318.

10. Somerville, C. C.; Knight, I. T.; Straube, W. L.; Colwell, R. R. *Appl. Environ. Microbiol.* **1989**, *55*, 548–554.

11. Tyler, T. L.; Watwood, M. E.; Colwell, F. S. *Characterization of Urease Genes in Subsurface Microbial Communities*, Annual Meeting of the American Society for Microbiology, abstract, 2002.

12. Muyzer, G.; Hottentrager, S.; Teske, A.; Wawer, C. *Denaturing Gradient Gel Electrophoresis of PCR-Amplified 16S rDNA-A New Molecular Approach to Analyse the Genetic Diversity of Mixed Microbial*

Chapter 7

^{13}C-Tracer Studies of Soil Humic Substructures That Reduce Heavy Metal Leaching

Richard M. Higashi[1], Teresa Cassel[2], Peter G. Green[3], and Teresa W.-M. Fan[4,*]

[1]Crocker Nuclear Laboratory and Center for Health and Environment, University of California, Davis, CA 95616
[2]Crocker Nuclear Laboratory and [3]Civil and Environmental Engineering, University of California, Davis, CA 95616
[4]Department of Chemistry, University of Louisville, Louisville, KY 40208

For stabilization of heavy metals at contaminated sites, the unavoidable biogeochemistry of soil organic matter and particularly humic substances (HS), not just their static properties, in relation to heavy metal binding is critically important for long-term sustainability. Yet this factor is poorly understood at the molecular level. Soil with ^{13}C-labeled HS was spiked with Cd-accumulated wheat roots—a biogeochemically relevant form of Cd—and amended with five different organic byproducts (cellulose, wheat straw, pine shavings, chitin, and bone meal), then allowed to humify for 299 days in column format. Metal ion mobility was examined periodically by leachate collection and inductively coupled plasma mass spectrometry analysis of leachates. The labeled molecular substructures in the HS were then examined by multinuclear 2-D nuclear magnetic resonance and pyrolysis-gas chromatography mass spectrometry. The residual isotopic abundances in different substructures of soil HS were used as turnover indicators for the humification process of various organic amendments. In this report, we describe the residual ^{13}C labeling of HS substructures, which revealed that turnover—not the abundance—of polysaccharidic and lignic HS substructures was associated with reduced leaching of Cd, Ni, and Cs.

Introduction

Diverse human activities including strategic and energy development at various U.S. Department of Energy (DOE) and military installations have resulted in contamination of soils and waterways that can seriously threaten the health of humans and ecosystems. Unlike organic pollutants which undergo degradation, detoxification, and even mineralization to CO_2, metal contaminants cannot be readily decomposed beyond their elemental form. Thus, the options are limited to removal and transformation into less toxic and/or less bioavailable form(s). Physical removal technologies such as excavation and soil washing are costly and destructive to the soil ecosystems. The much less invasive phytoextraction approach (i.e., extracting metals by plant roots and translocating them to shoots) is still being developed, often requiring the addition of synthetic chelators (1-3) while the costs associated with disposal of metal-laden plant matter and ecotoxic risk from the standing crop remains in debate.

In recent years, studies have explored the transformations of metal ions from toxic to less toxic or mobile form(s) such as from Cr(VI) to Cr(III) or from U(VI) to U(IV) (4, 5). This approach would be useful for redox-sensitive metals and metalloids but not applicable to metals in general. Meanwhile, in situ stabilization or bioavailability reduction via the addition of inorganic and organic amendments has received increasing attention due to the potential for low cost and general applicability to a number of metals and metalloids, with some approaches that are expected to have minimal disturbance to ecosystem functions. For example, the addition of inorganic amendments such as lime, zeolites, apatites, and oxides of Fe and Mn has been investigated for their efficacy in immobilizing various metals in laboratory settings (6-8). Transition metals, heavy metals such as Pb and Cd, as well as radionuclides such as uranium and cesium were shown to be immobilized or less bioavailable for plant uptake.

However, the immobilization efficiency of inorganic amendments depends on pH, soil organic content, and soil mineralogy (8). In some cases, the approach perturbs existing ecosystems since amendments have many active sites and must be incorporated physically into the soil matrix. Additionally, it is unclear how different vegetation, microbial processes, and the attendant humification processes may impact the long-term efficacy of the stabilization. Plants and microbes can have aggressive means for mining the soil, which could remobilize the sorbed metals. In addition, the interaction (e.g., "coating") of soil organic matter (OM) with mineral matrices could interfere with metal precipitation, cation exchange capacity, and/or fixation into mineral interlayers, which are the three common mechanisms for metal immobilization by mineral addition (6). Therefore, the long-term influence of soil biogeochemistry on soil OM will need to be considered to implement the approach of inorganic amendments.

Since soil OM is going to be unavoidable, some have attempted to engineer this complex system. Alkaline composted biosolids appear to be the preferred test materials for metal stabilization studies (6, 9, 10). Trace elements such as Cd, Cu, Pb, Ni, and Zn were shown to be less exchangeable in soils and/or less available to plants after addition of biosolids. However, biosolids can be a source of contaminant metals that may build up with repeated applications (6) while the long-term persistence in metal stabilization is unclear.

As the main product of plant-microbe activity and an important sink for both cationic and oxyanionic metals, the importance of the "recalcitrant" portion of soil OM—humic substances (HS)—and their genesis in metal ion interaction cannot be overstated. HS is a major factor in soil: the amount of C in the form of HS (ca. 1400-1500 Pg C) is thought to represent two-thirds of the terrestrial C pool (11-13), making it by far the most abundant natural OM in soil and sediments (14) and a primary driver of soil biogeochemical processes. HS can react with metal ions through complexation, ion exchange, sorption, or precipitation. Their polyanionic nature and high affinity for polyvalent cations (15, 16) makes them a major sink for heavy metal ions in soils and sediments. In addition, HS have much slower turnover rates than other OM, strongly interact with soil matrices (which makes them less mobile), and are readily manipulated through vegetation and/or organic amendments. These properties of HS appear highly desirable for stabilizing contaminant metals with minimal disturbance to the soil ecosystems. Other potential benefits of HS-based stabilization approach include stabilizing and/or enhancing biodegradation of organic pollutants since HS can support microbial degradation of organic pollutants and are known to interact with organic pollutants to reduce their toxicity to biota (17). Moreover, the HS-based approach can either be implemented alone or in combination with other in situ remediation techniques, and is well-suited for mixed waste sites where both metal and organic contaminants are of concern, which is a common, rather than exceptional, occurrence.

We have been exploring biostabilization of Cd via humification of organic byproducts from plant roots as well as various agricultural or industrial processes. This work builds on the knowledge and tools that we have generated relating root exudation and metabolism (18–21), as well as soil OM structures (22–24), to metal mobilization and accumulation by plants. For example, in recent studies, isolated soil HS appeared to facilitate metal sequestration while alleviating Cd-induced growth reduction in wheat plants (25). The ability of HS to protect plants from toxic effects of excess metals has also been reported previously (26, 27).

Due to the complex nature of HS, it is conceivable that different parts of the HS structure may mediate different functions regarding metal ion mobilization or immobilization. In addition, different HS substructures have different biogeochemical turnover rates, mainly due to physical or chemical protection by soil matrix and/or innate biochemical stability. For example, strongly bound metal ions can be released from HS if the appropriate substructures—say, the proteinaceous remnants—are degraded by microbes, weathering, or other factors. This multifunctional, biogeochemical cycling nature of HS is often underconsidered. Therefore, to gain a mechanistic understanding of HS functions in metal ion stabilization, the importance of HS substructures turnover on metal ion sequestration must be determined, in contrast to the more prevalent metal ion binding studies on biogeochemically static HS structures.

Thus, our present goal was to investigate HS structures and turnover that relates to reduced metal ion leaching in vegetated topsoils, in response to various organic amendments. We first prepared Cd-loaded wheat root tissues as a relevant form of biogeochemical Cd, since roots accumulate Cd extensively (e.g., *19, 20, 25*). We also performed ^{13}C-labeling soils from the DOE Savannah River Site (SRS); the rationale for this soil is given in the Methods section. These roots and soils were employed in soil column metal leaching experiments where different organic amendments of natural isotopic abundance were humified for 299 days. The residual isotopic abundance in different substructures of HS in these soils was used as turnover indicators for the humification process of various organic amendments. In this report, we describe the residual ^{13}C labeling pattern of HS which revealed that turnover of selected HS substructures was associated with reduced metal ion leaching.

Materials and Methods

Soil was collected from two horizons (surface and 0.5 m depth) in an uncontaminated area near the Tim's Branch stream system at the DOE SRS site, and stored at 4 °C in the dark until use. Our interest in working with this type of soil stems from the fact that Tim's Branch, a second order stream system flowing into a tributary of the Savannah River, experienced large influxes of depleted and natural uranium (U), nickel (Ni), and aluminum (Al) as well as other metals. Tim's Branch soils and vegetation have had a strong impact on the deposition of metals and metal-laden sediments (*28*).

A soil aging experimental system capable of stable operation for several months (see left half of Figure 1) was used for analytical method development and for an initial survey of the effect of organic amendments on soil OM transformation. This system was then used to produce ^{13}C-labeled SRS soils by incubating the soils with [U-^{13}C]-glucose for 34 weeks at 25 °C. In Figure 1, the left setup for labeling soil is a polysulfone, airtight, Buchner vacuum filtration system (Nalgene, Inc.), while the right setup for humification of organic amendments consists of parts for a polypropylene solid-phase extraction device (Alltech, Inc., Deerfield, IL). [U-^{13}C]-glucose together with KNO_3 dissolved in deionized H_2O was added to each pot to double the percent nitrogen in the soil and to increase the existing soil carbon by 10-33% (Table I). These additions were performed at week 0, 5, 10, and 23. Aging of labeled

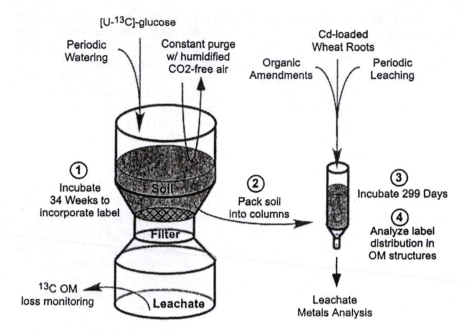

Figure 1. Schematic of soil labeling and humification setups. Experimental design is explained in the text. The left setup for labeling soil is a polysulfone airtight filtration system, while the right setup for humification of organic amendments consists of parts for a polypropylene solid-phase extraction device.

components was initiated with 100 grams each of field moist soil in four pots, two each with the surface and subsurface soils. Soil pots were irrigated monthly with a nitrogen free micronutrient solution, adapted from Bundy and Meisinger (29), containing 7mM $CaSO_4$, 2mM $MgSO_4$, and 5 mM KH_2PO_4 at pH = 6.0 applied under vacuum for approximately 30 minutes. Leachates were collected from the receivers, freeze-dried, and analyzed for loss of labeled supplements. The first four irrigations used 25 mL of solution, and the subsequent six were made with 15 mL of a 2X concentrated solution to minimize leaching of the labeled supplements.

Table I. Addition of Labeled (or Unlabeled) Carbon and Nitrogen to Soil Preparation

Soil type	Initial N	Initial C	N added		C added	
	%	%	%	KNO$_3$ (g)	%	Glucose (g)
Surface	0.03	0.80	0.03	0.23	0.11	0.50
Subsurface	0.01	0.15	0.03	0.23	0.05	0.25

Ten grams each of the enriched surface soils was used for HS extraction and 2.5 g of all soils was collected for other analyses. The remaining surface soil was combined with subsurface soil in a 75:95 ratio to create the experimental soil labeled with ^{13}C. This soil was mixed thoroughly with Cd-loaded wheat roots (grown in a Cd-spiked hydroponic media [19]) to yield nominal 3.5 mg Cd/g soil, and then divided into 20 g each aliquot. Each of the five different organic amendments (cellulose, wheat straw, pine shavings, chitin, and bone meal) was mixed with the aliquoted soil at 35 mg amendment/g soil. Cellulose was metal-free, X-ray fluorescence analytical grade from Spex Certiprep (Metuchen, NJ), wheat straw was obtained from a University of California–Davis agricultural field site, pine shavings and bone meal were purchased from a local gardening store, and chitin was obtained from Sigma-Aldrich (St. Louis, MO). A subsample of the aliquot was taken before the soil was divided and packed into two 15-mL polypropylene columns with soil retained by a 20-µm polyethylene frit and 0.7-µm glass fiber filter. These duplicate columns of the five amendments were kept at 25 °C in the dark and irrigated weekly alternated with 250-µL of deionized water or the nitrogen-free 2X nutrient solution described above to maintain microbial activities. Irrigation solutions were allowed to infiltrate for 30 min before the columns were centrifuged to collect the leachate for 1 min at approximately 100 xg.

Leachates were stored at -20 °C until analysis. Six leachates for each column were pooled to generate monthly samples which were centrifuged to remove particulates and 100-200 µl each aliquots were acid-digested, and analyzed by inductively coupled plasma mass spectrometry (ICP-MS) for a number of elements including transition metals, Cd, Pb, Hg, Cs, Sr, Ag, and Sn. An Agilent 7500i ICP-MS (Agilent, Inc., Palo Alto, CA) with autosampler was used to deliver the digests to a Babbington nebulizer housed in a cooled spray chamber. The plasma was robustly tuned to minimize matrix effects, doubly-charged ions (Ce^{+2}/Ce~1.5%) and oxides (CeO/Ce~0.4%). External standards were prepared in mixtures with elements near the anticipated concentrations to maximize precision, minimize carry-over, and take as full advantage of the dynamic range as possible to quantify major, minor, and trace elements simultaneously.

HS was extracted from the labeled SRS soils with 0.25 M NaOH following our previous procedure (23). The isolated HS was then subjected to multinuclear, one and two dimensional nuclear magnetic resonance spectroscopy (2-D NMR) and pyrolysis gas chromatography-mass spectrometry (pyro-GCMS) analysis, as described previously (*23, 24, 30*).

Results and Discussion

For stabilization of heavy metals at contaminated sites, the unavoidable biogeochemistry of HS—not just the static properties of HS—in relation to heavy metal binding is critically important for long-term sustainability, a factor that is poorly understood at the molecular level. Using a soil isotopic labeling system to produce material for soil column experiments (Figure 1), the humification process in relation to metal ion mobility was investigated. The soil HS was labeled with ^{13}C using ^{13}C-glucose as the precursor, progress of which was monitored by pyro-GCMS.

We found that pyrolyzed fragments that are of protein and polysaccharidic origins were highly enriched in ^{13}C. Figure 2 illustrates the salient information from this analysis. Pyro-GCMS thermolyzes peptidic groups to form indole (C_8NH_{10}) with expected all-^{12}C ion at mass-to-charge ratio (m/z) 117, and the smaller natural abundance ^{13}C ion at m/z 118. Depending on the number of carbons labeled, a distribution of peaks from m/z 117 –> 125 can be expected to be seen for this peptidic marker, with m/z 125 representing a fully ^{13}C-labeled structure. An example of non-labeled standard peptidic material, bovine serum albumin, is shown in the bottom panel. In the upper panel, there appears to be primarily two major pools of soil peptidic structures: non-labeled (m/z 117), and fully labeled (m/z 125). The non-labeled pool, persisting after 34 weeks of incubation with ^{13}C glucose, apparently represents a "recalcitrant" pool of peptidic material in soil, consistent with previous findings (e.g., *23, 24*).

Following 34 weeks of labeling, the soils were repacked into columns (Figure 1), and Cd-loaded roots were added. The addition of Cd-loaded root, from wheat plants grown hydroponically in elevated Cd, was to simulate vegetated conditions (e.g., natural or under phytoremediation) where metal-loaded roots are an inevitable part of the belowground system. Other metals—notably Ni, Sr, Cs, and Pb—were extant in the SRS soil, and provided an opportunistic study of the leaching of these metals in addition to Cd. Thus, it should be noted that none of these metals were experimentally spiked to the soil as solutions, as is commonly done, but rather were incidentally pre-aged with contaminant metals (Ni, Sr, Cs, Pb), to which a biogeochemically relevant form of Cd was added (bioaccumulated in roots).

The columns were also amended with five different organic byproducts and allowed to humify for 299 days in column format. The organic amendments used were cellulose (polysaccharide), wheat straw (lignocellulose/silicate), pine shavings (lignocellulose/phenolic), chitin (nitrogenous polysaccharide), and bone meal (divalent cation phosphate). The reasons for these amendments are as follows. Cellulose is a major single-substance terrestrial OM input to soil from plant matter; lignocellulose/silicate are representative of grassy and other plant material; lignocellulose/phenolic are representative of pine forest inputs;

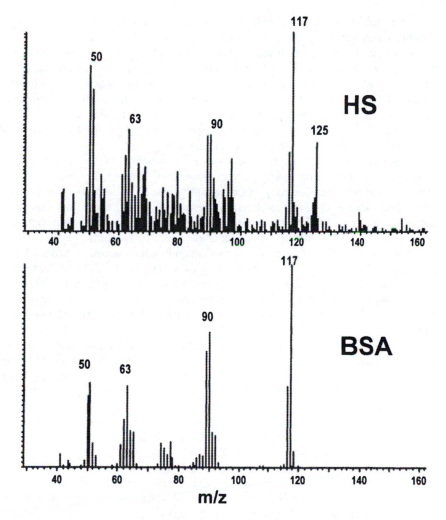

Figure 2. Pyro-GCMS output showing pattern of labeling peptidic groups in the soil incubated with ^{13}C glucose (expt. described in text). HS: ^{13}C-labeled soil humic material. BSA: Bovine serum albumin, a standard peptidic material.

chitin is representative of insect and fungal OM inputs to soil; and phosphate from bone meal is intended to provide a biological source of divalent cation binder for comparison. Metal ion mobility was examined periodically by leachate collection and ICP-MS analysis of leachates. The labeled molecular substructures in the HS of the starting and ending soil in columns were then examined by multinuclear 2-D NMR and pyro-GCMS.

The mobility of metal appeared dependent on both the type of metals and organic amendments applied, presumably reflecting the different humification processes and interaction of humified products with the metal. During the aging time course, distinct metal leaching patterns were observed as shown for Cd in Figure 3. Most notably, all but one (i.e., chitin) of the organic amendments eventually led to lower Cd leaching rates (relative to the unamended control) after 10 months of incubation. It is also interesting to note that bone meal and chitin amendments led to an initially high Cd leaching (up to 36 days of incubation), but dropped below the control level thereafter.

However, Figure 4 shows that, when the amount of metal leached is summed over the entire aging period, bone meal and chitin-amended soils showed a net increased leaching of Cd, in addition to Ni and Cs, over the control treatment. This is contrary to the simple expectation that these two materials may be good metal binders, being rich in phosphates (i.e., calcium phosphate for bone meal) as well as carbonyls and amines (N-acetyl glucosamine for chitin). In contrast, cellulose and pine shavings amendments showed a reduced net leaching of Cd, Sr, and Cs, while wheat straw amendment exhibited decreased leaching of Pb and Cs. These three materials are rich in lignin and cellulose, which suggests that humification of lignocellulosic materials may be beneficial for metal stabilization. A corresponding ^{15}N-labeled experiment showed the same results.

The humified product from some of the aged soils (after 299 days of incubation) has been analyzed by 1-D and 2-D NMR, as shown in Figures 5 and 6. Figure 5 compares the 1-D ^{1}H NMR spectra of soil humates extracted from soils amended and aged with cellulose, wheat straw, pine shavings, chitin, and bone meals for 299 days. These data reveal the relative increase in the spectral region of 3-4.8 ppm for soils amended with pine shavings, wheat straw, and cellulose, while a decreased intensity was observed for soil amended with bone meal, as compared with the unamended soil. As described below, this region is largely composed of resonances arising from amino acids and polysaccharides. It should be noted that the overall spectral appearance for these products is similar to that of humates isolated from natural soils, which suggests that the organic amendments have been adequately humified.

Further 2-D ^{1}H total correlation spectroscopy (TOCSY) analysis of the humate from the cellulose treatment indicates that this region is rich in αCH and CH_2 of amino acid residues plus CH/CH_2 of polysaccharides (Figure 6). These assignments were made based on the proton covalent connectivity and chemical shift information acquired from the TOCSY spectrum, and are in agreement with the 2-D [1]H-^{13}C heteronuclear single quantum coherence (HSQC) analysis (data not shown) and assignment of other soil humates (23, 31). In addition, based on the chemical shift information from TOCSY and/or HSQC, the amino acid residues are all in peptidic linkages (24), while the deoxyribose residue is most likely to be originated from nucleic acids. The TOCSY analysis also reveals the abundance of amino acid and saccharidic residues (Figure 6) while the HSQC data indicated the persistence of ^{13}C-enrichment in amino acids including Val, Leu, Ile, Thr, Ala, Asp, and Gly (data not shown). Moreover, the

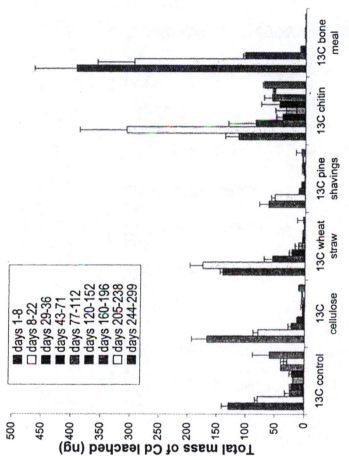

Figure 3. Cd leaching from ^{13}C-labeled SRS soil columns amended and aged with cellulose, wheat straw, pine shavings, chitin, and bone meal. Cd was leached from the soil columns periodically during the 299 days of incubation period. Initial Cd load (added as Cd-loaded wheat roots) in the soil averaged 2.4 mg/kg.

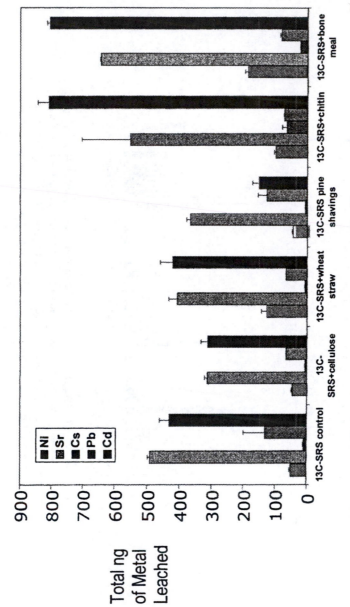

Figure 4. Total sum of metals leached from the amended ^{13}C labeled SRS-soils after 299 days of aging. The metals in all leachates from Figure 3 (average of two replicates) were summed over 0-299 days and plotted as a function of the amendment.

significant presence of Glu/Gln and Pro residues in the humified product, as the case for the starting soil (data not shown), could be attributed to a combination of slow turnover of unlabeled residues and some enrichment, as discussed above.

Complementary pyro-GCMS analysis of the pre-^{13}C labeled, organic-amended SRS soils provided quantitative information on the abundance of soil OM substructure and their % isotopic enrichment. In Figure 7, these parameters are plotted against total metals leached over 299 days (data from Figure 4). "Abundance" refers to the total lignic or polysaccharidic (PS) soil OM. "Incorporation" is the total minus the ^{13}C labeled soil OM, that is, displacement of ^{13}C from a given soil OM type, which is an indication of turnover. The analysis involved quantification of each partially labeled ^{13}C isotopomers for each OM constituent (e.g., a simple case was illustrated in Figure 2). "Isotopomer" analysis is the % [13]C enrichment in individual C within a structural fragment. Thus, in the present case, a single pyro-GCMS run took 12 passes of multi-peak data reduction to quantify all the isotopomers for six OM constituents.

As shown in Figure 7 top panel, a higher PS incorporation (closed squares) related well to the reduced Cd leaching, while the PS abundance (open squares) did not. This indicates that the turnover of PS was the controlling factor, not the total amount of PS accumulated in the system. In fact, for the pine shavings, there was very little accumulation of PS over 299 days, while the cellulose treatment actually lost PS (Figure 7, open squares). Yet these two were the most effective in reducing metal leaching (Figure 4), possibly because of their high turnover (Figure 7, closed squares). The lignic OM (circles) showed a trend similar to that for PS.

For Ni, the results for both PS and lignic OM showed a similar general trend as Cd (Figure 7, middle panel). The typically monovalent Cs behaved somewhat differently from the divalent metals Cd and Ni, in that amendments—including the control—that resulted in any turnover of PS (closed squares) or lignic OM (circles) resulted in lower leaching (Figure 7, bottom panel). As with Cd and Ni, the abundance of the PS OM (open squares) did not relate to Cs leaching. Other soil OM types (e.g., peptidic, alkanic, alkenic) and total soil OM showed little relation to leaching of any elements analyzed in this study (data not shown). However, this does not rule out the involvement of specific residues (e.g., peptidic Glu) in reducing metal leaching since pyro-GCMS data do not distinguish individual amino acids or saccharide monomers.

Thus, Figure 7 reveals a possible mechanism for the effectiveness of pine shavings and cellulose in reducing Cd, Ni, and Cs leaching, namely, the turnover of PS is important for reduced metal leaching. Figure 7 also reveals that both turnover and abundance of lignic OM may play a role, so that the combination of lignic + PS materials, such as natural plant materials, may be the most efficacious. Yet, this cannot be extrapolated to all plant materials. Figure 4

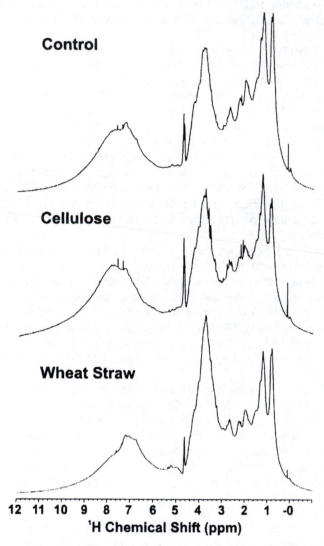

Control

Cellulose

Wheat Straw

¹H Chemical Shift (ppm)

Figure 5. 1-D ¹H NMR analysis of humified products from ¹³C labeled SRS soil amended with different byproducts. This spectral region is rich in functional groups originated from αCH and CH₂ of amino acid residues and CH/CH₂ of carbohydrates, as shown by the ¹H TOCSY analysis of humate from the cellulose treatment in Figure 6.

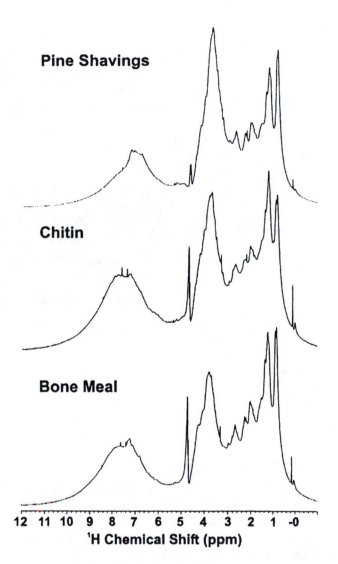

Figure 5. *Continued.*

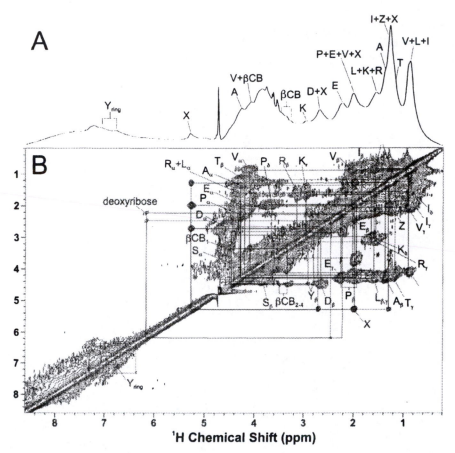

Figure 6. 2-D ^1H TOCSY analysis of soil humate isolated from ^{13}C-labeled SRS soil subsequently aged in cellulose. The spectra shown are 2-D TOCSY (B) along with the 1-D high-resolution ^1H spectra (A). Proton covalent connectivity is traced by rectangular boxes. βCB represents β-cellobiose while X and Z denote unassigned but abundant structures. Based on the proton connectivity, X should have the structure of CH_3-CH_2-CH_2-CH-X. The letters used for amino acids are as standard biochemical abbreviations: A, alanine; I, isoleucine; L, leucinel; V, valine; T, threonine; K, lysine; P, proline; E, glutamate; Q, glutamine; G, glycine; R, arginine; D, aspartate; N, asparagine; S, serine; and Y, tyrosine.

shows that wheat straw—a lignocellulosic material—resulted in increased leaching of Ni. Figure 7 may be revealing why this was so: for wheat straw, there was no PS incorporation (closed squares, middle panel) for 299 days. Once again, the net loss of PS (open squares, middle panel) over 299 days was not the factor for increased Ni leaching, as cellulose was effective in reducing Ni leaching but had considerably greater PS loss (Figure 7 middle panel, open square data point at lowest left).

These results suggest to us that, in order to reduce leaching and stabilize metals in soil, the starting materials (amendments, natural plant litter, etc.) may need to elicit biogeochemical changes that promote within the HS the turnover of PS and abundance plus turnover of lignic OM. Alternatively, "poor" starting materials with otherwise desirable characteristics, might stabilize metals if the biogeochemistry (vegetation, microbial community, etc.) of the site can be managed or coerced to promote turnover of PS and lignic OM into HS.

In summary, the combination of aging experiments, isotopic enrichment, and structure/dynamic characterization provided molecular-level information on humification processes in relation to metal mobility and binding that has eluded scientific inquiries in the past. We believe that much more novel mechanistic findings are yet to be uncovered for metal stabilization purposes, by continuing and expanding this approach.

Acknowledgments

This work was supported by the DOE Environmental Management Science Program, grant # DE-FG07-96ER20255, U.S. Environmental Protection Agency (EPA) grant #R825960010, and the Strategic Environmental Research and Development Program (SERDP) grant # DACA72-03-P-0013. Although the information in this document has been funded wholly or in part by the EPA, it may not necessarily reflect the views of the Agency and no official endorsement should be inferred. NMR spectra were recorded at the MRC Biomedical NMR Centre, London. The authors wish to thank Dr. Andrew N. Lane (currently at the J.G. Brown Cancer Center, University of Louisville) for NMR access, Dr. Robin Brigmon of the Savannah River Technical Center for the soils and discussion, and Dr. Thomas Young at University of California-Davis for providing access to the ICP-MS instrumentation.

References

1. Huang, J. W.; Cunningham, S. D. *New Phytol.* **1996**, *134*, 75–84.
2. Ebbs, S. D.; Kochian, L. V. *Environ. Sci. Technol.* **1998**, *32*, 802–806.
3. Lasat, M. M. *Journal of Environ. Qual.* **2002**, *31*, 109–120.

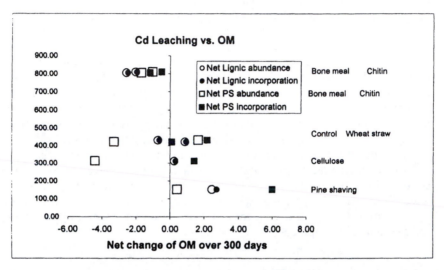

Figure 7. Results of pyro-GC-MS analysis of pre-¹³C labeled, organic amended
soils, plotted against data on ordinate from Figure 4 (total metals leached over
299 days). The abscissa is the net change (arbitrary units) of the OM parameter
over 299 days. "Abundance" refers to the total lignic or polysaccharidic (PS)
OM. "Incorporation" is the displacement of ¹³C from an OM type, effectively
constituent turnover. Data along a horizontal line are from a given treatment,
listed at right.

In the top panel, note that PS and lignic incorporation (closed symbols) related
well to the reduced Cd leaching. However, the PS abundance (open squares)
did not. This indicates that the <u>turnover</u>, e.g., via microbial processing, of PS is
the controlling factor.

For Ni (middle panel), the results show a similar general trend as Cd, while for
Cs (bottom panel), any amount of net turnover (closed squares) related to the
same level of reduced leaching; again, abundance of PS (open squares) did not
relate. Other OM types (e.g., peptidic, alkanic, alkenic) and total OM showed
no relation to metal leaching (data not shown). See text for further discussion.

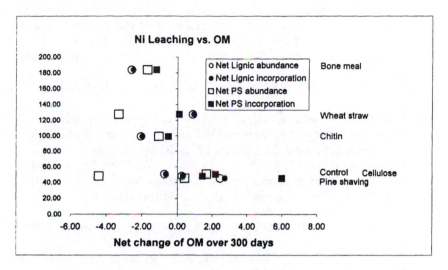

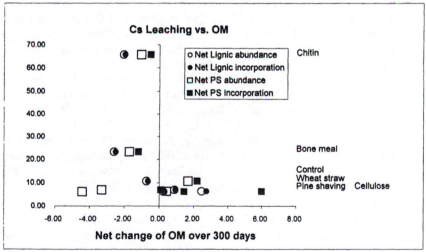

Figure 7. *Continued.*

4. Lovley, D. R.; Coates, J. D. *Curr. Opin. Biotechnol.* **1997**, *8*, 285—289.
5. Abdelouas, A.; Lutze, W.; Gong, W. L.; Nuttall, E. H.; Strietelmeier, B. A.; Travis, B. J. *Sci. Total Environ.* **2000**, *250*, 21–35.
6. Knox, A. S.; Seaman, J. C.; Mench, M. J.; Vangronsveld, J. Remediation of Metal- and Radionuclides-Contaminated Soils by In Situ Stabilization Techniques. In *Environmental Restoration of Metals-Contaminated Soils*; Iskander, I. K., Ed.; Lewis Publishers: Baca Raton, FL, 2001.
7. Seaman, J. C.; Meehan, T.; Bertsch, P. M. *J. Environ. Qual.* **2001**, *30*, 1206–1213.
8. Seaman, J. C.; Arey, J. S.; Bertsch, P. M. *J. Environ. Qual.* **2001**, *30*, 460–469.
9. Mench, M. J.; Vangronsveld, J.; Clijsters, H.; Lepp, N. W.; Edwards, R. In *Phytoremediation of Contaminated Soil and Water*; Terry, N., Banuelos, G., Eds.; Lewis Publishers: Boca Raton, FL, 2000, pp 325–358.
10. Pierzynski, G. M.; Schwab, A. P. *J. Environ. Qual.* **1993**, *22*, 247–254.
11. Schlesinger, W. H. In *Annual Review of Ecology and Systematics;* Johnston, R. F., Frank, P. W., Michener, C. D., Eds. Annual Reviews Inc.: Palo Alto, CA, 1977; Vol. 8, pp 51–81.
12. Eswaran, H.; Van Den Berg, E.; Reich, P. *Soil Sci. Soc. Am. J.* **1993**, *57*, 192–194.
13. Post, W. M.; Emanuel, W. R.; Zinke, P. J.; Stangenberger, A. G. *Nature (London)* **1982**, *298*, 156–159.
14. Hayes, M. H. B.; Malcolm, R. L. Considerations of Compositions and of Aspects of the Structures of Humic Substances. In *Humic Substances and Chemical Contaminants;* Clapp, C. E., Hayes, M. H. B., Senesi, N., Bloom, P. R., Jardine, P. M., Eds.; SSSA: Madison, WI, 2001.
15. Davies, G.; Ghabbour, E. A.; Cherkasskiy, A.; Fataftah, A. Tight Metal Binding by Solid Phase Peat and Soil Humic Acids. In *Humic Substances and Chemical Contaminants;* Clapp, C. E., Hayes, M. H. B., Senesi, N., Bloom, P. R., Jardine, P. M., Eds.; SSSA: Madison, WI, 2001.
16. Zhang, Y.-J.; Bryan, N. D.; Livens, F. R.; Jones, M. N. Complexing of Metal Ions by Humic Substances. In *Humic and Fulvic Acids: Isolation, Structure, and Environmental Role*; Gaffney, J. S., Marley, N. A., Clark, S. B., Eds.; American Chemical Society: Washington, DC, 1996.
17. Perminova, I. V.; Gretschishcheva, N. Y.; Petrosyan, V. S.; Anisimova, M. A.; Kulikova, N. A.; Lebedeva, G. F.; Matorin, D. N.; Venediktov, P. S. Impact of Humic Substances on Toxicity of Polycyclic Aromatic Hydrocarbons and Herbicides. In *Humic Substances and Chemical Contaminants*; Clapp, C. E., Hayes, M. H. B., Senesi, N., Bloom, P. R., Jardine, P. M., Eds.; SSSA: Madison, WI, 2001.
18. Shenker, M.; Fan, T. W. M.; Crowley, D. E. *J. Environ. Qual.* **2001**, *30*, 2091–2098.

19. Fan, T. W. M.; Lane, A. N.; Shenker, M.; Bartley, J. P.; Crowley, D.; Higashi, R. M. *Phytochemistry* **2001**, *57*, 209–221.
20. Fan, T. W.-M.; Baraud, F.; Higashi, R. M. Genotypic Influence on Metal Ion Mobilization and Sequestration via Metal Ion Ligand Production by Wheat. In *Nuclear Site Remediation*; Eller, P. G., Heineman, W. R., Eds.; American Chemical Society Symposium Series 778; American Chemical Society: Washington, DC, 2000; pp 417–431.
21. Fan, T. W. M.; Lane, A. N.; Pedler, J.; Crowley, D.; Higashi, R. M. *Anal. Biochem.* **1997**, *251*, 57–68.
22. Higashi, R. M.; Fan, T. W. M.; Lane, A. N. *Analyst* **1998**, *123*, 911–918.
23. Fan, T. W. M.; Higashi, R. M.; Lanes, A. N. *Environ. Sci. Technol.* **2000**, *34*, 1636–1646.
24. Fan, T.W-M.; Lane, A. N.; Chekmenev, E.; Wittebort, R. J.; Higashi, R. M. *J. Peptide Res.* **2004** *63*, 1–12.
25. Baraud, F.; Fan, T. W. M.; Higashi, R. M. In *Environmental Chemistry*; Robert, D., Ed.; European Association of Chemistry and the Environment, Springer Publishing: New York, in press.
26. Kinnersley, A. M. *Plant Growth Regulation* **1993**, *12*, 207–218.
27. Strickland, R. C.; Chaney, W. R.; Lamoreaux, R. J. *Plant Soil* **1979**, *52*, 393–402.
28. Batson, V. L. ; Bertsch, P. M. ; Herbert, B. E. *J. Env. Qual.* **1996**, *25*, 1129–1137.
29. Bundy, L. G.; Meisinger, J. J. Nitrogen Availability Indices. In *Methods of Soil Analysis;* Weaver, R. W., Ed.; Soil Science Society of America Book Series No. 5; SSSA: Madison, WI, 1994.
30. Olk, D. C.; Cassman, K. G.; Fan, T. W. M. *Geoderma* **1995**, *65*, 195–208.
31. Simpson, A. J.; Burdon, J.; Graham, C. L.; Hayes, M. H. B.; Spencer, N.; Kingery, W. L. *Eur. J. Soil Sci* **2001**, *52*, 495–509.

Characterization, Fate, and Transport of Subsurface Contamination

Chapter 8

Compositional Effects on Interfacial Properties in Contaminated Systems: Implications for Organic Liquid Migration and Recovery

Linda M. Abriola[1], Avery H. Demond[2], Denis M. O'Carroll[2], Hsin-lan Hsu[2], Thomas J. Phelan[1], Catherine A. Polityka[3], and Jodi L. Ryder[2]

[1]Department of Civil and Environmental Engineering, Tufts University, Medford, MA 02155
[2]Environmental and Water Resources Engineering, Department of Civil and Environmental Engineering, University of Michigan, Ann Arbor, MI 48109
[3]HSW Engineering, 605 East Robinson Street, Orlando, FL 32801

Aquifer contamination resulting from the release or disposal and subsequent subsurface migration of dense nonaqueous phase organic liquids (DNAPLs) poses a threat to potable water supplies in the United States. A complicating factor in the characterization and remediation of these sites is that DNAPLs were typically disposed in mixtures that included acids, bases, surfactants and many other constituents. This paper highlights results of an ongoing research project designed to explore the influence of solid and organic phase composition on DNAPL migration and entrapment in contaminated aquifers. The integrated research program includes small-scale laboratory investigations to explore the effects of chemical heterogeneity, in particular variations in mineralogy and surfactant composition, on interfacial properties. Column-scale experiments are being conducted to examine the dependence of organic contaminant constitutive relationships on wettability. Finally, mathematical models developed from these observations are being incorporated into a compositional multiphase simulator to facilitate prediction of DNAPL behavior under conditions representative of field sites. Two-dimensional sand box experiments are also ongoing to validate the modeling approach. Results from this research demonstrate the dramatic influence of interfacial property variation on DNAPL migration and retention. Incorporation of these effects into numerical simulators is required to accurately determine the extent of DNAPL contamination and to select appropriate remediation schemes.

Introduction

An understanding of the transport behavior of complex dense nonaqueous phase liquid (DNAPL) contaminants is a prerequisite for the accurate assessment of chemical exposure and the design of effective subsurface remediation strategies. DNAPL contaminants are common nationwide and, of particular interest to this research, are frequently encountered at U.S. Department of Energy (DOE) sites. The origin of these contaminants is typically waste mixtures containing metals, radionuclides, chlorinated hydrocarbons and/or fuel hydrocarbons, as well as organic acids and bases and other surface active compounds generated in various industrial processes at DOE facilities (1-3). The composition of such mixtures can substantially alter their environmental behavior in comparison with pure organic liquids. For example, the presence of surfactants in acidic and alkaline NAPL mixtures has been shown to render a water-wetting porous medium intermediate to organic-wetting and also to significantly decrease DNAPL/aqueous phase interfacial tension (4-10). In addition to variations in interfacial properties created by the waste's composition, naturally occurring subsurface components can exhibit a variety of wetting characteristics. For example, coal, graphite and talc are intermediate- to organic-wet, whereas more common aquifer materials such as quartz and carbonate are water-wet (11). These factors suggest that wettability variations and interfacial tension reductions may be common in the contaminated subsurface.

Wettability and interfacial tension play a critical role in defining the constitutive relationships governing the multiphase flow and entrapment of DNAPLs. Wettability refers to the "tendency of one fluid to spread on or adhere to a solid surface in the presence of another immiscible fluid" (12). The contact angle, a measure of wettability, is the angle the fluid-fluid interface makes with the solid (13). In water/NAPL/solid systems, the liquid which has the higher affinity for the solid surface coats the solid and is re̓ ̄ʳᵉd to as the wetting phase, while the remaining liquid is known as the nonwetti̇ᵢₙₐ ̗ₕₐₛe. As the contact angle, measured through the water phase in a two-liquid NAPL/water system, approaches $0°$, the surface is said to be strongly water-wet. Conversely, as the contact angle approaches $180°$, the surface is said to be strongly organic-wet. A surface is termed intermediate-wet if the contact angle ranges from approximately $75°$ to $120°$ (*14, 15*). Interfacial tension arises at the interface separating two immiscible fluids and is defined as the energy per unit interfacial area required to create a new surface (*13*). The presence of surface-active species adsorbed at the interface tends to reduce the interfacial tension (*13, 16*).

Although spatial variations in wettability conditions and interfacial tension may substantially influence DNAPL migration and entrapment in the saturated zone, the physics of DNAPL flow in chemically heterogeneous systems has not been adequately investigated. Multiphase flow simulators typically assume that subsurface soils are completely water-wet with a uniform surface composition and that subsurface fluids are chemically homogeneous and pure. This paper presents results of an ongoing research project designed to explore the influence of variable solid and organic phase composition on DNAPL migration and entrapment. This research program integrates experimental measurements and mathematical modeling at a variety of scales. Small-scale experiments are being conducted to quantify the influence of interfacial properties on organic contaminant constitutive relationships, larger "bench scale" experiments are being conducted to validate proposed constitutive models, and numerical simulations are being conducted to assess the impact of these phenomena at the field scale.

Small-Scale Laboratory Investigations

Interfacial Properties of DNAPL/Aqueous Systems Containing Organic Acids and Bases

One component of this research is designed to investigate the influence of organic acids and bases on the interfacial behavior of organic liquid wastes. Previous research has explored the effects of an organic acid or an organic base individually on organic waste interfacial properties (*8-10*). Organic wastes, however, are typically mixtures containing numerous constituents such as surfactants, oils, greases, and organic solvents. Thus, although research that has utilized simple mixtures of one chlorinated solvent and one surfactant can serve as a useful foundation, further work is needed to examine more representative and complex waste mixtures.

Research has been undertaken to examine more realistic mixed-waste systems that contain both an organic acid and an organic base. Organic acids, such as hexadecanoic acid and oxalic acid, have been identified at many DOE waste sites (2). Octanoic acid (OA) $(CH_3(CH_2)_6COOH)$ was chosen as a representative organic acid because it has the same functional group as those organic acids found at waste sites and its high solubility in water makes it easy to study its pH-dependent interfacial behavior. Organic bases, such as amines, are used as anti-oxidants in grease and corrosion inhibitors in fuels. Dodecylamine (DDA) $(CH_3(CH_2)_{11}NH_2)$, was selected as a representative organic base. Tetrachloroethene (PCE) was selected as the representative DNAPL due to its frequent use and disposal at many DOE sites (2). Waste mixtures were prepared using PCE containing either 1 mM DDA, 1 mM OA or both (Aldrich Chemical, Milwaukee, WI). The ionic strength of the aqueous phase in all systems was fixed through the addition of 0.01 M NaCl, and pH was adjusted by adding small amounts of sodium hydroxide and hydrochloric acid.

To quantify the effects of mixed waste composition on wettability and interfacial tension equilibrium, aqueous phase receding contact angle and interfacial tension were measured. Interfacial tension was measured via a spinning drop tensiometer Model 500 (University of Texas, Austin, TX) and contact angles were obtained using axisymmetric drop shape analysis (17) on quartz slides. Contact angles are reported through the aqueous phase.

The effects of OA and DDA on the equilibrium contact angle and interfacial tension, as a function of pH, are presented in Figures 1 and 2 and are compared with data from single solute systems (18). In waste systems containing 1 mM OA alone, the wettability and interfacial tension of the system are unchanged from the pure PCE case. This is attributed to the limited surfactant interfacial activity at this relatively low concentration. When the pH is above the isoelectric point of quartz (pH = 2 to 3.5) (19), the quartz surface is negatively charged and the deprotonated and neutral forms of OA do not adsorb to the quartz surface and therefore do not alter the contact angle. Conversely, in the systems containing 1 mM DDA only, both the interfacial tension and contact angle are observed to decrease as the pH decreases below DDA's pKa (the negative logarithm of the equilibrium constant) (pH = 10.6). This behavior is attributed to the conversion of DDA to its cationic form (18). The cationic form of DDA sorbs to the negatively charged quartz surface, rendering the solid surface more hydrophobic. As the solid surface becomes more hydrophobic, its affinity for the aqueous phase decreases, resulting in an increase in contact angle. In the absence of DDA the contact angle measured on the quartz slide is quite small (approximately 10°), whereas the maximum contact angle in the presence of 10^{-3} M DDA is larger (approximately 70°).

When both DDA and OA are present in the mixed waste system, enhanced interfacial activity is observed. For example, the maximum contact angle nearly doubles, from approximately 70° (water-wet) with DDA alone to 130° (organic-

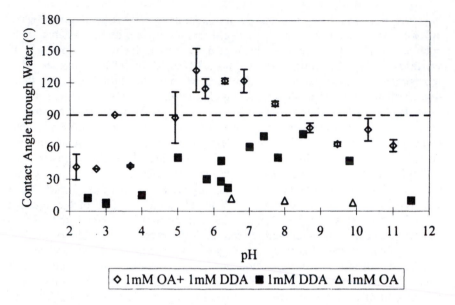

Figure 1. Contact angle as a function of pH for several OA & DDA concentrations (error bars, when available, represent 95% confidence intervals).

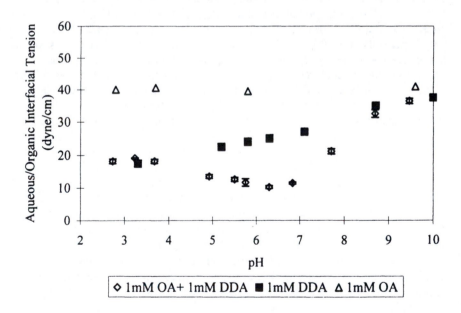

Figure 2. Interfacial tension as a function of pH for several OA & DDA concentrations (error bars, when available, represent 95% confidence intervals).

wet) when both DDA and OA are present. In addition, the interfacial tension decreases from approximately 25 dynes/cm with DDA alone, at pH = 6.3, to approximately 10 dynes/cm when both DDA and OA are present. These results suggest that interfacial activity has a greater dependence on pH in the mixed organic waste systems compared to that in the individual surfactant systems. In addition, the data suggest that the surface activity of mixed organic wastes cannot be predicted from the superposition of activity observed in the presence of individual waste components and a synergism among solutes may be expected in mixed organic wastes. In this system the synergism is at a maximum around neutral pH or within the range found in natural systems. Thus, trace quantities of surface active compounds, even at levels low enough such that, individually, no effect is observed, may have a significant impact on DNAPL migration and entrapment.

The Influence of Soil Surface Chemistry on Surfactant Sorption and Wettability Alterations

Experiments are also being conducted to assess variations in subsurface wettability that arise due to variations in solid surface composition.

Varying surface mineralogy is common to subsurface media, resulting in spatially varying wettability and surface chemistry properties. Three solid phases were selected for investigation, representative of a range of surface characteristics: quartz, quartz treated with a hematite surface coating (*20*), and quartz treated with Rhodorsil Siliconate 51T (a potassium methylsiliconate surface coating, Rhodia Silicones, Cranbury, NJ) (*21*). Bottle point experiments were conducted with these media and laboratory grade PCE, dyed with 0.25 g/L of Oil-Red-O to aid visualization (Fisher Scientific, Fair Lawn, NJ), as the representative DNAPL. DDA was selected as the representative organic base. Four bottle test systems were created: an untreated quartz sand/PCE/aqueous system, a hematite treated quartz/PCE/DDA/aqueous system, an untreated quartz/PCE/DDA/aqueous system, and a Rhodorsil-treated quartz/PCE/aqueous system. The untreated quartz sand/PCE/aqueous system served as a water-wet standard (contact angle = 10°), and the Rhodorsil-treated quartz/PCE/aqueous system served as the intermediate-wet standard (contact angle = 77°). The addition of DDA to the other systems permitted examination of the effect of the type of surface on surfactant adsorption. The aqueous phase used in the untreated quartz/PCE/DDA/aqueous and the hematite treated quartz/PCE/DDA/aqueous systems was buffered using an acetate solution (pH = 5.19). The aqueous phase in the two standard systems, the water-wet and intermediate-wet systems, was not buffered. It is likely that the addition of the Oil-Red-O dye and the DDA organic base to the PCE phase lowered the

PCE/aqueous phase interfacial tension (*10*, *49*). These reductions in IFT, however, are not large enough to significantly influence the visual experiments.

For the standard systems, dyed PCE was added to the water-saturated media, untreated quartz sand and Rhodorsil-treated sand. In the other systems, DDA was added to the dyed PCE phase and then the organic mixture was added to the water-saturated hematite-coated and untreated quartz sands. All bottles were capped and then vigorously shaken by hand.

Figure 3 shows a sampling of bottle point experiments for the four systems. In the untreated quartz/pure PCE/aqueous system (bottle 1), PCE is visible as a distinct phase and does not coat the sand grains, consistent with water-wet behavior. In contrast, in the Rhodorsil-treated sand/pure PCE/aqueous system (bottle 4) PCE coats the sand grains and is not visible as a separate phase, consistent with intermediate-wet behavior. In the systems containing DDA, observations indicate that PCE wets the quartz sand in the quartz/PCE/DDA/aqueous system (bottle 3). In this system, PCE coats the sand grains and is not visible as a separate phase, similar to the behavior observed in the Rhodorsil-treated sand/pure PCE/aqueous system. Since the aqueous phase pH in this system is 5.19 (above the point of zero charge), the quartz surface charge is negative and favors the adsorption of the cationic form of DDA. The adsorption of DDA onto the solid surface increases its hydrophobicity and renders the surface intermediate-wet. PCE does not wet the hematite-coated sand (bottle 2); it is visible as a distinct phase and is not coating the sand grains. These results are similar to the untreated quartz/pure PCE/aqueous system (bottle 1). The point of zero charge for hematite is pH = 6 (*20*). Since the aqueous phase pH in this system is 5.19, the hematite surface is positively charged and does not favor the adsorption of DDA. These results suggest that the hematite-coated sand in the PCE/DDA/aqueous system is water-wet and the quartz sand in the PCE/DDA/aqueous system is intermediate-wet. This experiment illustrates the importance of mineral coatings and the subsequent wettability alterations in subsurface systems. Thus the ability of DDA to change the wettability of the subsurface system will depend not only on pH and the presence of other surface active solutes, such as OA, but also on the presence of mineral coatings on quartz. Because mineral coatings are generally not uniform, the subsurface may, consequently, present an environment of spatially and temporally varying wettability.

Variability in the Wettability of Natural Soils

The experimental observations discussed above demonstrate that surfactant sorption to mineral surfaces results in wettability alterations. In natural porous media, wettability variations can also occur in the absence of surfactants due to variability in soil phase composition. For example, carbonaceous materials, which may be present in shales, coals, or as particle coatings, are believed to contribute to the non-water-wet character of some soil and mineral surfaces (*22–27*).

To examine the range of wettabilities among carbon containing solid-phase components, materials were selected to be representative of a range of carbon source environments and degree of aging. The following materials were selected: Silica sand (US Silica, Ottawa, IL), Ann Arbor II soil (Brookston series, Ann Arbor, MI), Utica shale (Bituminous shale, Ward Natural Science, Rochester, NY), Garfield oil shale (Kerogen shale, Ward Natural Science, Rochester, NY) and Waynesburg coal (Bituminous coal, Ward Natural Science, Rochester, NY). Ann Arbor II soil represents an agricultural soil composed of quartz or feldspar particles coated with humic acid recently formed under aerobic conditions (*28*). The Utica and Garfield shales are similar in that they contain insoluble, organic matter formed under anoxic sedimentary conditions and mixed with fine grained mineral particles (*29*). The Waynesburg coal, formed in anoxic wetland environments, has a higher organic matter to mineral matter ratio than the shales (*29*).

A series of bottle point experiments was conducted to assess the wettability of these materials. The solid materials were equilibrated with an aqueous 10^{-2} M NaCl solution for at least 24 hours. PCE, dyed red to improve visualization, was then added to the soil/aqueous phase slurry and the mixture was shaken. No surface active agents were added to these systems. Experimental results indicate a range of wetting conditions (Figure 4). The quartz sand system serves as a water-wet standard (bottle 1). The Ann Arbor II soil and Utica shale also exhibit water-wet behavior (bottles 2 and 3). In these three bottles, PCE is present as a distinct phase and does not coat the soil particles. In contrast, the Garfield oil shale and Waynesburg coal (bottles 4 and 5) systems appear organic-wet. In these systems, only the clear aqueous phase is visible above the soil and PCE coats the surfaces of these natural materials.

The presence of organic carbon has often been cited as a reason for the non-water-wet behavior of soils (*23, 26, 30–32*). However, these bottle point experiments also suggest that the nature of the soil organic carbon is important. For example the Ann Arbor II soil (bottle 2) has a natural humic acid coating but is water-wet. In addition, the Utica shale and Garfield oil shale (bottles 3 and 4) exhibit differences in wetting behavior even though they are both shale materials. Differences in wetting behavior may be attributed to the number of oxygen-containing surface-functional groups (*25, 33*). As a soil ages, the number of oxygen-containing functional groups decreases causing the soil surface to become less polar and less hydrophilic (*25*). NMR spectrum of the humic acid coating of the Ann Arbor II soil, a geologically young soil, indicates many oxygen-containing surface-functional groups. The soil surface, therefore, has a high degree of polarity, and is hydrophilic and water-wet. Soil aging and differences in the polar character may also explain the differences in the

Figure 3. Bottle point experiment illustrating the effect of mineral coatings on wettability alterations. From left to right: (1) quartz + pure PCE + H_2O, (2) hematite + DDA + PCE + H_2O @ pH = 5.19, (3) quartz + DDA + PCE + H_2O @ pH = 5.19, and (4) quartz treated with Rhodorsil 51T + pure PCE + H_2O.

Figure 4. Bottle point experiment illustrating the wettability of different natural carbonaceous materials. From left to right: (1) silica sand, (2) Ann Arbor II soil, (3) Utica bituminous shale, (4) Garfield oil shale, and (5) Waynesburg bituminous coal.

observed wetting behavior between the shale materials. If the soil ages in a reduced environment, the loss of oxygen containing surface-functional groups may cause the soil surface to transition from water-wet to organic-wet behavior. Given these observations, wettability variations are likely at many waste sites not only due to the presence of surfactants in the waste mixture but also due to heterogeneous solid phase composition.

The Effect of Aquifer Material Wettability on Capillary Pressure/Saturation Relationships

The previous sections described ongoing research to assess the range of wettability variations that may be encountered at contaminated sites. In addition, this project has evaluated the impact of wettability on constitutive relationships. Capillary pressure-saturation experiments were conducted for synthetic aquifer materials composed of F35/50/70/110 Ottawa sand (d_{50} = 0.024 cm and uniformity index = 3.06) coated with octadecyltrichlorosilane (OTS) (*34*) and Rhodorsil Siliconate 51T (*21*). The OTS sands were strongly organic-wet with contact angles approaching 165°, while the siliconate-coated sands were intermediate-wet. PCE was used as the representative DNAPL. The aqueous phase in all systems was Milli-Q water equilibrated with the organic and solid phases. Experiments were conducted in the absence of surface-active agents in order to decouple the effects of particle surface wettability and reduction in interfacial tension.

The capillary pressure/saturation experiments were conducted following the procedure of Salehzadeh and Demond (*35*). Each of the sand mixtures was dry packed into a column 1.27 cm long, and the column was then flushed with several pore volumes of carbon dioxide to displace air in the pore space. To saturate the column and completely displace and solubilize the carbon dioxide, the columns were then flushed with 100 pore volumes of PCE-equilibrated Milli-Q water. Teflon and nylon membranes (Pall Corporation, Ann Arbor, MI) were placed at the bottom and top of the column to act as a capillary barrier to water and PCE flow, respectively. Finally, the bottom of the column was connected to a PCE reservoir and the top was connected to a capillary tube that acted as a water reservoir. The capillary pressure in the system was adjusted by changing the height of the PCE reservoir. Equilibrium was assumed to be achieved when the water level in the capillary tube did not change over a two-hour period. The capillary pressure/saturation experiments started with the column 100% saturated with Milli-Q water. Primary drainage capillary pressures were obtained by incrementally raising the height of the PCE reservoir. When a residual water saturation was achieved, primary imbibition data were obtained by decreasing the capillary pressure by lowering the PCE reservoir.

In these experiments, the capillary pressure is defined as the difference between the organic and aqueous phase pressures. Drainage refers to water drainage from the column, and imbibition refers to PCE drainage from the column. Primary drainage and imbibition capillary pressure-saturation curves are shown in Figure 5 for water-, intermediate- and organic-wet F35/F50/F70/F110 sands. Apparent water saturation is defined as the sum of the effective water saturation and the entrapped organic saturation (36). The primary drainage capillary pressures for the water- and intermediate-wet sands are positive over all saturations (Figure 5a). In contrast, the organic-wet drainage capillary pressure is slightly negative over water saturations ranging from 100% to 45%, and becomes slightly positive thereafter. Based on these results, it is hypothesized that, for a strongly organic-wet material, PCE will not spontaneously displace water and that capillary forces are essentially negligible along the majority of the primary drainage path. Primary imbibition data indicate that the capillary pressure remains positive for the water-wet sand for all water saturations (Figure 5b). In contrast, for the intermediate- and organic-wet sands, significant water pressure is required to displace PCE from the pore space.

These measurements reveal that the PCE entry pressure for the intermediate- and organic-wet sands is significantly less than that for the water-wet sand. In addition, once the organic liquid is present in the intermediate- and organic-wet sands, significant water pressure is required to displace it. Thus, these data suggest that the presence of intermediate- and organic-wet sands in a contaminated aquifer would result in higher organic liquid phase saturations than those in a water-wet sand of the same size distribution, and in significantly higher water pressures necessary to displace the organic liquid from the pore space.

Wettability Effects on Relative Permeability/Saturation Relationships

Mathematical models of DNAPL subsurface movement require capillary pressure/saturation and relative permeability/saturation constitutive relationships for DNAPL/water/soil systems. Wettability alterations may have a considerable effect on capillary pressure/saturation relationships as discussed in the previous section. They may also have an impact on relative permeability/saturation relationships; yet this has received less attention in the literature in large part due to the experimental difficulties associated with measuring these relationships.

Predictive models based on the capillary pressure/saturation relationships are commonly used to estimate relative permeability in the absence of experimental data (37, 38). One such predictive model, that of Bradford et al. (37), is used in Figure 6 to illustrate the effects of varying wettability on relative

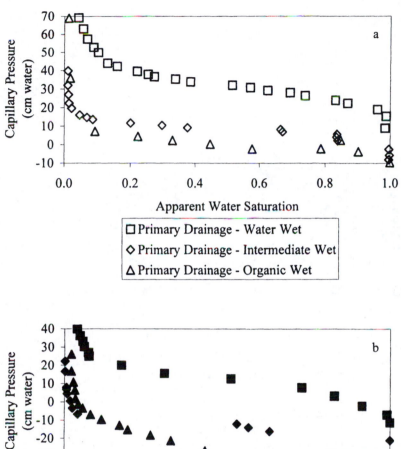

Figure 5. (a) Primary drainage and (b) primary imbibition two-phase (water/PCE) capillary pressure-saturation curves for water-, intermediate-, and organic-wet sand.

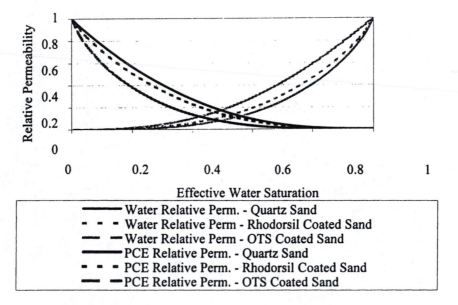

Figure 6. Predicted water-organic relative permeability relationships for sands of various wetting conditions for the F35/F50/F70/F110 size fraction.

permeabilities. In this example, relative permeability/saturation relationships are predicted for the synthetic sands used in the capillary pressure/saturation experiments, presented in the previous section. This model is based upon the assumption that the non-wetting fluid occupies the largest pores and the wetting fluid occupies the smaller pores. Thus a fluid's relative permeability is greater when it is the non-wetting fluid and smaller when it is the wetting fluid. For example, the relative permeability to water is greater in the OTS coated sand, where water is the non-wetting fluid, than in the quartz sand, where it is the wetting fluid. Relative permeabilities to both fluids in the Rhodorsil-coated sands always lie between the two extreme wettability cases. Application of this model suggests that relative permeability is sensitive to variations in solid phase wettability. Since the constitutive relationships that govern multiphase flow are functions of wettability, an assumption of a completely water-wet medium will lead to errors in the prediction of DNAPL migration.

Two-Dimensional Bench Scale Experiments

To observe the impact of wettability variations at a larger scale, PCE infiltration experiments were conducted in a "two-dimensional" sandbox (*39*). The box was constructed with an aluminum back and sidewalls, and a tempered glass front which facilitated visual observation of PCE migration. It was packed to a uniform porosity of 32.3% with clean 20/30 sand (d_{50} = 0.071cm and uniformity index = 1.21), incorporating three sand layers of different textures and wetting properties (Figure 7). Dyed PCE (47.33 ml) was injected with a syringe pump (Harvard Apparatus, South Natick, MA) for a 66.6-minute period at a constant rate (0.71mL/min). The injection location was at the midpoint between the glass and the aluminum backing, 28.7 cm above the aluminum base.

Upon release, the PCE migrated downward and pooled on top of the water-wet F70/F110 sand lens (d_{50} = 0.015 cm and uniformity index = 2.25). Further downward migration did not occur until a pool of sufficient size developed, after which the PCE cascaded over the sides of the lens. Similar pooling and cascading behavior occurred when the PCE reached the water-wet F35/F50 lens (d_{50} = 0.036 cm and uniformity index = 1.88) (Figure 8a). In contrast, a negligible entry pressure is required for PCE to enter the organic-wet F35/F50 lens located at the left of the tank (Figure 7). Continued downward migration of PCE below this organic-wet lens occurred only after the lens neared complete PCE saturation. PCE that migrated below the F35/F50 lenses was retained by the F20/F30 organic-wet layer near the bottom of the tank. No PCE migrated into the water-wet F20/F30 layer below the organic-wet layer. Thus, the organic-wet sands act as a capillary barrier to effectively inhibit further migration. These results are expected based on the capillary pressure/saturation results presented in the previous section (Figure 5). Both the

174

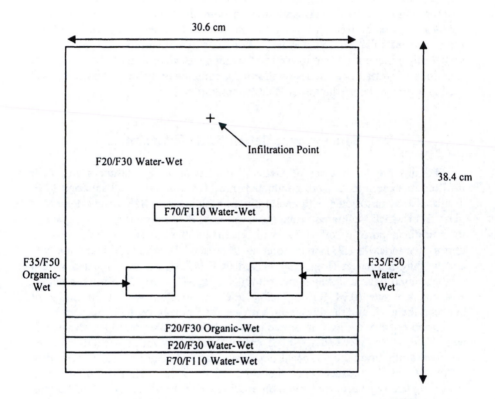

Figure 7. Schematic representation of two-dimensional sandbox packing (not to scale).

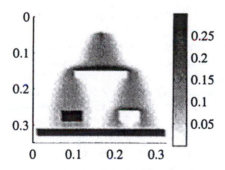

*Figure 8. Two-dimensional infiltration photo and simulation result of PCE
saturation at elapsed time = 60 minutes using Brooks-Corey/Burdine capillary
pressure/relative permeability/saturation constitutive relationships.*

infiltration and the primary drainage capillary pressure/saturation experiments suggested that a negligible entry pressure was required to displace water in organic-wet sands. In addition, the primary imbibition capillary pressure/saturation experiment suggested that a significant negative capillary pressure was required for water to reinvade the organic-wet sand; as a result, little redistribution of PCE is observed in the organic-wet lenses.

A multiphase numerical simulator, M-VALOR (40), modified to account for the influence of wettability variations on the constitutive relationships (39, 41), was used to simulate the sand box experiment. In the simulator, measured capillary pressure-saturation relationships were employed, and relative permeability relationships were estimated based on capillary pressure relationships and wettability (39, 42). Figure 8b presents the simulation results at 60 minutes using the Brooks-Corey (43)/Burdine (44) capillary pressure/relative permeability/saturation constitutive relationships. Simulations that accounted for varying wettability accurately predicted the observed PCE migration pathways, including the retention of PCE in the organic-wet layers. These experimental results and simulations indicate that contrasts of capillary properties at interfaces lead to higher NAPL saturations, increased lateral spreading, and decreasing depths of NAPL infiltration. The simulation results also indicate that the use of independently measured constitutive relationships result in good approximations of organic migration and entrapment under the conditions employed in this bench-scale experiment.

Field-Scale Simulations

The modified numerical simulator, M-VALOR, is also being used to explore the effect of coupled textural and chemical heterogeneity on predictions of organic liquid migration and entrapment at larger scales. While the impact of textural heterogeneity on field-scale migration and entrapment has been investigated (45, 46), comparatively less is known about the potential effect wettability variations may have on DNAPL migration. For this study, permeability (k) fields were generated using conditional, sequential Gaussian simulation based upon geostatistical parameters derived from an analysis of particle size distributions obtained from a PCE-contaminated site in Oscoda, MI (47). Representative realizations of permeability fields are shown in Figure 9. Here, the degree of porous media heterogeneity, based on the ln(k) variance, increases from left to right. Heterogeneous spatial distributions of wettability were generated by correlating the organic-wet mass fraction (F_o) of sand particles to permeability through the use of an exponential function (41, 48). Figure 10 shows two F_o distributions that are (perfectly) positively and negatively correlated to aquifer permeability. In the positively correlated case, the higher permeability zones are organic wetting, whereas in the negatively

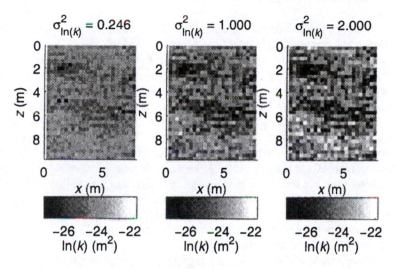

Figure 9. Representative permeability distributions. ln(k) variance increase from left to right.

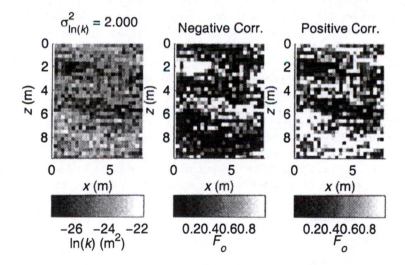

Figure 10. Correlation of solid phase wettability to permeability. In the negatively correlated system, low permeability zones are organic wetting and, in the positively correlated system, high permeability zones are organic wetting.

correlated case, the organic-wet soils are associated with the lower permeability zones.

A suite of simulations was conducted to examine the influence of different exponential permeability/wettability functions and different levels of textural heterogeneity on DNAPL entrapment and dissolution. In the simulations presented in Figure 11, 96 liters of PCE were introduced into a 7.92 m × 9.75 m domain over a period of 400 days and were then allowed to redistribute for an additional 330 days. The level of spatial discretization selected for these simulations was consistent with the scale of the experiments used to estimate the appropriate constitutive relationships. DNAPL saturation distributions representative of the variability of observed infiltration and entrapment behavior are summarized in Figure 11. Here, the degree of textural heterogeneity increases from the top to the bottom row and different wettability treatments are shown in the columns from left to right. The first two columns depict systems that are completely water- and completely organic-wet. The last two columns show systems where the organic-wet mass fraction (F_o) are positively and negatively correlated to permeability.

An examination of Figure 11 suggests that spatial variations in wettability can have an observable impact on the migration and entrapment of organic liquids in the subsurface. As observed in the two-dimensional sandbox experiments, the occurrence of organic-wet solids tends to reduce depths of organic liquid penetration and to increase the magnitude of the maximum entrapped organic liquid saturations. This behavior is consistent with the organic liquid's preferential tendency to reside in organic-wet media. Figure 11 also demonstrates that the positively and negatively correlated systems behave much more similarly to the completely organic-wet system than to the completely water-wet system. This observation suggests that only a small fraction of organic-wet material, regardless of whether it is associated with high or low permeability soil, is necessary to create significant differences in organic liquid migration and entrapment. The plots in the leftmost column of Figure 11 demonstrate the considerable effect that textural heterogeneity, as quantified by $\ln(k)$ variance, has on organic-phase distributions in a completely water-wet medium. In the higher variance systems, the similarity in penetration and vertical spreading of the organic liquid among the water-wet, correlated, and organic-wet systems is significant. It appears that, as soils become increasingly organic-wet, the influence of textural heterogeneity becomes less important.

Conclusion

The research presented above highlights research conducted at a variety of scales on the influence of wettability variations on DNAPL behavior. Such wettability variations can arise from the adsorption of surface-active DNAPL

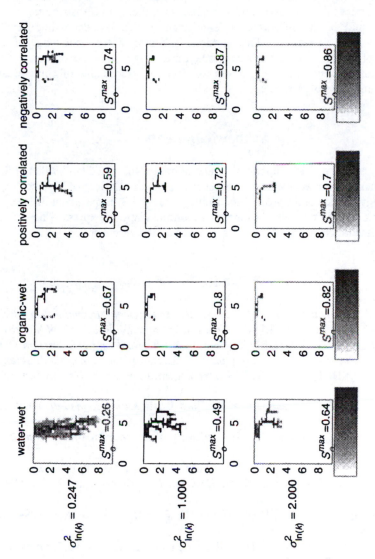

Figure 11. Representative organic liquid (PCE) saturation distributions. The maximum organic liquid saturation encountered in the domains is noted on each plot.

constituents onto mineral surfaces or from the composition of naturally occurring minerals in the subsurface. The assumption that a soil is water-wet when it is, in fact, organic-wet may lead to significant errors in delineating the extent of migration and persistence of contamination in the subsurface. Experimental and simulation results suggest that the spatial distribution of an organic liquid in the subsurface is significantly different depending on the wettability of the medium. This research indicates that a knowledge of the aqueous phase chemistry, the subsurface mineralogy, and the organic waste composition at a site will be required for accurate predictions of contaminant behavior and for risk assessment.

Acknowledgments

This research was supported under Grant No. DE-FG07-96ER14702, Environmental Management and Science Program, Office of Science and Technology, Office of Environmental Management, U.S. Department of Energy. Any opinions, findings, conclusions, or recommendations expressed herein are those of the authors and do not necessarily reflect the views of the DOE.

REFERENCES

1. Sloat, R. J. *Hanford Low Level Waste Management Reevaluation Study*. ARH-231 DEI; Atlantic Richfield Hanford Company: Richland, WA, 1967.
2. Riley, R. G.; Zachara, J. M.; Wobber, F. J. *Chemical Contaminants on DOE Lands and Selection of Contaminant Mixtures for Subsurface Science Research*; DOE/ER-0547T; U.S. Department of Energy: Washington, DC, 1992.
3. Jackson, R. E.; Dwarakanath, V. *Ground Water Monit. Remediat.* **1999**, *19*, 102–110.
4. Powers, S. E.; Anckner, W. H.; Seacord, T. F. *Journal of Environ. Eng.* **1996**, *122*, 889–896.
5. Powers, S. E.; Tamblin, M. E. *J. Contam. Hydrol.* **1995**, *19*, 105–125.
6. Barranco, F. T.; Dawson, H. E. *Environ. Sci. Technol.* **1999**, *33*, 1598–1603.
7. Demond, A. H.; Desai, F. N.; Hayes, K. F. *Water Resour. Res.* **1994**, *30*, 333–342.
8. Lord, D. L.; Demond, A. H.; Salehzadeh, A.; Hayes, K. F. *Environ. Sci. Technol.* **1997b**, *31*, 2052–2058.
9. Lord, D. L.; Hayes, K. F.; Demond, A. H.; Salehzadeh, A. *Environ. Sci. Technol.* **1997a**, *31*, 2045–2051.

10. Lord, D. L.; Demond, A. H.; Hayes, K. F. *Transp. Porous Media* **2000**, *38*, 79–92.
11. Anderson, W. G. *J. Pet. Technol.* **1987**, *39*, 1283–1300.
12. Craig, F. F., Jr. *The Reservoir Engineering Aspects of Waterflooding; Monograph Series v. 3*; Society of Petroleum Engineers: Dallas, TX, 1971.
13. Hiemenz, P. C.; Rajagopalan, R. *Principles of Colloid and Surface Chemistry*, 3rd ed.; Marcel Dekker: New York, 1997.
14. Treiber, L. E.; Archer, D. L.; Owens, W. W. *Soc. Pet. Eng. J.* **1972**, *12*, 531–540.
15. Morrow, N. R. *J. Can. Pet. Technol.* **1976**, *15*, 49–69.
16. Adamson, A. W.; Gast, A. P. *Physical Chemistry of Surfaces,* 6th ed.; Wiley: New York, 1997.
17. Cheng, P.; Li, D.; Boruvka, L.; Rotenberg, Y.; Neumann, A. W. *Colloids Surf.* **1990**, *43*, 151–167.
18. Lord, D. L. Ph.D. Dissertation, University of Michigan, 1999.
19. Domenico, P. A.; Schwartz, F. W. *Physical and Chemical Hydrogeology,* 2nd ed.; Wiley: New York, 1998.
20. Johnson, P. R.; Sun, N.; Elimelech, M. *Environ. Sci. Technol.* **1996**, *30*, 3284–3293.
21. Fleury, M.; Branlard, P.; Lenormand, R.; Zarcone, C. *J. Pet. Sci. Eng.* **1999**, *24*, 123–130.
22. DeBano, L. F. *J. Hydrol.* **2000**, *231*, 4–32.
23. Capriel, P.; Beck, T.; Borchert, H.; Gronholz, J.; Zachmann, G. *Soil. Biol. Biochem.* **1995**, *27*, 1453–1458.
24. Franco, C. M. M.; Clarke, P. J.; Tate, M. E.; Oades, J. M. *J. Hydrol.* **2000**, *231*, 47–58.
25. Fuerstenau, D. W.; Rosenbaum, J. M.; Laskowski, J. *Colloids Surf.* **1983**, *8*, 153–174.
26. Karnok, K. A.; Rowland, E. J.; Tan, K. H. *Agron. J.* **1993**, *85*, 983–986.
27. Horsely, R. M.; Smith, H. G. *Fuel* **1951**, *30*, 54–63.
28. Huang, W. Ph.D. Dissertation, University of Michigan, 1997.
29. Killops, S. D.; Killops V. J. *An Introduction to Organic Geochemistry*; Longman Group: Harlow, U.K., 1993.
30. Giovannini, G.; Lucchesi, S. *Soil Sci.* **1997**, *162*, 479–496.
31. Ma'Shum, M.; Tate, M. E.; Jones, G. P.; Oades, J. M. *J. Soil Sci.* **1988**, *39*, 99–110.
32. Savage, S. M.; Martin, J. P.; Letey, J. *Soil Sci. Soc. Am. Proc.* **1969**, *33*, 405–409.
33. Somasundaran, P.; Zhang, L.; Fuerstenau, D. W. *Int. J. Miner. Process.* **2000**, *58*, 85–97.
34. Anderson, R.; Larson, G.; Smith, C. *Silicon Compounds: Register and Review*; 5th ed.; Huls America Inc.: Piscataway, NJ, 1991.
35. Salehzadeh, A.; Demond, A. H. *J. Environ. Eng.* **1999**, *125*, 385–388.

36. Kaluarachchi, J. J.; Parker, J. C. *Transp. Porous Media* **1992**, *7*, 1–14.
37. Bradford, S. A.; Abriola, L. M.; Leij, F. J. *J. Contam. Hydrol.* **1997**, *28*, 171–191.
38. Oostrom, M.; Lenhard, R. J. *Adv. Water Resour.* **1998**, *21*, 145–157.
39. O'Carroll, D. M.; Bradford, S. A.; Abriola, L. M. *J. Contam. Hydrol.,* in press.
40. Abriola, L. M.; Rathfelder, K. M.; Maiza, M.; Yadav, S. "VALOR code version 1.0: A PC code for simulating immiscible contaminant transport in subsurface systems," EPRI TR-101018, 1992.
41. Bradford, S. A.; Abriola, L. M.; Rathfelder, K. M. *Adv. Water Resour.* **1998**, *22*, 117–132.
42. Bradford, S. A.; Leij, F. J. *J. Contam. Hydrol.* **1997**, *27*, 83–105.
43. Brooks, R. H.; Corey, A. T. *Hydraulic Properties of Porous Media;* Hydrology Paper No. 3; Civil Engineering Department, Colorado State University: Fort Collins, CO, 1964.
44. Burdine, N. T. *Transactions of the American Institute of Mining and Metallurgical Engineers* **1953**, *198*, 71–78.
45. Dekker, T. J.; Abriola, L. M. *J. Contam. Hydrol.* **2000**, *42*, 187–218.
46. Kueper, B. H.; Gerhard, J. I. *Water Resour. Res.* **1995**, *31*, 2971–2980.
47. Lemke, L. D.; Abriola, L. M.; Goovaerts, P. *Water Resour. Res.* **2004**, *40*, doi: 01510.01029/02003WR001980.
48. Phelan, T. J.; Bradford, S. A.; O'Carroll, D. M.; Abriola L. M.; Lemke, L. D. *Adv. Water Resour.,* **2004**, *27*, 411–427.
49. Tuck, D. M.; G. M. Iversen; and W. A. Pirkle. *Water Resour. Res.* **2003**, *39*, doi:1210.1029/2001WR001000.

Chapter 9

Chaotic Processes in Flow through Fractured Rock: Field and Laboratory Experiments Revisited

Boris Faybishenko

Earth Sciences Division, Ernest Orlando Lawrence Berkeley National Laboratory, Berkeley, CA 94720

Recent laboratory and field experiments have demonstrated a variety of flow processes in fractured rock, such as intrafracture film flow along fracture surfaces, coalescence and divergence of multiple flow paths along fracture surfaces, intrafracture water dripping, and flow and transport through a fracture network. Although direct measurements of variables characterizing the individual flow and chemical transport processes in fractured rock are not technically feasible, their cumulative effect can be characterized using a phase-space analysis of time-series data. The phase space analysis of time-series data gathered from laboratory and field infiltration tests show that flow in heterogeneous soils and fractured rock can exhibit the properties of a nonlinear dynamic system, and in particular, deterministic chaos with a certain random component. To better understand the physical processes leading to chaos, the author presents evidence and introduces models of chaotic advection for flow in unsaturated fractured rock, using the results of field and laboratory investigations. Chaotic advection for flow through fractured rock manifests itself through stretching and folding of material lines for flow along partially saturated fractures, chaotic temporal variations of infiltration and outflow rates, water pressure fluctuations, pulsing, and feedback processes. A better understanding of chaotic advection can be used for improving predictions of flow and transport in fractured rock and developing new, practical methods for remediation of contaminated sites, nuclear waste disposal in geological formations, and in other areas of earth sciences.

Introduction

The Environmental Management (EM) Program of the U.S. Department of Energy (DOE) is responsible for cleaning up more than 100 sites that were contaminated from research, development, production, and testing of nuclear weapons, as well as hundreds of sites with organic contaminants. Because remediation of contaminated soils and groundwater at many sites has not been as successful as expected, the Top-to-Bottom Review Team of DOE called for (a) an improved ability to predict contaminant fate and transport, and (b) the development of *in situ* techniques to access and treat subsurface contaminants (Report "Reinventing Cleanup: R&D Needs and Strategies," dated November 28, 2001). Characterization and prediction of flow and transport in both the unsaturated and saturated zones are key elements in designing effective remediation technologies. One of the most challenging problems is prediction of flow and transport within fractured rock in the vadose zone. This problem cannot be resolved without first characterizing the physics of unstable flow phenomena in unsaturated fractures.

Unstable flow and transport processes in heterogeneous fractured-porous media are generated by the interplay of such processes as intrafracture film flow along fracture surfaces, coalescence and divergence of multiple flow paths along fracture surfaces, intrafracture water dripping, and flow and transport through a fracture network. These processes are, in turn, coupled with intrafracture condensation and evaporation, fracture-matrix interaction, vertical fluid flow, and capillary barrier effects at the intersection between fractures. The phase space analysis of time-series data gathered from laboratory and field infiltration tests show that flow in heterogeneous soils and fractured rock can exhibit the properties of a nonlinear dynamic system, and in particular, deterministic chaos with a certain random component (*1, 2*). The presence of rock discontinuities at different scales appears to be one of the main reasons for failure of scaling relations developed for scaling of soil hydraulic parameters (*3*).

Until recently, flow and chemical transport processes in heterogeneous, subsurface porous and fractured media—with spatial and temporal irregularities of water pressure, saturation, and concentration—were assumed to be random. These processes are conventionally described using partial differential equations (e.g., the Richards' model for unsaturated soils), with volume and time averaged hydraulic parameters (*4*) whose spatial distribution is assigned stochastically (e.g., *5, 6*). Moreover, predictions of flow in fractured rocks are conducted using models for unsaturated hydraulic parameters originally developed for porous media (e.g., van Genuchten or Brooks and Cory models), which do not take into account complex fracture-flow or fracture-matrix interaction processes, occurring in fractured rock. These methods do not consider that one of the main reasons for the apparent randomness of the experimentally observed water-flow variables (e.g., water velocity and pressure) is the presence of nonlinear dynamic processes, and, particularly, deterministic chaos (*1, 2*).

To help explain the differences between regular (nonchaotic deterministic), random (turbulent), and deterministic chaotic systems, Figure 1 shows typical trajectories for each type of motion. Laminar (regular) flow is fluid motion with smooth and orderly adjacent layers or laminar slip, occurring with no mixing or insignificant mixing of fluid layers, or with a very slow molecular-diffusion process, up to several orders of magnitude slower than that under turbulence. Chaotic flow is fluid motion that occurs in a nonlinear dynamic system, and it exhibits random-looking, erratic behavior, but which can be predicted on a short-time scale. For a chaotic system, the flow trajectories diverge exponentially over time, so that chaotic flow becomes specifically unpredictable for later times, but whose range of fluctuations can be predicted using stochastic methods. In contrast to laminar and deterministic chaotic flow, turbulence is fluid motion with irregular (mostly random) fluctuations, characterized with three velocity components (in three dimensions). Because turbulent flow is characterized by a series of multiple spatial and temporal scales, turbulent motion is practically unpredictable at any given time, but volume and time-averaged characteristics of turbulent flow can be predicted.

Modern investigations into the deterministic chaotic properties of natural systems began with the pioneering work of Lorenz (*7*). Some examples of dynamic systems that display nonlinear deterministic-chaotic behavior with aperiodic and apparently random variability include atmospheric (*7, 8*), geologic (*9*), geochemical (*10*), and geophysical (*11, 12*) processes, avalanches resulting from the perturbation of sandpiles of various sizes (*13*), falling of water droplets (*14*), river discharge and precipitation (*15*), oxygen isotope concentrations (*16*), viscous fingering in porous media (*17*), oscillatory fluid release during hydrofracturing in geopressured zones buried several kilometers in actively subsiding basins (*18*), thermal convection in porous media at large Rayleigh numbers (*19*), and instabilities at fluid interfaces (*20*). Note that experimental and theoretical investigations have shown very similar (even universal) patterns of changes in variables characterizing different physical, biological, mechanical, and chemical systems, described using methods of nonlinear dynamics and chaos (*21*).

The study of deterministic chaotic processes involved in flow and transport through fractured rock, which has been supported by the DOE Environmental Management Science Program, has been conducted since 1996 with several resulting publications describing the results of laboratory and field infiltration tests (*1–3, 22–24*). The phase-space analysis of time-series data obtained from these experiments, including calculations of the diagnostic parameters of chaos, show that chaotic processes are typical for intrafracture flow in subsurface media, in combination with a certain random component.

To further advance knowledge about flow and transport in fractured porous media, this paper will present evidence of chaotic advection for flow and transport in unsaturated fractured rock, and describe subsurface flow processes with models of chaotic advection, using the results of field and laboratory investigations.

This paper is structured as follows: a review of the main processes affecting unsaturated flow dynamics in fractured rock; an outline of the fundamental concepts of deterministic chaos and a discussion of the physics of chaotic processes relevant for flow and transport through fractured rock: water dripping, chaotic diffusion, chaotic advection, chaotic mixing, and feedback processes; presentation of examples demonstrating evidence of chaotic advection based on the reanalysis of the results of the following experiments: (1) field small-scale ponded infiltration tests conducted at the Hell's Half Acre site in Idaho (*24*), and (2) laboratory fracture replica flow experiments conducted by Persoff and Pruess (*25*), Tokunaga et al. (*26*), Su et al. (*27*), and Geller et al. (*28*); and finally concluding remarks and a brief suggestion on developing engineered remediation technologies, based on enhancing subsurface chaotic processes, for cleaning up contaminated sites.

Processes Affecting Flow Dynamics in Fractured Rock

Flow and transport in a fractured rock system are affected by superposition, feedback, and competition of processes related to two key elements:

1. *Complex geometry of preferential flow paths* (as affected by rock discontinuity and heterogeneity on all scales—from a rough fracture surface to an irregular fracture network).

2. *Nonlinear dynamic processes* such as episodic and preferential flow, funneling and divergence of flow paths, transient flow behavior, nonlinearity, film flow along fracture surfaces, intrafracture water dripping, entrapped air, fracture-matrix interaction, and pore-throat effects.

The complex geometry of flow paths in fractured rock results, primarily, from rock discontinuities that are present on all scales, extending from the microscale of microfissures (among the mineral components of the rock) to the macroscale of various types of joints and faults (*29, 30*). The complexity of the fracture-network geometry can cause either divergence or convergence of localized and nonuniform flow paths in different parts of fractured media, as well as capillary barrier effects at the intersection of flow paths.

One of the most important features of flow in fractured rock is fast, preferential flow, which could be caused by several processes, such as film flow (*31*), water channeling (*4, 27, 32*), and fingering (*33*). Pore-throat and preferential flow effects, variable surface wettability, fracture roughness, and asperity contacts could also potentially cause spatial and temporal flow instability and chaos. As a result, the flow rate (*24, 34*) and capillary pressure (*33*) exhibit significant high-frequency temporal fluctuations under constant boundary conditions during infiltration into the subsurface. Air compression ahead of the wetting front creates a pulsation of water pressure at the wetting front (*35*). Heterogeneous fracture asperities are possible causes for episodic flow events, even under steady-infiltration boundary conditions (*36*). Intrafracture water dripping is affected by the viscosity, surface tension, and phase changes along an irregular surface, along with sticking, spreading, tortuosity, accumulation and episodic flow of water droplets, impurities, roughness, temperature, and the surface slope.

The intrafracture flow processes are affected by fracture aperture, fracture-surface roughness, asperity contacts, and the fracture-matrix interaction (*36, 37*). A combination of various intrafracture flow processes may cause intrafracture water dripping, film flow, and the presence of stagnant flow zones in dead-end pores or even along fracture surfaces (Figure 2). The chemicals retained in these zones can only diffuse into the flow-through zone, which is a slow (rate-limited) process, unless special measures are applied to enhance the diffusion process (*38*). Figure 2 demonstrates a schematic of the intrapore (or intrafracture) velocity field, including: (1) a central flow-through channel, and (2) stagnant regions (within dead-end pores or fractures), with vorticity flow zones (*39, 40*). Fluid flow through a channel causes a mass exchange through molecular diffusion between the channel and the surrounding porous media, as well as fluid mixing due to the vorticity of flow.

Several attempts were made to explain fast water seepage in unsaturated fractured rocks (under negative water pressure) using concepts of film flow (*31*). Sticking, spreading, and flowing of water droplets on solid surfaces may

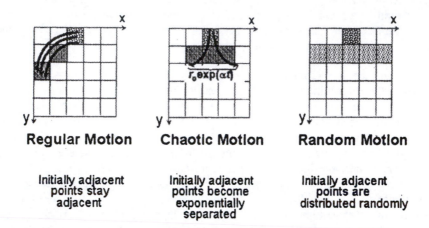

Figure 1. Comparison of regular (i.e., nonchaotic deterministic), deterministic chaotic, and random behavior—modified from (52). The rate of divergence of neighboring trajectories for deterministic chaotic behavior is determined by the Lyapunov exponent, λ.

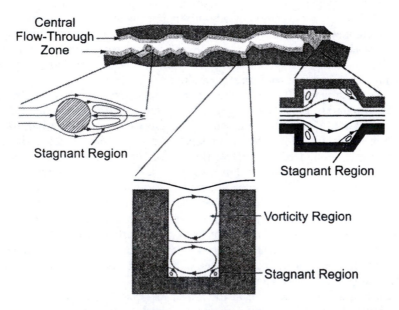

Figure 2. Schematics of film flow and stagnant zones in a fracture (the illustrations of stagnant and vorticity zones are shown according to [38]).

complicate film flow and water dripping. These processes depend on numerous factors, such as traces of impurities, roughness, temperature, and the contact angle of a drop (41, 42). Liquid flow in fractures is affected by a combination of surface tension, gravity, and inertia (43). For two-phase (water and air) flow in a fracture, the liquid layer is bounded on one side by the supporting solid matrix and on the other side by a surrounding fluid interface. If the surrounding fluid is gas, the film has a free surface, identical to flow along inclined planes. Recent theoretical and computational research (44, 45) on films flowing down inclined planes indicated highly ordered patterns that can spontaneously appear in some driven dissipative systems, which can be described by a 4th order partial differential equation (see following section). Tokunaga et al. (26) and Tokunaga and Wan (31) considered the film flow on fracture surfaces as analogs to water movement in partially saturated porous media, which is described by the Richards equation.

Physics of Chaotic Processes

Fundamentals of Deterministic Chaos

The term *deterministic chaos* is used to describe dynamic processes with random-looking, erratic data, in which random processes are not a dominant part of the system (46). Deterministic chaos is developed either in a dissipative system (the motion irreversible, as in systems with friction or systems exchanging energy with external media), such as the unsaturated zone, or in a conservative system (a Hamiltonian system that conserves energy), which is typical for the saturated zone. In order for a physical system to exhibit deterministic-chaotic behavior, the following conditions must be met:

- *The system is sensitive to initial conditions*, in that small changes in initial conditions significantly influence the system behavior over time, such that the system may exhibit nonperiodic behavior and not repeat its past behavior. The presence of nonlinear dynamics implies that small changes in initial and boundary conditions and parameters will ultimately lead to significant changes in the results of predictions.
- *The system is nonlinear*, but while nonlinearity is a necessary condition, it is not sufficient for a system to be chaotic (47).
- *Intrinsic properties of the system*, not random external factors, cause the irregular, chaotic dynamics of system components.

One of the most powerful techniques for identifying the presence of chaos is the phase-space reconstruction of time-series data. The *phase space* of a dynamic system is defined as an *n*-dimensional mathematical space, with orthogonal coordinates representing the *n* variables needed to specify the instantaneous state of the system (*48, 49*). The trajectories of the system's vector in the *n*-dimensional phase-space evolve in time from initial conditions onto the geometrical object called an *attractor*. The attractor is a set of points in a phase space towards which nearly all trajectories converge, and the attractor describes an ensemble of states of the system. Different variables of the phase space can be used as coordinates to graphically construct the attractor. Examples of the variables are:

1. The relationship between different system parameters (for example, directly measured physical variables such as capillary pressure, moisture content, and flow rate); it is said that the attractor is plotted in the parameter space.
2. The one-dimensional scalar array (*50*), $X_i(t)$, of one of the physical variables (e.g., time-series of pressure, temperature, velocity, or saturation array) or its first and second derivatives.
3. The scalar data, $X_i(t)$, and the values $X_i(t+\tau)$ and $X_i(t+2\tau)$, i.e., the values of X_i separated by a time-delay, τ, between successive measurements (this procedure is called a pseudo-phase-space reconstruction).

The bounds of the attractor characterize the range of long-term variations in system parameters. If the nonlinear dissipative dynamic system converges towards an attractor with aperiodic trajectories (i.e., chaotic) trajectories, the attractor is called a strange, or chaotic, attractor. A minimum of three dimensions in the phase space is required for a chaotic attractor to exist. Note that a correlation dimension of 5 or less characterizes a low-dimensional deterministic chaotic system (*51*). Trajectories on the attractor are not closed, i.e., a single trajectory will never return to its initial point, but will visit all points of the attractor in infinite time. Adjacent trajectories in the phase-space of the chaotic attractor diverge exponentially with time. Chaotic attractors have geometric structures that are fixed, despite the fact that the trajectories moving within them appear unpredictable. The geometric shape of the chaotic attractor can be used to identify the processes involved in the system's chaotic behavior.

Another important property of a chaotic system is the feature of stretching (divergence of neighboring trajectories) and folding (confinement to the bounded space, so that the system cannot explode to infinity) of a system trajectory in a phase space. A simple example of stretching and folding is as follows: In making bread, kneading a block of dough involves multiple stretching and folding over the dough in different directions. After a series of kneading, two nearby points of the dough will move away from each other in a complex manner. In other words, these two points 'lose memory' of their initial locations (52). A process of dough kneading resembles the formation of a geometric shape of a chaotic attractor, as the local separation of phase-space trajectories corresponds to stretching the dough, and their attraction corresponds to folding the stretched dough back onto itself.

Although chaotic fluctuations could theoretically be purely deterministic and be described using simple ordinary differential equations, real physical processes usually contain a stochastic (or noise) component (11, 53, 54). Note that a combination of many deterministic-chaotic processes could lead to an apparent randomness of experimentally observed data and a transition to high-dimensional chaos (with a correlation dimension more than 5 [51]), which is difficult to distinguish from randomness.

Water Dripping

A simple and common example of chaotic dynamical phenomena is that of water dripping from an apparently simple device—the leaking faucet. Shaw (55) showed the development of deterministic chaos for water dripping from a single faucet as a combination of gravitational, external, and surface tension forces. The source of instability is surface tension. Most of the models used to simulate water dripping are based on a relaxation-oscillator model. For example, to describe the time variations of dripping frequency, Shaw (55) developed a simple dynamical model to represent a coupling of a variable mass with a spring. This mass-spring model describes oscillations of a mass point, with mass increasing linearly with time at a given flow rate Q, and for a fixed value of the spring constant k. When the spring extension exceeds a certain threshold, which describes the detachment (snapping) of a falling drop, a part of the total mass m is removed. Although this empirical model provides a mechanistic description of the system and displays the presence of a chaotic regime, it does not provide a physical explanation of the dripping water phenomena in a continuous state fluid-dynamic system.

A fluid mechanics model to describe the dynamics of a dripping faucet is given by Coulett et al. (56), who described the dynamics of flow instability using hydrodynamical equations linearized about the stationary pendant drop. They also used a lubrication model for the fluid, embodied in a Lagrangian approach, with the following assumptions: (1) during its motion, the drop remains axisymmetric, (2) the radial component of the fluid velocity is negligible compared to the axial component, which depends on the vertical coordinate alone, and (3) no overturning of the interface occurs. Kiyono et al. (57) showed experimentally and numerically that a chaotic attractor of a dripping faucet is low-dimensional in a continuous state space. They also demonstrated that the dripping faucet dynamics could be described using a potential function with only two variables, such as the mass of the pendant drop and the position of the center of mass. Néda et al. (*58*) found that dynamic behavior of the faucet can be described based on the drop size and drip interval statistics. Coullett et al. (*56*) show that for small flow rates, the dripping may be periodic; for intermediate flow rates, it may be chaotic. As the flow rate increases, such behavior displays a so-called boundary crisis. The physical aspects of water dripping from a faucet will be compared with those observed during flow visualization tests in fracture replicas (*27, 28*).

Chaotic Diffusion

Diffusion is conventionally defined as the Brownian motion of a particle and is determined from

$$x' \sim \varepsilon(t) \tag{1}$$

where x' is a particle velocity along a coordinate x, and $\varepsilon(t)$ are random forces changing with time t. For a Gaussian-correlated random function $\varepsilon(t)$, it follows that

$$\langle x^2 \rangle \sim t \tag{2}$$

indicating that the squared distance from the origin increases linearly with time (52, p 33).

For motion in a dissipative deterministic chaotic system, deterministic chaotic diffusion along a coordinate x is developed according to the following condition (*59*):

$$\langle x^2 \rangle \sim t^\alpha \tag{3}$$

where α is the power factor, which could vary with time (see below the section on the laboratory experiments).

According to the deterministic-chaotic concept, macroscopic transport coefficients, such as diffusion coefficient and reaction rate, will exhibit irregular behavior as a function of a control-system parameter (*60*). The physics behind deterministic chaotic diffusion is that the trajectory in the phase space is generated by chaotic motion as a result of losses of its memory, but not generated by random forces. The deterministic-chaotic diffusion-reaction process (for assessing the reaction rate in chemical systems) replaces the old stochastic transport models (*52, 60*). Examples of using eq 3 will be given in the section on the analysis of laboratory experiments.

Chaotic Advection

Chaotic advection of laminar flows was recognized early in the 1980s by Aref (*61*). Chaotic advection is defined as the complex behavior of a passive scalar, i.e., idealized particles that are so small that they do not perturb the flow and move along the flow; but large enough not to generate Brownian motion (i.e., not to diffuse), and do not influence the hydrodynamics of the flow field. It is assumed that (*62*)

$$V_{particle} = V_{fluid}$$

$$V_{particle} = (dx/dt, dy/dt, dz/dt).$$

(4)

The dynamics of passive particle (or quantities, e.g., temperature or concentration) advection in the incompressible, laminar flow may be described by the Lagrangian representation of the fluid element (for low Reynolds numbers < 1) with a time-dependent (unsteady) or time-independent (steady) velocity, U, given by (*63, 64*)

$$dx/dt = u_x(x, y, z, t); \quad dy/dt = u_y(x, y, z, t); \quad dz/dt = u_z(x, y, z, t) \qquad (5)$$

where u_x, u_y, and u_z are the Cartesian components of the velocity field $u(u_x, u_y, u_z)$ along coordinates x, y, and z, and $\nabla \cdot U = 0$.

To demonstrate how the geometry of flow could cause the development of three-dimensional chaotic advection, Figure 3 illustrates a flow progression through a twisted pipe model (as an analog for the flow through a fracture). This flow consists of a sequence of consecutive half-tori rotated with respect to each other by a pitch angle. In this simplified model, at the interface between two half-tori, the flow pattern adjusts itself instantaneously to the pipe twist. Based on this simplification, a passive scalar evolves, following the streamlines of the half torus through which it is passing at any given time, and these streamlines are the same as for an isolated torus (*60*, pp. 8, 9). This example demonstrates that chaotic advection in flow processes in even laminar flows can follow extremely complicated, chaotic trajectories.

Chaotic advection for an incompressible fluid in a restricted final volume (assuming the validity of volume conservation) is developed when the fluid is attracted to a stagnation point, presenting a hyperbolic fixed point, along one direction and repelled along another. Dripping water along an unstable flow path at the fracture surface represents phenomena of repeated stretching and folding, which is reminiscent of a strange attractor and chaotic advection for a dynamic chaotic system.

Velocity variations (perturbations) at the fracture intersections could cause chaotic advection, as well. For example, Figure 4a schematically depicts

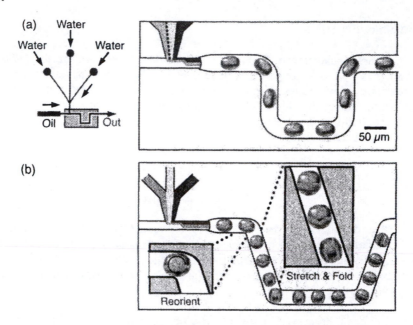

Figure 3. Stretching and folding of streamlines and mixing by the baker's transformation in plugs moving through winding microfluidic channels: (a) schematic of the experiment using the microfluidic network (left), and microphotograph of plugs (right), and (b) schematic of the stretching and folding (87).

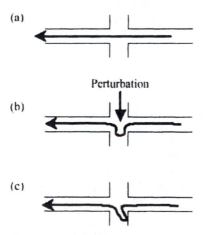

Figure 4. Schematics of (a) horizontal flow along the main flow channel, with a straight streamline showing the direction of flow; (b) streamline folding affected by perturbation, and (c) stretching of a streamline after switching off a perturbation, which is favorable to chaotic mixing.

horizontal flow along the main flow channel, with a straight streamline showing the direction of flow under a constant flow velocity. Figure 4b shows a streamline folding in the crossing flow paths, as affected by perturbation, and Figure 4c shows stretching of a streamline after switching off a perturbation, which is favorable to chaotic advection, diffusion, and mixing. Deterministic chaotic advection (without turbulent fluctuations of the flowfield) of a passive scalar, through a two-dimensional or three-dimensional time-dependent flow field (taking place along chaotic trajectories), may enhance both chaotic diffusion and chaotic mixing (*40, 65, 66*).

Chaotic Mixing

As chaotic regimes are associated with stretching and folding of material lines, the advection of blobs of markers in the fluid tend to create complex structures, which, in turn, facilitate molecular diffusion and mixing. Basic processes involved in chaotic mixing are illustrated in Figure 5 (*40*). Figure 5a shows the case of stretching and folding of streamlines for a passive blob with two similar fluids having a negligible interfacial tension with no interdiffusion. Figure 5b shows an active blob when the boundaries between fluids become diffusive, enhancing the mixing process because of drastic differences in the concentration (right-hand side schematic of Figure 5b). Figure 5c shows the case in which the difference in interfacial tension forces causes the blob to break up into small fragments, which may, in turn, stretch and break, producing smaller fragments.

According to Ottino (*40*), mixing reduces length scales, involving thinning of material volumes and dispersion throughout the space, possibly involving breakup and (for miscible fluids) uniformity of concentrations, and enhancing chemical diffusion and transport. Note that chaotic mixing processes may not be related to kinematic processes (*40, 61, 67*).

The following situations could be typical for flow in fractured rock. Mixing and dispersion of fluids result from two types of complex interactions between flow events (even for small Reynolds numbers) on different scales (1) drop length-scales, such as breakup, coalescence, and hydrodynamic interactions, and (2) agglomerate length-scales, such as surface erosion, fragmentation, and aggregation (*68*). Chaotic regions could coexist with nonchaotic zones (see below the section on laboratory experiments for examples from laboratory visualization experiments). The fact of no significant separation between processes on different scales leads to nonasymptotic properties of transport and mixing—in other words, it is not possible to characterize dispersion in terms of asymptotic quantities such as average velocity and diffusion coefficients (*64*).

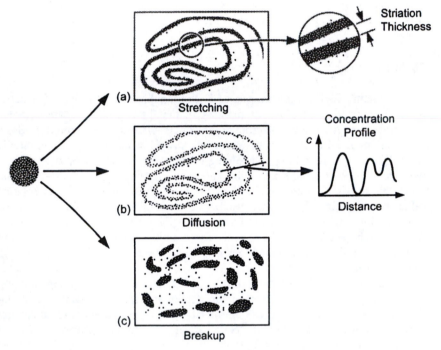

Striation
Thickness

(a) Stretching

Concentration
Profile

(b) Diffusion

Distance

(c) Breakup

Figure 5. Basic processes involved in chaotic mixing (40, page 2, Figure 1.1.1): (a) stretching and folding of streamlines for a passive blob with two similar fluids having a negligible interfacial tension with no interdiffusion; (b) evolution of the active blob when the boundaries between fluids become diffusive, enhancing the mixing process caused by drastic differences in the concentration (right-hand side schematic of Figure 5b); and (c) the case where the difference in interfacial tension forces causes the blob to break up into small fragments.

Feedback Phenomena

Open systems (e.g., geological, ecological, biological) respond to changes in system variables through either positive or negative feedback, or both. Feedback is well understood and widely studied in the fields of electrical engineering and biology. For a positive feedback, the system responds in the same direction as that variable, accelerating the system behavior. A positive feedback can potentially run a nonlinear system out of control, ultimately resulting in the system's collapse. For a negative feedback, the system responds in the opposite direction as that variable, slowing down the system behavior. While positive feedback tends to accelerate the system and enhance nonlinear effects, negative feedback stabilizes the response of an open system and reduces nonlinear effects.

Flow and transport in the subsurface could be affected by a feedback between chemical, microbiological, and water flow processes. For example, the water flow process may generate the development of biofilms around sediment particles, which, in turn, reduce the sediment permeability. At the same time, reduced permeability causes the flow rate to decrease, so that the microbial activity diminishes. The formation of the near surface low-permeability layer causes the hydraulic gradient in the near-surface zone to increase, and may cause the atmospheric air to enter the subsurface, reducing the flow rate from the surface (*69*). In addition, the colloidal dynamics in fractured porous media are complicated by the electrokinetic and hydrodynamic interaction between colloids, nonequilibrium adsorption, nonsorptive interactions of bacteria and colloids with particles, growth and grazing by protozoa, and detachment from solid surfaces, which are different from the dynamics in an open space (*70*). These processes can be described by a set of nonlinear, coupled electrokinetic and convective diffusion equations for ion densities, in combination with Navier-Stokes equations for the mass current, indicating that colloidal dynamics are nonlinear (*71–72*).

Intrafracture flow, including film flow and dripping (which, for example, causes time variations of water pressure—see the section on laboratory experiments), represents the type of a system with both positive and negative feedbacks. A combination of forward and backward capillary pressure waves for two-phase flow in a fracture results from a superposition of positive and negative feedback mechanisms—see the section on the two-phase fracture flow experiment.

Conventionally used Darcy's law and Richards' equation (2nd order partial differential equations) are examples of a system with a positive feedback, but without a negative feedback component. Negative and positive feedback mechanisms are taken into account by using the difference-differential equation for soil-moisture balance (*73*) and the Kuramoto-Sivashinsky equation (*74*) (see below).

Examples of Chaotic Advection from Field and Laboratory Tests

Field Infiltration Tests

Site Characterization

Several small-scale ponded infiltration tests were conducted in fractured basalt at the Hell's Half Acre field site (near Idaho National Engineering and Environmental Laboratory). A small water reservoir (40 cm × 80 cm) was constructed on the surface exposure of a vertical fracture at an overhanging basalt ledge (*24*). A constant water level was maintained in the water reservoir. The vertical fracture extended from the top to the bottom of the basalt ledge, intersecting a horizontal fracture that was exposed at the face of the basalt column. The infiltration rate was measured in the surface water reservoir, and the outflow rate (of dripping water) was measured using a grid of pans installed underneath the overhanging ledge. To infer the physics of the flow processes from time-series data, we provided a phase-space reconstruction of the infiltration and outflow rates, and determined diagnostic parameters of chaos for dripping intervals.

Infiltration and Outflow Rates

Despite the fact of maintaining the constant-head ponded water level, both infiltration and outflow rates exhibited a general three-stage trend of temporal variations identical to that observed during the infiltration tests in soils in the presence of entrapped air (*69, 75*). Gradual (low-frequency) time variations in the flow rate were accompanied by high-frequency oscillations (Figure 6a). Using noise-reduced trends of the infiltration and outflow rates, we plotted 3D pseudo-phase space attractors shown in Figures 6b and 6c. We can see that the attractors have spiral shapes and a few saddle points, with a smaller size of the outflow-rate attractor, implying the possibility of a deterministic-chaotic process and a dispersion process in flow (*76*) along a 1-m fractured basalt column.

The relation between the local outflow rate and the total outflow rate is nonlinear. For example, the plots in Figure 7 show a semi-logarithmic relation between the local outflow rate (for two pans located outside of the perimeter of the infiltration gallery) and the total outflow rate. The plots also show the threshold for a significant discharge from these fractures—20 mm/min for Pan 8, and 30 mm/min for Pan 12. In addition, the discharge from fractures causes the changes in infiltration into the subsurface, which is an indication of the feedback phenomenon, a typical feature of a dynamic chaotic system.

Phase-Space Analysis of Water Dripping Intervals

The examination of time-series data using the phase space analysis shows that water-dripping behavior from a fracture exhibits different types of chaos (*2*). Figure 8a presents an example of time series data for water dripping intervals. For these data, Figures 8b and 8c display the results of the time-domain analysis—Fourier transformation and autocorrelation function, respectively. Figures 8d-8g present the results of the phase-space analysis of water dripping intervals, including the diagnostic parameters of deterministic chaos, such as time delay, global embedding dimension, local embedding dimension, and Lyapunov exponents. The determination of diagnostic parameters of chaos from time-series data was provided using the methods described by Abarbanel (*50*).

Figure 9 shows several segments of the time-series and corresponding 2D projections of the attractors for one of the dripping points. The beginning of the test is characterized by quasi-periodic, almost double-cycling fluctuations around a constant mean value (Points 1–500). Then, a slight increase in the mean value accompanies a shift in the attractor (Points 500–1,100), which persists until the fluctuations gradually die out (Points 1,100–2,500). The next segment (Points 2,500–4,400) represents a gradual increase in the periodicity of fluctuations, followed by quasi-periodic fluctuations, with a now-inverted attractor. An apparently increased amplitude modulation in Figure 9a resembles the solution of a pendulum perturbed by both parametric and external periodic forces—typical for spatio-temporal interactions (*77*). These interactions are likely taking place within the fracture.

The analysis of long-term data sets shows the presence of both high-frequency and low-frequency fluctuations. The analysis of these data indicates the presence of different time scales involved in flow processes through a fracture, which are associated with the intrafracture flow and capillary barrier processes. Based on the presence of two time scales in the pattern of the water-dripping frequency observed at the fracture exit, we hypothesize that the observed time series is generated by a combination of intrafracture flow and a capillary barrier effect at the exit of the fracture. We assume that dripping from the fracture generates mostly high-frequency fluctuations, while intrafracture flow generates mostly low-frequency fluctuations.

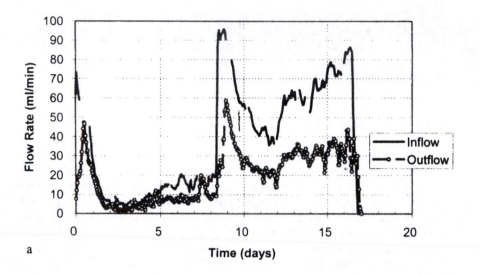

a

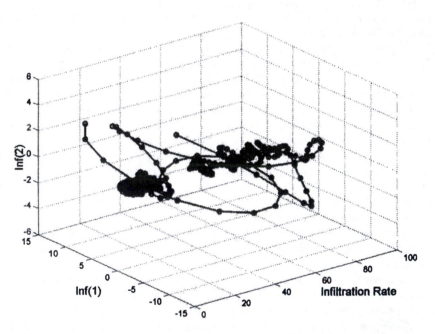

Figure 6. The results of the Hell's Half Acre infiltration test: (a) Infiltration and total outflux rates; (b) and (c) 3D pseudo-phase space attractors [in coordinates q, q(t-1), and q(t-2)] of infiltration and total outflow rates, respectively. All flow rates are area averaged. The time series data are from the Hell's Half Acre Site, ID, Test 8, 1998 (24).

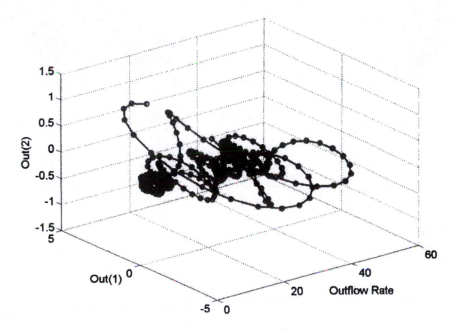

Figure 6. *Continued.*

Pan 8 Flux vs. Total Outflux

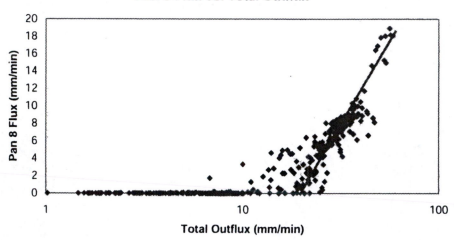

Pan 12 Flux vs. Total Outflux

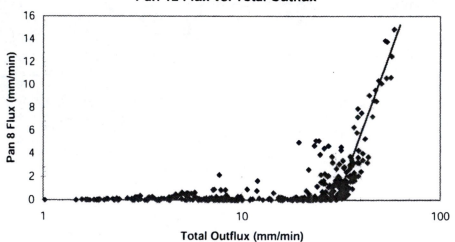

Figure 7. Relations between the total outflux recorded during the infiltration test (Test 8) and local outflux rates from fractures outside the perimeter of the infiltration gallery, collected in Pans 8 and 12 (time series data are shown in Figure 7b).

Thus, our analysis of water-dripping intervals reveals that different timescales of water dripping are not related to changes in boundary conditions, because the tests were conducted at a constant water level in the water-supply gallery, but to variations in intrafracture flow processes and capillary barrier effects.

Feedback

The Hell's Half Acre infiltration tests also show the dependence of the infiltration rate upon the outflow rate. The dependence of infiltration upon conditions below and above the fracture was also shown based on the results of the series of larger-scale ponded infiltration tests in fractured basalt at the Box Canyon site in Idaho (*22*).

Laboratory Experiments to Study Intrafracture Flow Processes

Intrafracture Flow Paths' Coalescence, Divergence, Film Flow, and Dripping

Test designs. This section includes the discussion of the laboratory tests conducted using fracture models (*26, 27, 28*).

Su et al. (*27*) conducted a series of laboratory tests using dyed water supplied into two different setup models: (1) a transparent replica (about 15-cm wide by 30-cm long) of a natural, rough-walled rock fracture from the Stripa Mine, Sweden, for inlet conditions of constant pressure and flow rate, over a range of inclination angles, and (2) parallel glass plates with three types of apertures. In these experiments water was supplied through the whole width of a fracture model.

Geller et al. (*28*) conducted a series of laboratory experiments in which water was injected at a constant flow rate (from 0.25 to 20 mL/hr) into fracture models (smooth, parallel glass plates separated by 350 µm, and textured glass plates, inclined 60° from the horizontal) through a single capillary tube that terminated at the entrance to the fracture model. (In these experiments, we also investigated the effects of the size and material of the capillary tube and the type of contact between the capillary tube and fracture model.) Liquid pressure was monitored upstream of the entrance to the fracture.

Tokunaga et al. (*26*) measured the propagation of the wetting front along the horizontal glass surface, which was assumed to be a single fracture model.

Analysis of fracture-flow dynamics. The results of investigations by Su et al. (*27*) demonstrated phenomena of intermittent flow under unsaturated

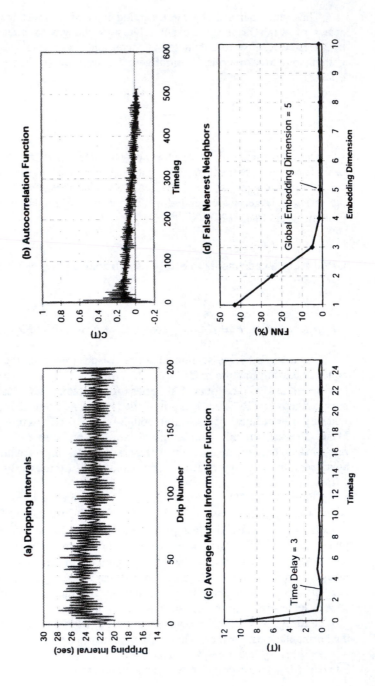

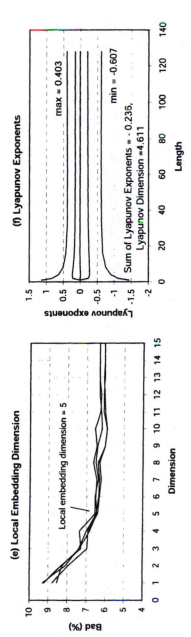

Figure 8. *Example of the phase-space analysis of water dripping intervals from the Hell's Half Acre infiltration test (Test 8, Dripping Point 6): (a) time-series of dripping water intervals, (b) autocorrelation function, (c) mutual information function used to determine the time delay, (d) false nearest neighbors used to determine the global embedding dimension, (e) local embedding dimension, and (f) Lyapunov exponents and Lyapunov dimension.*

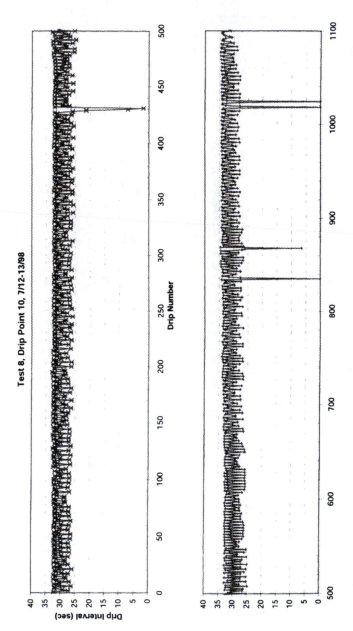

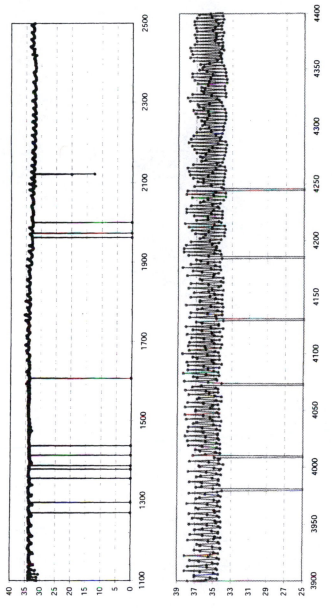

Figure 9. Water dripping intervals and corresponding 3D attractors showing the transient (nonstationary) character of deterministic chaos in flow through a fracture (Dripping Point 10. Test 8, 1998).
Continued on next page.

b

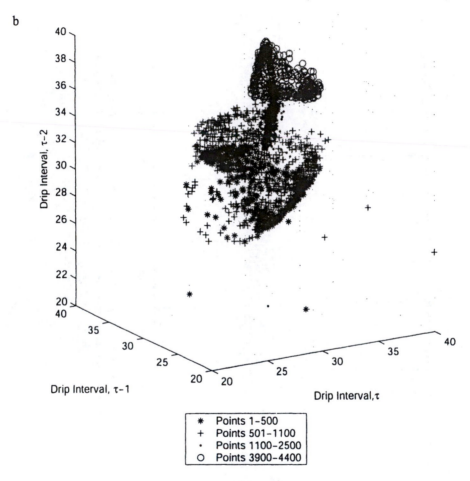

Figure 9. *Continued.*

conditions along two main pathways (Figure 10). The photograph in Figure 10 demonstrates the coexistence of a relatively stable (nonchaotic) zone and unstable (chaotic) flow along the individual flow pathways with divergent and coalescent flow paths, as well as the location of intrafracture water dripping, and intermittent film flow. Despite the local flow geometry changed rapidly over time, the average coverage of these seeps was practically stable over time.

Geller et al. *(78)* observed similar behavior in experiments conducted using a transparent replica of a natural rock fracture with a variable aperture. The presence of established flow paths along the fracture surface is consistent with the concept of a "self-organized" critical state (*79*). It is apparent from the flow experiments in unsaturated fractures that the surface coverage has a critical value (despite supplying additional water to the surface), and the system organizes itself in such a way that the excess water is removed through streams.

Su et al. (*27*) also quantified the combined effect of gravity, capillary, and viscous forces on flow in a fracture model, using Bond (*Bo=gravity force/capillary force*) and capillary (*Ca=viscous force/capillary force*) numbers. Figure 11 demonstrates the effect of capillarity and gravity driven flow, and shows that for the Bond number from 0.004-0.027, the volume of water per a dripping event remains insensitive to the Capillary number; and for the Bond number from 0.053 to 0.057, the volume of water per a dripping event increases drastically as the Capillary number increases.

To understand the physics of the propagation of preferential flow zones (or water fingers) along the fracture surface, including the contribution of the molecular diffusion and gravity effect, this author analyzed the data from Su et al. (*27*). The results of measurements of the length of fingers with time, presented in Figure 12a, were analyzed using a model of deterministic chaotic diffusion given by eq 3 (see the section on physics of chaotic processes). The results of calculations of coefficient α for eq 3 are shown in Figure 12b. One can see from Figure 12b that this coefficient is higher than unity, indicating a complex interplay of diffusion, gravity, and surface tension. Moreover, this coefficient decreases over time during the finger (film flow) propagation until water dripping starts, and then it abruptly increases for the fracture model inclination of 46 and 81 degrees. Coefficient α increases gradually for the fracture model inclination of 19 degrees. Note from Figure 12b that coefficient α becomes greater as the inclination increases.

Geller et al. (*28*), using a fracture replica, observed that water seeped through the fracture in discrete channels that undergo cycles of snapping and reforming, and liquid drips detached at different points along the water channel. Pressure fluctuations upgradient of the pressure sensor could be correlated to the growth and detachment of drips in the interior of the fracture observed directly and recorded with a video camera (Figure 13). This figure demonstrates the sequence of images for dripping phenomena within a fracture, indicating that after the break-up of the drop (thread snaps), the system does not return to the

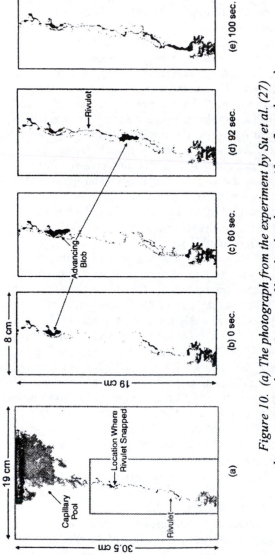

Figure 10. (a) The photograph from the experiment by Su et al. (27) demonstrating the coexistence of localized and non-uniform flow channels, which are originated from water-filled regions (or small, relatively stable, capillary pools), connected by unstable (chaotic) rivulets of liquid, and (b)-(e) A sequence of intermittent events along the flow channel.

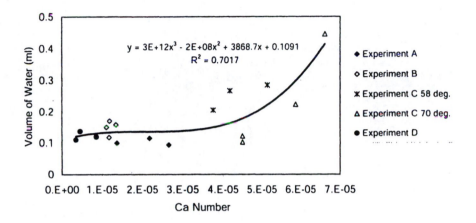

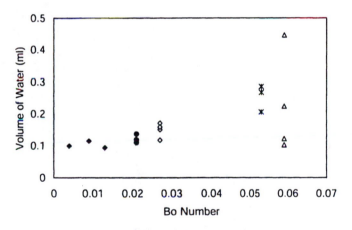

Figure 11. Effect of the capillary and Bond numbers on the volume of water between intermittent flow events (27).

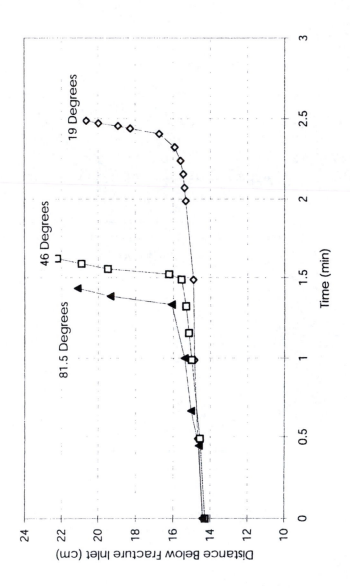

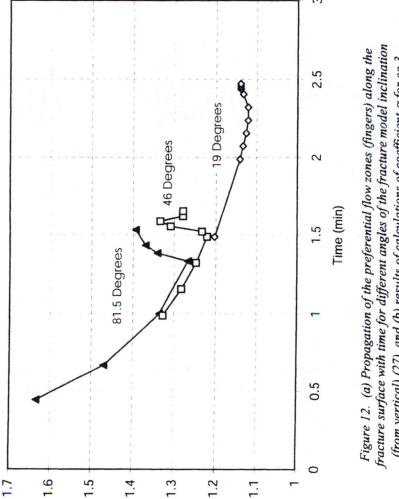

Figure 12. (a) Propagation of the preferential flow zones (fingers) along the fracture surface with time for different angles of the fracture model inclination (from vertical) (27), and (b) results of calculations of coefficient α for eq 3, which describes the process of chaotic diffusion, using the results of the Su et al. (27) and Tokunaga et al. (26) experiments.

214

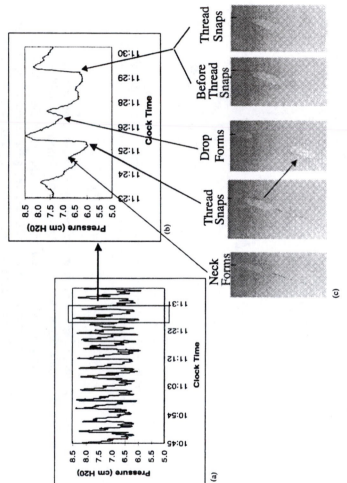

Figure 13. Correspondence between pressure time-trend and drip behavior for 0.25 mL/hr, needle point source within smooth glass plates separated by 0.36 mm shim. (a) Pressure data. (b) Expansion of the boxed section shown in (a). (c) Frames from video tape recording of experiment showing drip behavior (28).

same condition it was in before snapping, resulting in a different breaking point. (This is another indication of the sensitivity of dripping in a fracture to initial conditions, which is a necessary criterion for a deterministic chaotic system.) Comparison of the physical processes of intrafracture water dripping and of a single leaky faucet shows that these processes appear to be qualitatively different in that the intrafracture dripping is associated with the development of unstable (chaotic) rivulets of liquid formed along the fracture surface, while this effect is nonexistent for dripping from a single faucet.

This author also analyzed the data on the propagation of the wetting front along a horizontal roughed glass surface, assumed to be one side of fracture, from Tokunaga et al. (*26*). The results of calculations of a coefficient α for eq 3 are also given in Figure 12b. These data indicate a stronger process of diffusion at the beginning of the test than thereafter. The fact that α is not equal to 1 indicates that chaotic diffusion could have affected flow along the fracture model surface.   It is likely that the wetting front propagation (from the experiments by Su et al. [*27*] and Tokunaga et al. [*26*]) is affected by molecular van der Waals forces (*80*).   The effect of molecular van der Waals forces was also supported by the physico-mathematical analysis given by Faybishenko et al. (*81*).

Two-Phase Fracture Flow

Persoff and Pruess (*25*) conducted a series of two-phase flow experiments by simultaneously injecting water and nitrogen gas (representing the wetting and nonwetting phases) in replicas of natural rough-walled rock fractures in granite (from the Stripa mine in Sweden) and tuff (from Dixie Valley, Nevada). In these experiments, the water and gas flow rates were changed stepwise, and the gas and liquid pressures were measured at the inlet and outlet edges of the fracture for each of the constant flow rates. Persoff and Pruess (*25*) explained that instabilities in the liquid and air pressures (observed under constant liquid and gas injection rates) resulted from recurring changes in phase occupancy between liquid and gas at a critical pore throat in the fracture, as well as competition between fluid pressures caused by injection and capillary effects driving the liquid to the critical throat.

The results of the phase-space analysis of time-series data for capillary pressure for one of the tests show that the time intervals (x) between pressure spikes (pulses) can be described using a simple exponential equation, given as a difference equation by:

$$x_{n+1} = A\, x_n \exp(-\alpha x_n) \qquad (6)$$

with an additional small noisy component (α). In eq 6, n denotes the number of a time interval between pressure spikes, and A is a coefficient. This type of equation was used to describe the time-series data in population dynamics (*82, 83*).

For another experiment, temporal variations in capillary pressure exhibit quasi-periodic cycling with relatively short periods of laminar flow, which are interrupted by chaotic fluctuations (most likely indicative of a liquid breakthrough at a pore throat) (*25*). However, the inlet and outlet cycling patterns are different, as shown in Figure 14a. Theoretically, the forward and return waves of pressure must decay in the direction of flow, implying the dispersion of flow (*76*, p. 228). A larger magnitude of fluctuations and a longer duration of the laminar phase for the outlet capillary pressure than those at the inlet end are probably caused by a capillary barrier (pore-throat) effect near the exit from the fracture. We hypothesize that the observed quasi-periodic pressure oscillations at both inlet and outlet ends of the fracture result from a superposition of the forward and return capillary-pressure waves. Moreover, the effect of the outlet pressure fluctuation on the inlet pressure can be considered as a mechanism of negative feedback.

The phase space analysis of both inlet and outlet capillary pressures produce a zero Lyapunov exponent (*1*), implying that this dynamic system can be described by differential equations (*50*). A comparison of the pseudo-phase-space three-dimensional attractors for the inlet and outlet capillary pressures shows (Figures 14b and 14c) that these attractors are analogous to those described using the solution of the Kuramoto-Sivashinsky equation (Figures 14d and 14e) discussed below.

Kuramoto-Sivashinsky Equation for Chaotic Flow through Fractures

A 4th-order partial differential equation, called the Kuramoto-Sivashinsky (K-S) equation, is given in a canonical form by (*74*)

$$\frac{\partial \phi}{\partial \tau} + \phi \frac{\partial \phi}{\partial x} + \frac{\partial^2 \phi}{\partial x^2} + \frac{\partial^4 \phi}{\partial x^4} = 0 \qquad (7)$$

where ϕ, x and τ are dimensionless film thickness, length, and time, respectively. The gravitational, capillary, and molecular (van der Waals) forces included in the derivation of the K-S equation are identical to those occurring in fractures. In the K-S equation, the second term is a nonlinear term; the third and fourth are the destabilizing and stabilizing terms, on the same order of magnitude, that describe dissipative processes (*84, 85*). For small Reynolds numbers, typical for flow through fractured rock, the instability described by the K-S equation is

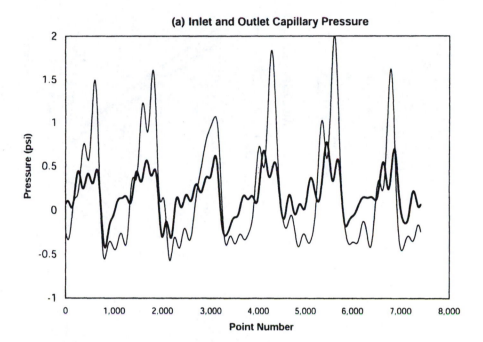

Figure 14. (a) Temporal variations of inlet and outlet capillary pressures from Experiment C of Persoff and Pruess (25) using Stripa natural rock under controlled gas-liquid volumetric flow ratio of two (the data are noise-reduced using Fourier transformation; (b) and (c) 3D attractors of inlet (time delay, τ=12) and outlet (τ=7) capillary pressures; and (d) and (e) 3D attractors of the solution of the Kuramoto-Sivashinsky equation (7) for the upper boundary of the flow domain (τ=6) and the lower boundary of the flow domain (τ=7). Time delays were determined using the average mutual information function (50). Time series data for Figures 14d and 14e are shown in Figure 15.

Continued on next page.

218

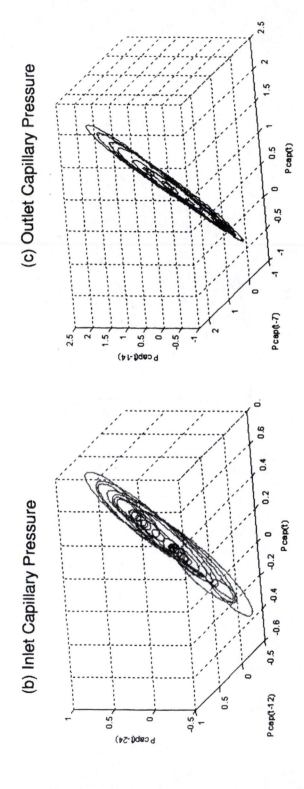

(c) Outlet Capillary Pressure

(b) Inlet Capillary Pressure

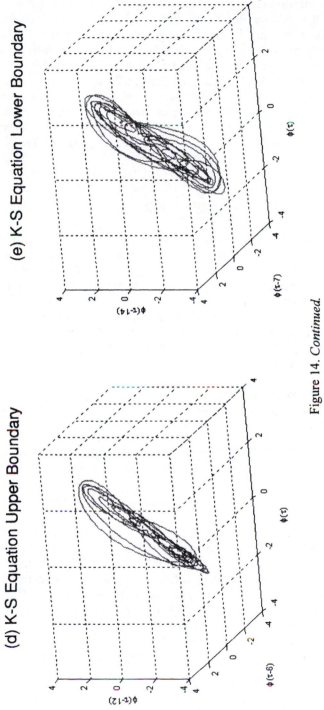

(d) K-S Equation Upper Boundary

(e) K-S Equation Lower Boundary

Figure 14. *Continued.*

created by molecular forces (81). Figure 15 shows examples of the solution of a one-dimensional K-S equation, illustrating the possibility of irregular but deterministic variations in film thickness along the fracture surface. The solution of eq 7 generates unusual, highly-ordered, but nonperiodic patterns of trajectories on attractors describing the evolution of the film thickness (45).

Thus, the complex water-flow behavior observed under laboratory and field conditions can represent the cumulative effect of the many degrees of freedom involved in water flow. For the fracture flow process described by the K-S equation, we can reasonably hypothesize that on a local scale, the linear relationship between the pressure head and the flow rate (i.e., Darcy's law) is invalid.

Concluding Remarks

Laboratory and field experiments show the presence and interplay of such processes as intrafracture film flow along fracture surfaces, coalescence and divergence of multiple flow paths along fracture surfaces, and intrafracture water dripping. The nonlinear dynamics of flow and transport processes in unsaturated fractured porous media arise from the dynamic feedback and competition between various nonlinear physical processes along with the complex geometry of flow paths. The apparent "randomness" of the flow field does not prohibit the system's "determinism" and is, in fact, described by deterministic chaotic models using deterministic differential or difference-differential equations.

Although direct measurements of variables characterizing the individual flow and chemical transport processes under field conditions are not technically feasible, their cumulative effect can be characterized by the phase-space analysis of time-series data for the infiltration and outflow rates, capillary pressure, and dripping-water frequency. The time-series of low-frequency fluctuations (assumed to represent intrafracture flow) are described by three-dimensional attractors similar to those from the solution of the Kuramoto-Sivashinsky equation. These attractors demonstrate the stretching and folding of fluid elements, followed by diffusion.

The analysis of experimental laboratory and field data shows evidence of chaotic processes, including chaotic advection, diffusion, and mixing, for unsaturated flow in fractured rock. These flow processes are unsteady, with two slow (fracture-matrix interaction) and one fast (flow along the fracture surface) velocity components. Stretching and folding of the fluid elements result from the deconvolution and convolution of individual flow paths, as well as intrafracture water dripping.

The evidence of chaos from experimental field and laboratory tests, and from mathematical models, poses an interesting challenge for scientists. The

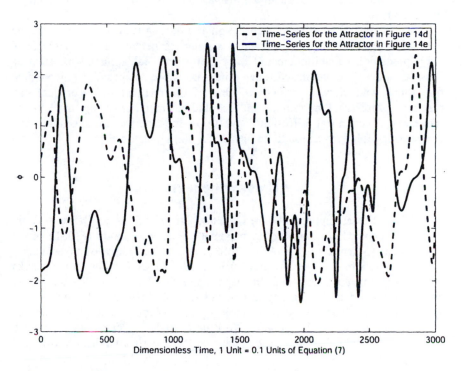

Figure 15. Illustration of the solution of the one-dimensional K-S equation showing time series data shown as 3D attractors in Figures 14d and 14e (numerical simulations were provided by David Halpern of the University of Alabama, Tuscaloosa, AL).

concepts of nonlinear dynamics and chaos can be used to provide an alternative explanation for erratic behavior of variables characterizing infiltration processes. The presence of chaos can make the use of classical deterministic and stochastic methods impossible for short-term prediction, while conventional stochastic methods can be used for long-term predictions. It is equally important (or probably more important) to understand what cannot be predicted than to understand what can be. Using nonlinear dynamic models is an important step forward from classical statistical approaches, but they are at an early stage of development.

An understanding of chaotic advection and mixing is also important for the prediction of chemical transport within subsurface heterogeneous media (86). Such prediction capabilities could have many practical applications, such as remediation of contaminated sites, nuclear waste disposal in geological formations, and climate predictions. One of the challenging theoretical and practical problems that remains to be studied involves using the chaotic processes of chemical diffusion, advection, and mixing in designing and engineering effective remediation schemes for contaminated sites.

Acknowledgments

This work was supported by the Director, Office of Science, Office of Basic Energy Sciences, Environmental Management Science Program (EMSP) of the DOE under Contract No. DE-AC03-76SF00098. The results of investigation were partially supported by the grant from EMSP of DOE. The author thanks C. Doughty for the review of the paper, and G. Su and T. Tokunaga for providing their data for the analysis included in this paper. The author also appreciates very much the collaboration with T. Wood and R. Podgorney of INEEL and J. Geller, S. Borglin, K. Pruess, and P. Persoff of Lawrence Berkeley National Laboratory, who conducted the experiments that are analyzed in this paper.

References

1. Faybishenko, B. Chaotic Dynamics in Flow through Unsaturated Fractured Media. *Adv. Water Res.* **2002,** *25* (7), 793–816.
2. Faybishenko, B. Nonlinear Dynamics in Flow through Unsaturated Fractured-Porous Media: Status and Perspectives. *Rev. Geophysics* **2004,** *42* (2).
3. Faybishenko, B.; Bodvarsson, G. S.; Hinds, J.; Witherspoon, P. A. Scaling and Hierarchy of Models for Flow Processes in Unsaturated Fractured

Rock. In *Scaling Methods in Soil Physics;* Pachepsky, Y. A., Radcliffe, D. E., Selim, H. M., Eds.; CRC Press: Boca Raton, FL, 2003; pp 373–417.

4. Pruess, K.; Faybishenko, B.; Bodvarsson, G. S. Alternative Concepts and Approaches for Modeling Flow and Transport in Thick Unsaturated Zones of Fractured Rocks. *J. Contam. Hydrol.* **1999**, *38* (1-3), 281–322.

5. Gelhar, L. W. *Stochastic Subsurface Hydrology*; Prentice-Hall: Englewood Cliffs, NJ, 1993.

6. Neuman, S. P.; Di Frederico, V. Correlation, Flow, and Transport in Multiscale Permeability Fields. In *Scale Dependence and Scale Invariance in Hydrology*; Sposito, G., Ed.; Cambridge University Press: New York, 1998; pp 354–397.

7. Lorenz, E. N. Deterministic Nonperiodic Flow. *J. Atmos. Sci.* **1963**, *20* (2), 130–141.

8. Nicolis, C. Predictability of the Atmosphere and Climate: Towards a Dynamical View. In *Space and Time Scale Variability and Interdependencies in Hydrological Processes*; Feddes, R. A., Ed.; Cambridge University Press: New York, 1995.

9. Turcotte, D. L. *Fractals and Chaos in Geology and Geophysics*; Cambridge University Press: New York, 1997.

10. Ortoleva, P. J. *Geochemical Self-Organization*; Oxford University Press: New York, 1994.

11. Dubois, J. *Non-Linear Dynamics in Geophysics*; Wiley: New York, 1998.

12. Read, P. L. Editorial: Nonlinear Processes. *Geophysics* **2001**, *8*, 191–192.

13. Rosendahl, J.; Vekic, M.; Kelley, J. Persistent Self-Organization of Sandpiles. *Phys. Rev. E: Stat. Phys., Plasmas, Fluids, Relat. Interdiscip. Top.* **1993**, *47* (2), 1401–1447.

14. Cheng, Z.; Redner, S.; Meaking, P.; Family, F. Avalanche Dynamics in a Deposition Model with "Sliding." *Phys. Rev. A: At., Mol., Opt. Phys.* **1989**, *40* (10), 5922–5935.

15. Pasternack, G. B. Does the River Run Wild? Assessing Chaos in Hydrological Systems. *Adv. Water Res.* **1999**, *23* (3), 253–260.

16. Nicolis, G.; Prigogine, I. *Exploring Complexity: An Introduction*; W. H. Freeman: New York, 1989.

17. Sililo, O. T. N.; Tellam, J. H. Fingering in Unsaturated Zone Flow: A Qualitative Review with Laboratory Experiments on Heterogeneous Systems. *Groundwater* **2000**, *38* (6), 864.

18. Dewers, T.; Ortoleva, P. Nonlinear Dynamical Aspects of Deep Basin Hydrology – Fluid Compartment Formation and Episodic Fluid Release. *Am. J. Sci.* **1994**, *294* (6), 713–755.

19. Himasekhar, K.; Bau, H. H. Large Rayleigh Number Convection in a Horizontal, Eccentric Annulus Containing Saturated Porous Media. *Int. J. Heat Mass Transfer* **1986**, *29*, 703–712.

20. Moore, M. G.; Juel, A.; Burgess, J. M.; McCormick, W. D.; Swinney, H. L. Fluctuations in Viscous Fingering. *Phys. Rev. E: Stat. Phys., Plasmas, Fluids, Relat. Interdiscip. Top.* **2002,** *65,* 030601(R).

21. Haken, H. Visions of Synergetic. *J. Franklin Institute* **1997,** *334B* (5-6), 759–792.

22. Faybishenko, B.; Doughty, C.; Steiger, M.; Long, J. C. S.; Wood, T. R.; Jacobsen, J. S.; Lore, J.; Zawislanski, P. T. Conceptual Model of the Geometry and Physics of Water Flow in a Fractured Basalt Vadose Zone. *Water Resour. Res.* **2000,** *37* (12), 3499–3520.

23. Faybishenko, B.; Witherspoon, P. A.; Doughty, C.; Geller, J. T.; Podgorney, R. R. Multi-Scale Investigations of Liquid Flow in a Fractured Basalt Vadose Zone. In *Flow and Transport through Unsaturated Fractured Rock*, 2nd ed.; Evans, D. D., Nicholson, T. J., Rasmussen, T. C., Eds.; Geophysical Monograph Series 42; American Geophysical Union: Washington, DC, 2001; pp 161–182.

24. Podgorney, R.; Wood, T.; Faybishenko, B.; Stoops, T. Spatial and Temporal Instabilities in Water Flow through Variably Saturated Fractured Basalt on a One-Meter Scale. In *Dynamics of Fluids in Fractured Rock;* Faybishenko, B., Whitherspoon, P. A., Benson, S. M., Eds.; Geophysical Monograph Series 122; American Geophysical Union: Washington, DC, 2000; pp 129–146.

25. Persoff, P.; Pruess, K. Two-Phase Flow Visualization and Relative Permeability Measurement in Natural Rough-Walled Rock Fractures. *Water Resour. Res.* **1995,** *31* (5), 1175–1186.

26. Tokunaga, T.; Wan, J.; Sutton, S. R. Transient Film Flow on Rough Fracture Surfaces. *Water Resour. Res.* **2000,** *36* (7), 1737–1746.

27. Su, G. W.; Geller, J. T.; Pruess, K.; Wen, F. Experimental Studies of Water Seepage and Intermittent Flow in Unsaturated, Rough-Walled Fractures, *Water Resour. Res.* **1999,** *35* (4), 1019–1037.

28. Geller, J. T.; Borglin, S. E.; Faybishenko, B. A. *Water Seepage in Unsaturated Fractures: Experiments and Evaluation of Chaotic Behavior of Dripping Water in Fracture Models*; Lawrence Berkeley National Laboratory: Berkeley, CA, 2001.

29. *Scale Effects in Rock Masses,* Proceedings of the Second International Workshop on Scale Effects in Rock Masses, Lisbon, Portugal, 1993; da Cunha, A. P., Ed.; A. A. Balkema: Brookfield, VT, 1993.

30. Priest, S. D. *Discontinuity Analysis for Rock Engineering*; Chapman and Hall: London, 1993.

31. Tokunaga, T. K.; Wan, J. Water Film Flow along Fracture Surfaces of Porous Rock. *Water Resour. Res.* **1997,** *33* (6), 1287–1295.

32. Johns, R. A.; Roberts, P. V. A Solute Transport Model for Channelized Flow in a Fracture. *Water Resour. Res.* **1991,** *27* (8), 1797–1808.

33. Selker, J. S.; Steenhuis, T. S.; Parlange, J. Y. Wetting-Front Instability in Homogeneous Sandy Soils under Continuous Infiltration. *Soil Sci. Soc. Am. J.* **1992**, *56* (5), 1346–1350.

34. Prazak, J.; Sir, M.; Kubik, F.; Tywoniak, J.; Zarcone, C. Oscillation Phenomena in Gravity-Driven Drainage in Coarse Porous Media. *Water Resour. Res.* **1992**, *28* (7), 1849–1855.

35. Wang, Z.; Feyen, J.; van Genuchten, M. Th.; Nielsen, D. R. Air Entrapment Effects on Infiltration Rate and Flow Instability. *Water Resour. Res.* **1997**, *34* (2), 213–222.

36. Ho, C. K. Asperity-Induced Episodic Percolation through Unsaturated Fractured Rock. Presented at the Geological Society of America Annual Meeting, Boston, MA, Nov 5-8, 2001; Abstract/Paper 56-0.

37. Gentier, S.; Hopkins, D.; Riss, J. Role of Fracture Geometry in the Evolution of Flow Paths under Stress. In *Dynamics of Fluids in Fractured Rock;* Faybishenko, B., Whitherspoon, P. A., Benson, S. M., Eds.; Geophysical Monograph Series 122; American Geophysical Union: Washington, DC, **2000;** pp 168–169.

38. Bresler, L.; Shinbrot, T.; Metcalfe, G.; Ottino, J. M. Isolated Mixing Regions: Origin, Robustness, and Control. *Chem. Eng. Sci.* **1997**, *52* (10), 1623–1636.

39. Akselrud, G. A.; Altshuler, M. A., *Introduction to Capillary Chemical Technology;* Khimia: Moscow, Russia, 1983.

40. Ottino, J. M. *The Kinematics of Mixing: Stretching, Chaos, and Transport*; Cambridge University Press: New York, 1989.

41. Deryagin, B. V.; Zorin, Z. M.; Churaev, N. V. Water Films on Solid Hydrophilic Surfaces. In *Water in Disperse Systems;* Deryagin, B.V., Ovcharenko, F. D., and Churaev, N. V., Eds.; *Chemistry:* Moscow, Russia, 1989; pp 210–228.

42. Middleman, S. *Modeling Axisymmetric Flows: Dynamics of Films, Jets, and Drops*; Academic Press: New York, 1995.

43. Glass, R. J.; Nicholl, M. J.; Rajaram, H.; Wood, T. R. Unsaturated Flow through Fracture Networks: Evolution of Liquid Phase Structure, Dynamics, and the Critical Importance of Fracture Intersections. *Water Resour. Res.* **2003**, *39* (12), 1352.

44. *Hydrodynamic Instabilities and the Transition to Turbulence,* 2nd ed.; Swinney, H. L., Gollub, J. P., Eds.; Springer-Verlag: Berlin, Germany, 1985.

45. Indereshkumar, K.; Frenkel, A. L. Wavy Film Flows Down an Inclined Plane. Part I: Perturbation Theory and General Evolution Equation for the Film Thickness. *Phys. Rev. E: Stat. Phys., Plasmas, Fluids, Relat. Interdiscip. Top.* **1999**, *60*, 4143–4157.

46. Moon, F. C. *Chaotic Vibrations: An Introduction to Chaotic Dynamics for Applied Scientists and Engineers*; Wiley & Sons: New York, 1987.

47. Acheson, D. *From Calculus to Chaos: An Introduction to Dynamics.* Oxford University Press: New York, 1997.
48. Tsonis, A. A. *Chaos: From Theory to Applications*; Plenum Press: New York, 1992.
49. Baker, G. L.; Gollub, J. P. *Chaotic Dynamics: An Introduction*; Cambridge University Press: New York, 1996.
50. Abarbanel, H. D. I. *Analysis of Observed Chaotic Data;* Springer: New York, 1996.
51. Sprott, J. C.; Rowlands, G. *Chaos Data Analyzer,* The Professional Version (2.1); Physics Academic Software: Raleigh, NC, 1995.
52. Schuster, H. G. *Deterministic Chaos: An Introduction;* Physik-Verlag: New York, 1989.
53. Kapitanyak, T. *Chaos in Systems with Noise;* World Scientific: New York, 1988.
54. Williams, G. P. *Chaos Theory Tamed;* Joseph Henry Press: Washington, DC, 1997.
55. Shaw, R. *The Dripping Faucet as a Model Chaotic System;* Aerial Press: Santa Cruz, CA, 1984.
56. Coullet, P.; Mahadevan, L.; Riera, C. Return Map for the Chaotic Dripping Faucet. *Prog. Theor. Phys.* **2000** (Suppl. 139), 507.
57. Kiyono, K.; Katsuyama, T.; Masunaga, T.; Fuchikami, N. Picture of the Low-Dimensional Structure in Chaotic Dripping Faucets. *Phys. Lett. A* **2003** *320*, 47–52.
58. Néda, Z.; Bako, B.; Rees, E. The Dripping Faucet Revisited. *Chaos* **1996,** *6* (1), 59–62.
59. Taitelbaum, H.; Koza, Z. Reaction–Diffusion Processes: Exotic Phenomena in Simple Systems. *Physica A,* **2000,** *285* (1-2), 166–175.
60. Gaspard, P.; Klages, R. Chaotic and Fractal Properties of Deterministic Diffusion–Reaction Processes. *Chaos* **1998,** *8* (2), 409–423.
61. Aref, H. Stirring by Chaotic Advection. *J. Fluid Mech.* **1984,** *143*, 1-21.
62. Aref, H. The Development of Chaotic Advection. *Phys. Fluids* **2002,** *14*, 1315–1323.
63. Cartwright, J. H. E.; Feingold, M.; Piro, O. An Introduction to Chaotic Advection. In *Mixing: Chaos and Turbulence;* Chat'e, H., Villermaux, E., Chomez, J. M., Eds.; Kluwer Academic Publishers: Norwell, MA, 1999; pp 307–342.
64. Boffetta, G.; Celani, A.; Cencini, M.; Lacorata, G.; Vulpiani, A. Nonasymptotic Properties of Transport and Mixing. *Chaos* **2000,** *10* (1), 50–60.
65. *Chaos Applied to Fluid Mixing*; Aref, H., Ed.; Pergamon: New York, 1995.
66. Lamberto, D. J.; Alverez, M. M.; Muzzio, F. J. Computational Analysis of Regular and Chaotic Mixing in a Stirred Tank Reactor. *Chem. Eng. Sci.* **2001,** *56*, 4887–4899.

67. Ottino, J. M.; Souvaliotis, A.; Metcalfe, G. Chaotic Mixing Processes: New Problems and Computational Issues. *Chaos, Solitons & Fractals* **1995**, *6*, 425–438.

68. Ottino, J. M.; DeRoussell, P.; Hansen, S.; Khakhar, D. V. Mixing and Dispersion of Viscous Liquids and Powdered Solids. *Adv. Chem. Eng.* **2000**, *25*, 105–204.

69. Faybishenko, B. Comparison of Laboratory and Field Methods for Determination of Unsaturated Hydraulic Conductivity of Soils. *Proceedings of the International Workshop on Characterization and Measurement of the Hydraulic Properties of Unsaturated Porous Media*; van Genuchten, M. Th., Leij, F., Wu, L., Eds.; 1998; pp 279-292.

70. Harvey, R. W.; Garabedian, S. P. Use of Colloid Filtration Theory in Modeling Movement of Bacteria through a Contaminated Sandy Aquifer. *Environ. Sci. Technol.* **1991**, *25*, 178–185.

71. Pagonabarraga, I.; Frenkel, D. Dissipative Particle Dynamics for Interacting Systems. *J. Chem. Phys.* **2001**, *115* (11), 5015–5026.

72. Horbach, J.; Frenkel, D. Lattice-Boltzmann Method for the Simulation of Transport Phenomena in Charged Colloids. *Phys. Rev. E: Stat. Phys., Plasmas, Fluids, Relat. Interdiscip. Top.* **2001**, *6406* (6 Part 1), 1507.

73. Rodriguez-Iturbe, I.; Entekhabi, D.; Lee, J. S.; Bras, R. L. Nonlinear Dynamics of Soil Moisture at Climate Scales: 2. Chaotic Analysis. *Water Resour. Res.* **1991**, *27* (8), 1907.

74. Sivashinsky, G. I.; Michaelson, D. M. *Prog. Theor. Phys.* **1980**, *63*, 2112.

75. Faybishenko, B. A. Hydraulic Behavior of Quasi-Saturated Soils in the Presence of Entrapped Air: Laboratory Investigations. *Water Resour. Res.* **1995**, *31* (10), 2421.

76. Rabinovich, M. I.; Trubetskov, D. I. *Oscillations and Waves in Linear and Nonlinear Systems*; Kluwer Academic Publishers: Norwell, MA, 1994.

77. Gumowski, I. *Oscillatory Evolution Processes: Quantitative Analyses Arising from Applied Science*; Manchester University Press: New York, 1989; Chapter 11, p 177.

78. Geller, J. T.; Su, G. In *Preliminary Studies of Water Seepage through Rough-Walled Fractures*; Pruess, K., Ed.; Lawrence Berkeley National Laboratory: Berkeley, CA, 1996.

79. Janosi, I. M.; Horvath, V. K. Dynamics of Water Droplets on a Windowpane. *Phys. Rev. A: At., Mol., Opt. Phys.* **1989**, *40* (9), 5232–5237.

80. Pismen, L. M.; Rubinstein, B. Y.; Bazhlekov, I. Spreading of a Wetting Film under the Action of van der Waals Forces. *Phys. Fluids* **2000**, *12* (3), 480.

81. Faybishenko, B.; Babchin, A. J.; Frenkel, A. L.; Halpern, D.; Sivashinsky, G. I. A Model of Chaotic Evolution of an Ultrathin Liquid Film Flowing Down an Inclined Plane. *Colloids Surf., A* **2001**, *192* (1-3), 377-385.

82. *Theoretical Ecology: Principles and Applications*; May, R. M., Ed.; Sinauer Associates: Sunderland, MA, 1981.

83. Sparrow C. *The Lorenz Equations, Bifurcations, Chaos, and Strange Attractors;* Springer-Verlag : New York, 1982.

84. Babchin, A. J.; Frenkel, A. L.; Levich, B. G.; Sivashinsky, G. I. Flow-Induced Nonlinear Effect in Thin Film Stability. *Ann. N.Y. Acad. Sci.* **1983**, *404*, 426.

85. Frenkel, A. L.; Babchin, A. J.; Levich, B.; Shlang, T.; Sivashinsky, G. I. Annular Flows Can Keep Unstable Films from Breakup: Nonlinear Saturation of Capillary Instability. *J. Colloid Interface Sci.* **1987**, *115*, 225.

86. Weeks, S. W.; Sposito, G. Mixing and Stretching Efficiency in Steady and Unsteady Groundwater Flows. *Water Resour. Res.* **1998**, *34* (12), 3315–3322.

87. Song, H.; Bringer, M. R.; Tice, J. D.; Gerdts, C. J.; Ismagilov, R. F. Experimental Test of Scaling of Mixing by Chaotic Advection in Droplets Moving through Microfluidic Channels. *Applied Physics Letters* **2003**, *83* (22), 4664–4666.

Chapter 10

Coupled Hydrological and Geochemical Processes Governing the Fate and Transport of Sr and U in the Hanford Vadose Zone

Melanie A. Mayes[1], Molly N. Pace[1], Philip M. Jardine[1],
Scott E. Fendorf[2], Norman D. Farrow[1], Xiangping L. Yin[1],
and John M. Zachara[3]

[1]Environmental Sciences Division, Oak Ridge National Laboratory,
Oak Ridge, TN 37831
[2]Department of Geological and Environmental Sciences, Stanford
University, Stanford, CA 94305
[3]Environmental and Molecular Science Laboratory, Pacific Northwest
National Laboratory, Richland, WA 99352

The goal of our research is to provide an improved conceptual and quantitative understanding of coupled hydrological and geochemical processes that are responsible for the migration of radionuclides and toxic metals in the vadose zone beneath the Hanford tank farms. Large, intact sediment cores were collected in the vertical and horizontal directions from the three relevant sedimentary units. Laboratory-scale saturated and unsaturated transport experiments were conducted using a suite of nonreactive and reactive tracers. The creation of capillary air barriers under unsaturated conditions in sedimentary layering promoted bedding-parallel flow in horizontal cores and preferential flow in vertical cores. Diffusional exchange with regions of immobile water was inferred by separation of multiple nonreactive tracers. The metal-chelate complex $SrEDTA^{2-}$ was not stable in the presence of the Hanford sediment. Sr^{2+} sorption in an intact

Hanford core was significant and not suggestive of accelerated transport. The transport of U(VI) was influenced by rate-limited interfacial reactions, as inferred by greater sorption in equilibrium isotherms versus miscible displacement in repacked sediments. The transport of U through an intact caliche core was less reactive than expected, which was attributed to greater surface area, reactivity, and sorption site availability of crushed disturbed sediments versus a consolidated intact core. Our results suggest that the transport of radionuclides and toxic metals in the Hanford far-field vadose zone will be dependent upon coupled hydrological and geochemical processes, which are a function of water content, sedimentary structure, and the rates and mechanisms of interfacial geochemical adsorption.

Introduction

At the U.S. Department of Energy (DOE) Hanford Reservation, near Richland, WA, plutonium production during the cold war era resulted in the generation of large quantities of high level radioactive and toxic waste. The primary method of disposal was underground burial within high-capacity single- and double-shelled steel tanks ("tank farms") in the 200 area (1). Approximately 1 million gallons of the single-shelled tank wastes are known to have leaked into the deep (70 m) vadose zone of this semi-arid region. The characteristics of the waste stream (high temperature, high ionic strength, caustic) promote dissolution and reprecipitation of subsurface minerals near the tank farms, but these effects are expected to decrease with depth and distance (2–4). Metal chelation by organics is also suspected to contribute to metal mobility (2, 3). Average annual precipitation is approximately 16 cm y^{-1}, though past waste releases have increased the natural moisture content (5-15%) of the region above the water table (3). Observed migration of radionuclides has previously been characterized as "accelerated," because the distribution of more mobile radionuclides (e.g., ^{99}Tc, ^{129}I, U, ^{3}H) was significantly wider and deeper than expected (5, 6). This suggests that there is uncertainty regarding the mechanisms of contaminant transport in unsaturated, unconsolidated sediments; however, historical unknowns in the waste stream and timing of leaks also contribute to the level of uncertainty (7). In contaminated boreholes in the 200 area, layered sediments appear to be conducive to anisotropic lateral flow, particularly in the finer-grained sediments of higher moisture content, but contaminants are intermittently found in such sediments (2, 3, 7). It is likely that a combination of physical and chemical processes is responsible for the complex spreading of radionuclides observed in the "far-field" vadose zone.

Hydrological mechanisms resulting in preferential contaminant transport in unsaturated sediments include finger, funnel, and lateral flow. Textural differences can promote the development of an unstable wetting front, which is characterized by rapid vertical flow within wet "fingers" which are separated by regions of much lower water content (8–14). In addition, funneling along lithologic heterogeneities may concentrate flow into or around sedimentary heterogeneities (15, 16). Vertical flow may be impeded by fine scale textural differences between sediment layers, resulting in local-scale perched water and lateral diversion (17–19). In addition, diffusion of contaminants into relatively immobile flow regimes can result in the creation of a long-term source of contamination (7).

Rate-limited diffusional exchange between relatively mobile and relatively immobile flow regimes is referred to as physical nonequilibrium (20, 21). Nonequilibrium may result from the interaction of preferential flow mechanism(s) with less mobile pore regions. Physical nonequilibrium has been quantified using multiple nonreactive tracers in structured (21–23) and heterogeneous porous (24, 25) media. This technique takes advantage of the rate of diffusion of an ion, which is inversely proportional to its size. This allows for identification of the contributions of matrix diffusion as inferred by tracer separation, and quantification of the rate of exchange between the relatively mobile and immobile flow regimes (26).

Accelerated contaminant transport in the far-field may be related to kinetic (rate-limited) geochemical reactions; e.g., chemical nonequilibrium. In addition, hydrologic transport may enhance chemical nonequilibrium when the flow velocity exceeds the rate of geochemical reaction and/or reduces the availability of sorption sites (27–30). This condition is referred to as coupled physical and chemical nonequilibrium (21). Unsaturated conditions, in particular, may influence geochemical reactions because the flow regime can become sequestered to more fine-grained beds with higher water content, thus bypassing coarse-grained, less saturated beds. Further, general relationships between sediment mineralogy and grain size suggest that bulk and unsaturated sediment reactivity may not be equivalent (31, 32).

The goal of this paper is to provide an overview of coupled hydrological and geochemical processes observed in large intact cores of Hanford sediments. A multiple nonreactive tracer technique will be utilized to quantify physical hydrology under saturated and unsaturated conditions. Adsorption isotherms will quantify the sorption of $SrEDTA^{2-}$ and U(VI) under kinetic and equilibrium conditions. Small-scale miscible displacement experiments will quantify the retardation of these contaminants in disturbed sediments under saturated, flowing conditions. Finally, these contaminants will be introduced into two intact, unsaturated cores to determine the effects of coupled physical and geochemical processes upon contaminant transport. The results of our research are expected to contribute to the conceptual and quantitative understanding of hydrological and geochemical contaminant transport in the "far-field" region of the Hanford subsurface.

Materials and Methods

Large undisturbed sediment cores (0.25-m diameter x 0.25-m length) were acquired from within the Hanford, Plio-Pleistocene, and Upper Ringold units using a rotary coring apparatus equipped with diamond bit core barrels attached to a track-mounted hydraulic motor (Figure 1). This modification allowed collection of undisturbed cores parallel and perpendicular to sedimentary bedding, in order to investigate hydrologic transport processes in horizontal and vertical directions. The Upper Ringold Formation, which comprises the lower ~11 m of the vadose zone, consists of fine-grained, semi-consolidated, laminated silts and sands deposited within a fluvial/lacustrine environment (33, 34). Two units of the Ringold were obtained, one horizontally-bedded and silty, and the other cross-bedded and sandy (24). The Plio-Pleistocene, or Cold Creek, Unit (thickness ~ 12 m) is a consolidated, calcite-cemented caliche composed of fluvial lithic and ash fragments (Figure 1) (35, 36). The Ringold and Plio-Pleistocene samples were obtained from the White Bluffs immediately north of the Hanford Reservation. The Hanford formation (informal designation) which comprises the upper ~45 m of the vadose zone, is a coarse- to fine-grained, heterogeneous, unconsolidated sand which was deposited as a result of cataclysmic floods during the most recent glacial period (Figure 2) (33, 35). The Hanford samples were obtained from the Environmental Restoration Disposal Facility (ERDF) in the 200W area of the Hanford Reservation (25). Detailed characterization efforts have shown that the core materials have similar physical and geochemical characteristics as sediments beneath the 200-West area of the Hanford tank farms (37).

Transport experiments were conducted on the intact cores at a variety of water contents using saturated and unsaturated flow experimental techniques (24, 25, 38). The intact cores were epoxied into PVC pipe and emplaced into fabricated Plexiglas endcaps. The cores were hydraulically connected to a membrane inside the endplate to facilitate either saturated or unsaturated flow. Under saturated conditions, influent solution was ponded (1-2 cm) at the upper boundary to facilitate delivery into the cores, and effluent was collected in a high-capacity fraction collector. Under unsaturated conditions, a vacuum-

Figure 1. Air-rotary coring apparatus for obtaining intact cores of the Plio-Pleistocene Unit, White Bluffs, Hanford Reservation. Diamond-bit core barrel (10-inch diameter) was attached to a track-mounted hydraulic motor. The entire apparatus was affixed to a backhoe bucket in order to obtain cores at any angle.

Figure 2. Excavation of the intact Hanford formation cores from the surrounding sediments. The annulus created by the coring apparatus was filled with foam or wax, and the surrounding material was carefully excavated, revealing the intact core.

regulated chamber at the lower boundary was utilized to maintain tension within the core (Figure 3). Solution was delivered to the upper boundary using a multi-channel peristaltic pump, and collected in a fraction collector inside the vacuum chamber. This unsaturated flow technique reproduced fate and transport in select pore regimes, thereby isolating hydrological mechanisms operative at different water contents (21).

A multiple nonreactive tracer technique was utilized to identify physical nonequilibrium processes, e.g., coupled preferential flow and matrix diffusion (21, 24, 25). Three nonreactive tracers, Br⁻, pentafluorobenzoic acid (PFBA), and piperazine-1, 4-bis(2-ethanesulfonic acid) (PIPES), were used, which differ only in their free water molecular diffusion coefficient, thus physical nonequilibrium may be inferred by observed separation of the tracers. Influent concentrations of the nonreactive tracers were 0.5 mM (Br, PFBA) and 1.0 mM (PIPES). Only Br⁻ was utilized in the repacked column experiments, because physical nonequilibrium as a result of sedimentary structure is expected to be minimal in repacked media. Analyses were accomplished using UV detection (190 nm) and low pressure liquid chromatography (Model DX-600, Dionex Corp, Sunnyvale, CA).

Equilibrium adsorption isotherms and kinetic batch experiments were conducted to quantify the interaction of U(VI) and SrEDTA^{2-} with subsurface media. Disturbed sediments were crushed (Plio-Pleistocene Unit only) and sieved < 2 mm. U(VI) isotherms were carried out in 30-cm^3 polypropylene centrifuge tubes using 2 g of geologic media and 4 ml of solution containing $0 - 8.4 \times 10^{-5}$ M U (as $UO_2(NO_3)_2$) in 0.1 M $NaNO_3$. Kinetic experiments utilized 8 g of geologic media and 16 ml of solution containing 1×10^{-3} M SrEDTA^{2-} (as $Sr(NO_3)_2$ and Na_2H_2EDTA) in 0.1 M NaCl. Geochemical modeling confirmed the formation of the SrEDTA^{2-} complex. Duplicate isotherm samples were placed on a shaker for 72 h, and centrifuged for 15 min at ~2300 RPM. Kinetic samples were conducted in a similar manner, except they were collected as a function of time for 30 d. The samples were then decanted into acidified scintillation vials for analysis and into glass tubes for pH measurements. Blanks (no soil) were also collected into acidified vials and saved for analyses.

Miscible displacement experiments were conducted on repacked sediments to determine relevant transport parameters (retardation coefficients) for U(VI) and SrEDTA^{2-} in the absence of the effects of sedimentary structure and unsaturated flow. The experiments were conducted in a glass column 1 cm in diameter and 4.5 cm in length (U(VI)) or 10 cm in length (SrEDTA^{2-}). The influent solutions consisted of $NaNO_3$ or NaCl matrix, nonreactive Br⁻ tracer $(2 \times 10^{-4}$ M), and the highest concentration of U and Sr utilized in the equilibrium batch experiments. A medical pump was used to deliver solution to the bottom of the column at an average pore water velocity of 2.1 cm hr^{-1} for

Figure 3. Oak Ridge National Laboratory unsaturated flow facility. The intact cores are contained in large polyvinyl chloride pipes, and a filter membrane was affixed to the sediments within a fabricated acrylic endcap at the lower boundary. Tension was maintained through the large white vacuum chambers, which also housed a fraction collector for samples of column effluent. Influent was delivered by a multi-channel pump at the upper boundary. Tensiometers, visible in the side of each core, were used to measure matric potential on a daily basis.

U(VI) and 0.33 cm h^{-1} for SrEDTA^{2-}. Effluent solution was collected as a function of time in a fraction collector.

All transport experiments were conducted in the following manner. A "matrix" solution was utilized to initiate steady-state flow, the "influent" containing (non)reactive tracers was injected for a finite volume, and then the matrix was utilized to displace the influent injection. Experiments were terminated when the effluent concentration was 2% of influent. The ionic strength (I) of the matrix and influent solutions was equivalent. Reactive injections were conducted at I = 0.1 M using NaNO$_3$ or NaCl to mimic far-field pore water conditions. The concentration of NaNO$_3$ or NaCl was adjusted to account for added NaOH and/or NaHCO$_3$ used in pH buffering. The pH of influent and matrix solutions was equivalent to the natural sediment pH (~ pH 8). For U experiments, pH adjustment and U solubility were controlled using the chemical equilibrium between 0.02 M NaHCO$_3$ and 1% CO$_2$(g). This equilibrium resulted in the formation of a uranyl-carbonate complex, UO$_2$(CO$_3$)$_3$$^{4-}$, which was dominant at pH 8 (39, 40). Uranium analyses were performed on a Kinetic Phosphorescence Analyzer (KPA) (Chemchek Instruments, Richland, WA) and/or an Elan 6100 inductively coupled plasma mass spectrometer (ICP-MS), which was also used for analyses of Sr and other metals (Perkin-Elmer, Shelton, CT). Total organic carbon (TOC) by combustion was utilized to quantify EDTA (Model TOC-5000, Shimadzu, Japan).

The transport of reactive tracers was sometimes modeled using the convective-dispersive equation (CDE) and/or the mobile/immobile (MIM) version of the CDE (41). This modeling is designed to provide simple parameters to quantitatively describe reactive transport, such as the retardation (R) coefficient. Note that this modeling effort is not designed to provide a mechanistic understanding of reactive transport. Description of this widely-used modeling strategy is available (24, 25), and the reader is referred to these previously published papers for additional details concerning the modeling effort.

Results and Discussion

Physical Hydrology

A propensity for lateral flow was suggested where sediments are fine-grained and strongly horizontally-bedded. The observations of similar, overlying breakthrough curves (BTC) and no tracer separation, regardless of saturation, in a horizontal core of the Ringold horizontally-bedded unit support this assertion (Figure 4a) (*24*). The absence of tracer separation suggested that the influence of physical nonequilibrium was minimal. This was likely a result of desaturation of coarse-grained beds and the subsequent restriction of flow to finer-grained beds. It is plausible that desaturation resulted in the formation of multiple capillary barriers, which minimized diffusional tracer exchange between adjacent beds. This mechanism may lead to lateral flow in the subsurface and contribute to enhanced transverse spreading of wastes released from the tanks (*2, 3, 7, 19*).

When flow was required to cross sedimentary discontinuities, however, the formation of capillary barriers in desaturated coarse layers inhibited stable flow and resulted in the development of preferential flow. The observation of early tracer breakthrough with decreasing water content in the Ringold cross-bedded core suggests that some portion of the media was bypassed by preferential flow (Figure 4b). An increase in the water content immediately preceding arrival of the tracers in the effluent suggested that perched water contributed to the development of preferential flow in the Ringold cross beds. Separation of the nonreactive tracers (Figure 4b) suggested an interaction between regions of relatively mobile and immobile regimes which can be quantified using tracer diffusion coefficients (*26*). This pattern of early tracer breakthrough and separation was observed in all unsaturated vertical Ringold cores and in all cross-bedded Ringold cores (*24*). The observance of immobile water as inferred by tracer separation is significant, because pockets of immobile water may impede vertical migration of waste released from the Hanford tanks, or serve as later sources of contamination. Further, relatively immobile regions provide an opportunity for increased contact time between contaminants and subsurface media, which may promote adsorption and/or precipitation.

The preferential flow results, observed only under unsaturated conditions, are consistent with other published unsaturated transport experiments in disturbed media (*42, 43*) in that early tracer breakthrough and the development of immobile water increased in importance as water content decreased. These results are also consistent with experiments in repacked sediments that have demonstrated the importance of the mechanism of finger flow in porous media (*8, 9, 12–14*). All of these previous studies, however, utilized disturbed or repacked sediments, and our results clearly demonstrate the influence of natural sedimentary layering upon unsaturated flow. Horizontal core (Figure 4a) and vertical core (Figure 4b) experiments appear to be divergent, but are consistent in that the formation of capillary barriers in layered sedimentary deposits influenced the transport of solutes under unsaturated conditions. Our results support field-scale observations of anisotropic flow in the subsurface (*10, 11, 19*), and suggest that such anisotropy is a function of both water content and layered sedimentation. However, similar experiments performed in the sandier Hanford flood deposits were less suggestive of preferential flow, as evidenced by similar breakthrough curves under saturated and unsaturated conditions,

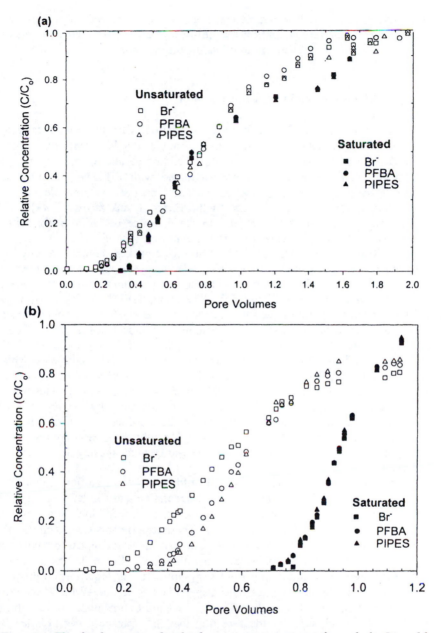

Figure 4. The displacement of multiple nonreactive tracers through the Ringold Formation under saturated and unsaturated conditions. (a) Displacement through a horizontal core of the horizontally-bedded Ringold Formation. (b) Displacement through a vertical core of the cross-bedded Ringold Formation. (Reproduced with permission from reference 24. Copyright 2002 Elsevier Science B.V.)

although an immobile flow regime was suggested by observed tracer separation (25). Overall, our results confirm the utility of using intact media samples to quantify hydrologic mechanisms of contaminant transport.

Coupled Hydrology and Geochemistry of Sr

The transport of Sr^{2+} through the Hanford flood deposits was investigated by performing kinetic adsorption and miscible displacement experiments through repacked sediments and an intact bedding-parallel core. Kinetic and repacked column experiments were conducted with $SrEDTA^{2-}$ as previous research has suggested that organic chelates may be responsible for the "accelerated" transport of Sr^{2+} beneath the Hanford tank farms (7, 44). The kinetic experiment indicated that Sr^{2+} in the presence of EDTA sorbs within 30 min (K_d = 2.7 ml g^{-1}) and remains stable over time (data not shown).

Transport of $SrEDTA^{2-}$ through the repacked column resulted in delayed breakthrough of Sr^{2+} (R = 75.4) relative to the nonreactive tracer Br$^-$ (Figure 5a). EDTA eluted from the column at approximately the same time as the nonreactive tracer Br$^-$ (Figure 5b). Therefore, $SrEDTA^{2-}$ (log K = 10.45) (44) must have rapidly dissociated (residence time = 30 h), forming a more stable complex with other cations available in the soil such as $CaEDTA^{2-}$ (log K = 12.41), Fe(III)EDTA$^-$ (log K = 27.70), and/or AlEDTA$^-$ (log K = 19.07) (45, 46). This is consistent with experiments on ORNL sediments which demonstrated that $SrEDTA^{2-}$ was not stable in the presence of Al^{3+} and Fe^{3+} (38, 47). Unfortunately, we were not able to speciate the resultant metal-EDTA complex in the effluent because we currently do not have the capabilities (48). Regardless, these results strongly suggest that EDTA does not contribute to the accelerated transport of Sr^{2+} under far-field conditions. Consequently, the transport experiment in the undisturbed core was conducted using Sr^{2+} instead of $SrEDTA^{2-}$.

Transport of Sr^{2+} through the intact Hanford core was significantly retarded as shown in Figure 5c (R = 73.9). Tracer separation and similar breakthrough relative to saturated conditions characterized the physical hydrology of this Hanford horizontal core (data not shown). These experiments suggested that the observed transport of Sr^{2+} through intact sediments does not appear to have been significantly influenced by physical nonequilibrium. However, the observed Sr^{2+} BTC asymmetry and tailing were suggestive of a nonequilibrium process, since the Sr^{2+} isotherm was linear and dispersion was accounted for. Modeling the observed Sr^{2+} BTC with the two-site nonequilibrium model produced an excellent fit, which supports the assertion that Sr^{2+} transport was affected by geochemical nonequilibrium process(es) (Molly N. Pace, Oak Ridge National Laboratory, unpublished data).

Coupled Hydrology and Geochemistry of U(VI)

The adsorption of U in the Plio-Pleistocene caliche sediment was investigated by performing equilibrium isotherm and miscible displacement experiments on repacked sediments. The adsorption of U(VI) onto the caliche material was observed to be Freundlich over the concentration range of these experiments (Figure 6a). Miscible displacement through repacked Plio-Pleistocene sediments (Figure 6b) resulted in decreased retardation of U (R = 3.9) compared to that predicted using the linear portion of the isotherm (R = 11.4). This probably reflects the longer residence time of the batch equilibrium (72 h) versus the column experiments (2 h), a kinetic limitation which has been confirmed with time-dependent batch experiments (data not shown). Therefore it is likely that geochemical nonequilibrium in the repacked columns inhibited U sorption, which is in contrast to geochemical equilibrium observed during Sr^{2+} transport. These results are supported by other experiments in Hanford sediments which suggested that the adsorption of U(VI) was rate-limited under flowing conditions (29, 30).

The physical hydrology of the Plio-Pleistocene Unit was characterized by performing saturated and unsaturated transport experiments using multiple nonreactive tracers (Figure 6c). Physical nonequilibrium was not significant, as evidenced by the co-elution of the nonreactive tracers under saturated and unsaturated conditions. The breakthrough of tracers was slightly more rapid under saturated conditions, which suggests that some preferential flowpaths (e.g., fractures) were eliminated as the water content decreased (21-24, 38). Minimal physical nonequilibrium is most likely related to the sedimentary depositional characteristics of the Plio-Pleistocene (Figure 1), which are different from the surrounding unconsolidated Hanford (Figure 2) (25) and Ringold sediments (24) that were conducive to physical nonequilibrium. First, our samples of Plio-Pleistocene were extensively cemented and thus lacked distinct sedimentary bedding observed in the Hanford and Ringold sediments. In addition, extensive secondary precipitation of $CaCO_3$ was observed in thin section (Figure 7), which would have the overall effect of reducing the porosity by infilling voids, bedding planes, and fractures. This observation is consistent with the depositional environment of the Plio-Pleistocene in which sediment supply was limited and periodic subaerial exposure resulted in carbonate precipitation (35, 36). However, it should be noted that the unit in the subsurface is quite heterogeneous due to these complex sedimentary depositional processes (35, 36), which suggests the need for additional studies.

Uranium transport through the undisturbed, unsaturated Plio-Pleistocene core resulted in similar retardation (R = 3.1) (Figure 6c) to that observed in the repacked sediments (Figure 6b). This is surprising considering the longer residence time in the intact core (280 h) versus the shorter residence time (2 h) and observed geochemical nonequilibrium in the repacked sediments. The

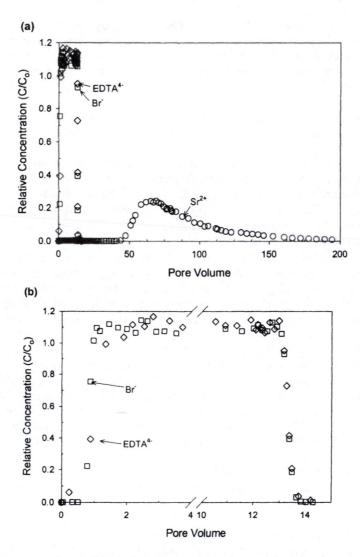

Figure 5. Coupled hydrology and geochemistry of the adsorption and displacement of Sr^{2+} and $SrEDTA^{2-}$ through the Hanford flood deposits. (a) Observed displacement of nonreactive tracer Br^-, and reactive $SrEDTA^{2-}$ through repacked sediments. (b) Enlarged view of (a) to show the co-elution of nonreactive tracer Br^- and $EDTA^{4-}$ formerly complexed with Sr^{2+}. (c) Observed displacement of Br^- and Sr^{2+} through the intact unsaturated, horizontal core.

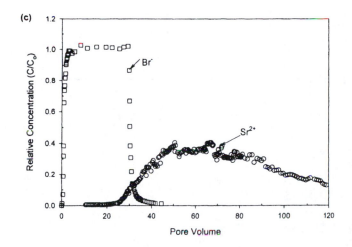

(c)

Figure 5. *Continued.*

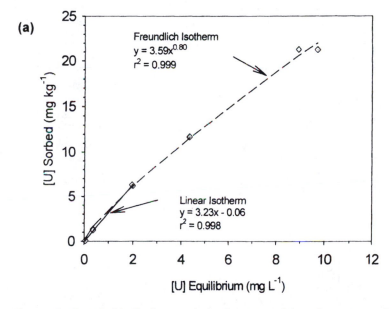

Figure 6. Coupled hydrology and geochemistry of the adsorption and displacement of U(VI) through the Plio-Pleistocene Unit. (a) Equilibrium adsorption isotherm showing both Freundlich and linear fits. (b) Observed (points) and modeled (lines) of the displacement of nonreactive tracer Br⁻ and U through repacked sediments. (c) Observed (points) and modeled (lines) of the displacement of multiple nonreactive tracers through the intact core, under saturated and unsaturated conditions. The displacement of U occurred under unsaturated conditions.

Continued on next page.

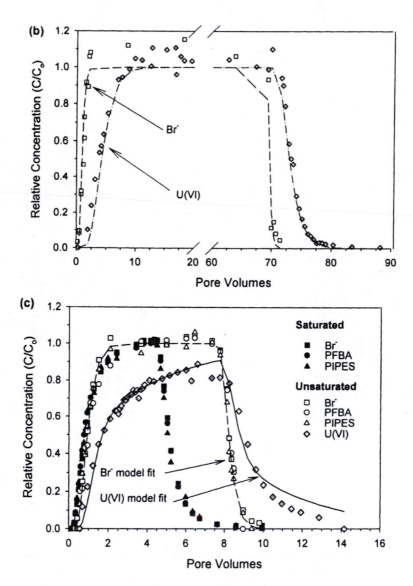

Figure 6. *Continued.*

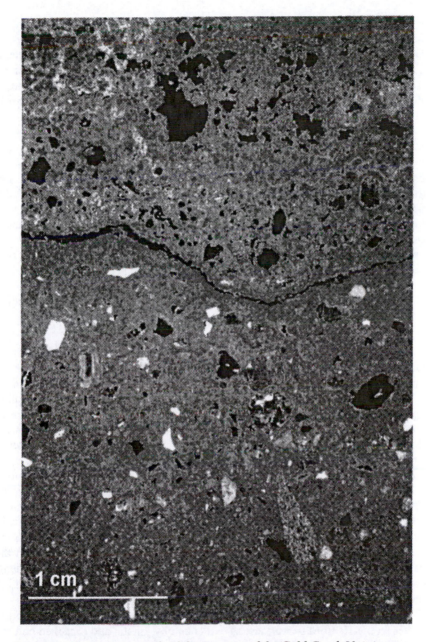

Figure 7. Photomicrograph of thin section of the Cold Creek Unit in cross-polarized light. The upper portion is dominated by voids (in black) and "popcorn-shaped" growth of secondary calcium carbonate. The lower portion is dominated by lithic fragments of different mineral compositions and few voids.

nonreactive tracer data imply that there were no active hydrologic mechanisms (e.g., physical nonequilibrium) that would have inhibited U sorption (Figure 6c). Decreased retardation of U in the intact core may be related to differences in surface area between the intact core and the crushed sediments used for the isotherms and repacked experiments. This would suggest a fundamental difference between using intact and reconstituted materials to make inferences about hydrologic and or geochemical behavior.

Adsorption of U onto Hanford and Ringold sediments occurred by the formation of an inner-sphere, ternary U-CO_3 complex on surfaces of Fe-oxide minerals, as determined using X-ray adsorption spectroscopy (XAS) (*49, 50*). Fe-oxide content in the caliche (1.6 g Fe kg^{-1} sediment), however, was quite low (*37*). In the intact Plio-Pleistocene, Fe minerals appear to exist as isolated lithic fragments, thus they may be located within extensive pore fillings of $CaCO_3$ precipitate (Figure 7). Sediment preparation for isotherms and repacked columns may have increased the surface area, reactivity, and availability of all minerals, including Fe-oxides. This hypothesis may account for the observed greater reactivity of the disturbed versus the undisturbed caliche material.

Conclusions

Coupled physical and chemical processes governing the transport of reactive and nonreactive tracers were investigated in intact cores and disturbed samples of sedimentary media from the Hanford subsurface. The nonreactive tracer experiments confirmed the importance of water content and sedimentary layering in unsaturated transport. Physical nonequilibrium processes such as preferential (finger or funnel) flow and matrix diffusion were manifested when flow was required to cross sedimentary bedding with different water contents, such as in cross-bedded or rippled sediments. The observed development of diffusionally-controlled immobile regimes may promote contaminant adsorption and/or precipitation, or may serve as a long-term source of contamination in the Hanford subsurface. Horizontally-bedded sediments exhibited minimal physical nonequilibrium due to the formation of capillary barriers in coarse-grained beds, which prevented diffusional solute exchange between pore regimes. Such sedimentary deposits may be conducive to anisotropic lateral spreading in the field. Such effects, however, were more important in fine-grained sediments with distinct sedimentary bedding (Ringold), and were less important for highly cemented (Plio-Pleistocene) or sandier (Hanford) sediments.

Organic complexation of Sr^{2+} is not expected to contribute to the mobility of ^{90}Sr in the Hanford subsurface, as evidenced by kinetic and repacked column experiments in which SrEDTA^{2-} was rapidly dissociated in the presence of Hanford sediments. Strontium adsorption was significant during transport in unsaturated, intact Hanford core in which preferential flow was minimal and geochemical nonequilibrium was suggested. This suggests that ^{90}Sr will be adequately contained in the Hanford far-field vadose zone.

Transport experiments through repacked samples of the Plio-Pleistocene caliche suggested that adsorption of U(VI) was kinetically-limited under short residence times. The retardation of U through an intact caliche core, however, was less than expected despite a long residence time. Physical nonequilibrium was minimal in the caliche, suggesting that the decreased reactivity of U was not related to the physical hydrology of the core. A difference in reactive surface area between the intact rock core and crushed repacked samples may account for these observations. Because U adsorption is generally associated with Fe-oxides, sample preparation may have increased the surface area, availability, and reactivity of Fe minerals. This suggested that such disturbance of consolidated media resulted in geochemical parameters which were not representative of geochemical reactions in intact samples. This conclusion is also supported by our unsaturated flow experiments, which suggests the physical characteristics of the intact sediments can influence unsaturated flow.

Acknowledgments

The authors would like to dedicate this publication to the late Norman D. Farrow who designed the intact core retrieval apparatus. His contributions to our research and his constant friendship over many years will both be deeply missed. This research was sponsored by the Office of Science Environmental Remediation Sciences Division, the Environmental Management Science Program (EMSP), DOE, under contract DE-AC05-00OR22725 with Oak Ridge National Laboratory, which is managed by the University of Tennessee-Battelle, LLC. The authors appreciate the efforts of Roland Hirsch and Mark Gilbertson, the program managers for EMSP who supported this work. This manuscript was improved by incorporating the comments of an anonymous reviewer.

References

1. Agnew, S. F.; Boyer, J.; Corbin, R. A.; Duran, T. B.; Fitzpatrick, J. R.; Jurgensen, K. A.; Ortiz, T. P.; Young, B. L. *Hanford Tank Chemical and Radionuclide Inventories: HDW Model;* Los Alamos National Laboratory: Los Alamos, NM, 1997; LA-UR-96-3860, Rev. 4.
2. Serne R. J.; Last, G. V.; Schaef, H. T.; Lanigan, D. C.; Lindenmeier, C. W.; Ainsworth, C. C.; Clayton, R. E.; LeGore, V. L.; O'Hara, M. J.; Brown, C. F.; Orr, R. D.; Kutnyakov, I. V.; Wilson, T. C.; Wagnon, K. B.; Williams, B. A.; Burke, D. S. *Characterization of Vadose Zone Sediment: Slant*

Borehole SX-108 in the S-SX Waste Management Area; Pacific Northwest National Laboratory: Richland, WA, 2002; PNNL-13757-4.

3. *Field Investigation Report for Waste Management Area S-SX;* U.S. Department of Energy: Richland, WA, 2002; RPP-7884, Rev. 0.

4. Pruess, K.; Yabusaki, S.; Steefel, C.; Lichtner, P. *Vadose Zone J.* **2002,** *1,* 68–88.

5. Dirkes, R. L.; Hanf, R. W. *Hanford Site Environmental Report for Calendar Year 1996;* Pacific Northwest National Laboratory: Richland, WA, 1997; PNNL-11472.

6. Hartman, M. J.; Dresel, P. E. *Hanford Site Groundwater Monitoring for Fiscal Year 1996;* Pacific Northwest National Laboratory: Richland, WA, 1997; PNNL-11470.

7. Knepp, A. *Field Investigation Report for Waste Management Area B-BX-BY;* CH2M HILL Hanford Group: Richland, WA, 2002; RPP-10098, Rev. 0.

8. Glass, R. J.; Steenhuis, T. S.; Parlange, J.-Y. *J. Contam. Hydrol.* **1988,** *3,* 207–226.

9. Glass, R. J.; Steenhuis, T. S.; Parlange, J.-Y. *Soil Sci.* **1989,** *148,* 60–70.

10. Ritsema, C. J.; Dekker, L. W.; Hendrickx, J. M. H.; Hamminga, W. *Water Resour. Res.* **1993,** *29,* 2183–2193.

11. Ritsema, C. J.; Dekker, L. W.; Nieber, J. L.; Steenhuis, T. S. *Water Resour. Res.* **1998,** *34,* 555–567.

12. DiCarlo, D. A.; Bauters, T. W. J.; Darnault, J. G.; Steenhuis, T. S.; Parlange, J.-Y. *Water Resour. Res.* **1999,** 35, 427–434.

13. Bauters, T. W. J.; DiCarlo, D. A.; Steenhuis, T. S.; Parlange, J.-Y. *J. Hydrol.* **2000,** *231,* 244–254.

14. Sililo, O. T. N.; Tellam, J. H. *Ground Water* **2000,** *38,* 864–871.

15. Kung, K.- J. S. *Geoderma* **1990,** *46,* 51–58.

16. Kung, K.-J. S. *Geoderma* **1990,** *46,* 59–71.

17. Fluhler, H.; Durner, W.; Flury, M. *Geoderma* **1996,** *70,* 165-183.

18. Newman, B. D.; Campbell, A. R.; Wilcox, B. P. *Water Resour. Res.* **1998,** *34,* 3485–3496.

19. Gee, G. W.; Ward, A. L. *Vadose Zone Transport Field Study: Status Report.* U.S. Department of Energy: Richland, WA, 2001; PNNL-13679.

20. Brusseau, M. L.; Hu, Q.; Srivastava, R. *J. Contam. Hydrol.* **1997,** *24,* 205–219.

21. Jardine, P. M.; O'Brien, R.; Wilson, G. V; Gwo, J.-P. In *Physical Nonequilibrium in Soils: Modeling and Application;* Selim, H.M., Ma, L., Eds.; Ann Arbor: Chelsea, MI, 1998; pp 243–271.

22. Langner, H. W.; Gaber, H. M.; Wraith, J. M.; Huwe, B.; Inskeep, W. *Soil Sci. Soc. Am. J.* **1999,** *63,* 1591–1598.

23. Mortensen, A. P.; Jensen, K. H.; Nilsson, B.; Juhler, R. K. *Vadose Zone J.* **2004,** *3,* 634–644.

24. Mayes, M. A.; Jardine, P. M.; Mehlhorn, T. L.; Bjornstad, B. N.; Ladd, J. L.; Zachara, J. M. *J. Hydrol.* **2003**, *275*, 141–161.
25. Pace, M. N.; Mayes, M. A.; Jardine, P. M.; Mehlhorn, T. L.; Zachara, J. M.; Bjornstad, B. N. *Vadose Zone J.* **2003**, *2*, 664–676.
26. Gwo, J. P.; Jardine, P. M.; Wilson, G. V.; Yeh, G. T. *J. Hydrol.* **1995**, *164*, 217–237.
27. Akratanakul, S.; Boersma, L.; Klock, G. O. *Soil Science* **1983**, *135*, 331–341.
28. Mayes, M. A.; Jardine, P. M.; Larsen, I. L.; Brooks, S. C.; Fendorf, S. E. *J. Contam. Hydrol.* **2000**, *45*, 243–265.
29. Gamerdinger, A. P.; Kaplan, D. I.; Wellman, D. M.; Serne., R. J. *Water Resour. Res.* **2001**, *37*, 3147–3153.
30. Gamerdinger, A. P.; Kaplan, D. I.; Wellman, D. M.; Serne., R. J. *Water Resour. Res.* **2001**, *37*, 3155–3162.
31. Maraqa, M. A.; Wallace, R. B.; Voice, T. C. *J. Contam. Hydrol.* **1999**, 36, 53–72.
32. Fesch, C.; Lehmann, P.; Haderlein, S. B.; Hinz, C.; Schwarzenbach, R. P.; Fluhler, H. *J. Contam. Hydrol.* **1998**, 33, 211–230.
33. Bjornstad, B. N. 1990. *Geohydrology of the 218-W-5 Burial Ground, 200-West Area, Hanford Site;* U.S. Department of Energy: Richland, WA, 1990; PNNL-7336.
34. Lindsey, K. A.; Gaylord, D. R. *Northwest Science* **1990**, *64*, 165–180.
35. *Standardized Stratigraphic Nomenclature for Post-Ringold Formation Sediments within the Central Pasco Basin;* U.S. Department of Energy: Richland, WA, 2002; USDOE/RL-2002-39, Rev. 0.
36. Slate, J. L. *Nature and Variability of the Plio-Pleistocene Unit in the 200 West Area of the Hanford Site;* U.S. Department of Energy: Richland, WA, 2000; BHI-01203, Rev. 0.
37. Roh, Y. Presented at the Clay Mineralogy Society Annual Meeting, Boulder, CO, June 2002.
38. Jardine, P. M.; Jacobs, G. K.; O'Dell, J. D. *Soil Sci. Soc. Am. J.* **1993**, *57*, 954–962.
39. Waite, T. D.; Davis, J. A.; Payne, T. E.; Waychunas, G. A.; Xu, N. *Geochim. Cosmochim. Acta* **1994**, *58*, 5465–5478.
40. Barnett, M. O.; Jardine, P. M.; Brooks, S. C.; Selim, H. M. *Soil Sci. Soc. Am. J.* **2000**, *64*, 908–917.
41. Toride, N.; Leij, F. J.; van Genuchten, M. Th. *The CXTFIT code for estimating transport parameters from laboratory or field experiments, version 2.1;* U.S. Department of Agriculture, US Salinity Laboratory: Riverside, CA, 1999; Res. Rep. 137.
42. Padilla, I.Y.; Yeh, T.-C.; Conklin, M.H. *Water Resour. Res.* **1999**, *35*, 3303–3313.
43. Gamerdinger, A.P.; Kaplan, D.I. *Water Resour. Res.* **2000**, *36*, 1747-1755.

44. Felmy, A.R.; Mason, M.J. *J. of Solution Chem.* **2003,** *32,* 283-300.
45. Zachara, J.M.; Gassman, P.L.; Smith, S.C.; Taylor, D.L. *Geochim. Cosmochim. Acta.* **1995,** *59,* 4449-4463.
46. Brooks, S.C.; Taylor, D.L.; Jardine, P.M. *Geochim. Cosmochim. Acta.* **1996,** *60,* 1899-1908.
47. Jardine, P.M.; Taylor, D.L. *Geoderma* **1995,** *67,* 125-140.
48. Ammann, A. *J. Chrom. A.* **2002,** *947,* 205-216.
49. Bargar, J.R.; Reitmeyer, R.; Davis, J.A. *Environ. Sci. Technol.* **1999,** *33,* 2481-2484.
50. Bostick, B.C.; Barnett, M.O.; Jardine, P.M.; Brooks, S.C.; Fendorf, S.E. *Soil Sci. Soc. Am. J.* **2002,** *66,* 99-108.

Chapter 11

Development of Accurate Chemical Equilibrium Models for the Hanford Waste Tanks: The System Na-Ca-Sr-OH-CO$_3$-NO$_3$-EDTA-HEDTA-H$_2$O from 25 to 75 °C

Andrew R. Felmy, Marvin Mason, Odeta Qafoku, and David A. Dixon

The Pacific Northwest National Laboratory, Richland, WA 99352

This manuscript describes the development of an accurate aqueous thermodynamic model for predicting the speciation of Sr in the waste tanks at the Hanford site. A systematic approach is described that details the studies performed to define the most important inorganic and organic complexation reactions as well as the effects of other important metal ions that compete with Sr for complexation reactions with the chelates. By using this approach we were able to define a reduced set of inorganic complexation, organic complexation, and competing metal reactions that best represent the much more complex waste tank chemical system. A summary is presented of the final thermodynamic model for the system Na-Ca-Sr-OH-CO$_3$-NO$_3$-EDTA-HEDTA-H$_2$O from 25 to 75 °C that was previously published in a variety of sources. Previously unpublished experimental data are also given for the competing metal Ni as well for certain chemical systems, Na-Sr-CO$_3$-PO$_4$-H$_2$O, and for the solubility of amorphous iron hydroxide in the presence of several organic chelating agents. These data were not used in model development but were key to the final selection of the specific chemical systems prioritized for detailed study.

Introduction

The high-level radioactive waste tanks at the U.S. Department of Energy (DOE) storage site at Hanford are extremely complex solutions containing high concentrations of electrolytes, radioactive species, other metal ions, and, in selected tanks, high concentrations of organic chelating agents. Developing accurate thermodynamic models for such chemical systems is an extremely daunting task not only because of the complex chemical nature of the solutions and the very strong aqueous complexes that can form but also since several potentially very insoluble phases can also form in these systems. This balancing of strong aqueous complexation reactions with insoluble precipitate formation means that even small errors in the thermodynamic models can result in large differences in the predicted solubilities for several constituents.

[90]Sr is a key radionuclide in the processing of the current high priority tanks at Hanford (AN-107 and AN-102) and is also an important radionuclide that may have leaked from the waste tanks currently being investigated by the Hanford Vadoze Zone Science and Technology (HVZS&T) program. The importance of ^{90}Sr can be seen in the recent data of Urie et al. (*1*) for tank AN-102 (Table I). [90]Sr is the radionuclide present at the highest concentration in the tank supernatant behind only [137]Cs. Current strategies for removal of these radionuclides from the supernatants call for ^{137}Cs removal via ion-exchange and the Sr removal (Sr/TRU) via solid phase precipitation induced by the addition of stable Sr and $KMnO_4$. The ^{90}Sr is removed by isotopic substitution into the precipitated phases. Accurate models for predicting the aqueous complexes in solution and the solid phases that can form during the Sr/TRU separation process are thus of high importance for treating not only these tank wastes of immediate importance but also tank wastes that will be processed later in the cleanup program. These same thermodynamic models are also useful for predicting the chemical species present in the initial tank waste leaks, and hence are useful for defining the source term in reactive transport models of subsurface contaminant migration.

Unfortunately, developing accurate thermodynamic models, which describe these aqueous complexation and solubility reactions, is extremely difficult. This difficulty is readily apparent from the chemical analysis of the supernatants or the feed solutions expected to enter the Waste Treatment Plant (WTP), see Table II.

These solutions contain high concentrations of electrolytes (Na, NO_3, NO_2, ...) inorganic complexants (OH, CO_3, F, ...) as well as organic chelates (EDTA, HEDTA, NTA, ...). Accurate models for these solutions must be able to describe all of these features including the effects of high ionic strength, inorganic and organic complexation, and the temperature. The latter feature is important since the processing solutions could range in temperature from 25 to

Table I. Radionuclide Concentrations in Hanford Waste Tank AN-102 (*1*)

Analyte	Supernatant ($\mu Ci/ml$)	Wet Centrifuged Solids ($\mu Ci/g$)
^{60}Co	8.49E-02	5.71E-02
^{90}Sr	5.72E+01	1.44E+02
^{99}Tc	1.48E-01	9.88E-02
^{125}Sb	NM	2.E-01
^{137}Cs	3.69E+02	2.16E+02
^{152}Eu	NM	1.E-02
^{154}Eu	2.31E-01	5.12E-01
^{155}Eu	1.00E-01	3.20E-01
^{238}U	NM	2.18E-05
^{237}Np	1.20E-04	9.21E-04
^{238}Pu	1.65E-03	1.19E-02
^{239}Pu	6.47E-03	5.56E-02
^{240}Pu	2.01E-03	1.50E-02
$^{239/240}Pu$	5.90E-03	4.17E-02
^{241}Am	1.65E-01	4.21E-01
^{242}Cm	6.29E-04	2.E-03
$^{243/244}Cm$	6.71E-03	1.72E-02

Note: NM, not measured.

75 °C. In addition to these reactions with the radionuclides, several metal ions in the solution are also present in high enough concentration to tie up significant fractions of the organic chelates. For example, the metal ions Fe (0.024 m), Ca (0.013 m), and Ni (0.0079 m) are present at high enough concentration to tie up significant fractions of the chelators such as EDTA (0.024 m), HEDTA (0.0094 m) or gluconate (0.022 m) if they form strong metal-chelate complexes under these basic high ionic strength conditions. Thus the solubility reactions for Sr can be significantly impacted by the presence of these other metal ions. Finally, the presence of strongly interacting inorganic ligands often results in a corresponding strong interaction in the solid phase and the resulting formation of very insoluble solid phases. This is certainly the case for Sr that can form very insoluble carbonates ($SrCO_3$) or phosphates ($Sr_3(PO_4)_2$). Unraveling all of these features has been the focus of this project.

In the remainder of this paper, we describe our overall approach for addressing this problem in chemical thermodynamics and the results obtained. The manuscript contains a complete list of the thermodynamic data and the original references describing the parameter development. However, previously unpublished results are also presented for several chemical systems that were not used in the development of the published thermodynamic data, yet are key

Table II. Waste Tank AN-107 Diluted Feed Composition (2)

Major Compounds	Concentration (m)	Minor Components	Concentration (m)	Organic Ligands	Concentration (m)
Na^+	8.9	Al	1.7×10^{-1}	Glycolate	.30
NO_3^-	3.1	Ba	3.4×10^{-5}	Gluconate	.022
NO_2^-	1.3	Ca	1.3×10^{-2}	Citrate	.055
CO_3^{2-}	1.6	Ce	2.3×10^{-4}	EDTA	.024
OH^-	0.84	Cd	4.9×10^{-4}	HEDTA	.0094
SO_4^{2-}	0.1	Cr	3.3×10^{-3}	NTA	.037
PO_4^{3-}	0.037	Cs	1.1×10^{-4}	IDA	.056
$^a F^-$	0.39	Cu	3.9×10^{-4}		
Cl^-	0.046	Fe	2.4×10^{-2}		
		K	3.8×10^{-2}		
		La	1.9×10^{-4}		
		Mn	2.3×10^{-3}		
		Nd	5.8×10^{-4}		
		Ni	7.9×10^{-3}		
		Pb	1.45×10^{-3}		
		Sr	3.5×10^{-5}		
		U	3.6×10^{-4}		
		Zn	3.4×10^{-4}		
		Zr	5.6×10^{-4}		

Note: IC analysis probably includes formate and acetate.

for establishing the rationale for the selection of the specific chemical systems for which detailed thermodynamic data were developed. The appendix presents a complete listing of the model parameters developed as part of this research. Summarizing these results in one manuscript may enhance the usefulness of these data by eliminating the need for researchers to obtain several different references.

Experimental Methods

This section describes the experimental methods used for studies of the solubility of sodium phosphate $Na_3PO_4(c)$ in Na_2CO_3 solutions, the solubility of $Fe(OH)_3(am)$ in solutions of organic chelates, and the solubilities of $Ni(OH)_2$ and $Ca(OH)_2$ in NaOH. Information on the experimental methods for all other chemical systems is given in the original references.

$Na_3PO_4(c)$ in Na_2CO_3

The solubility of $Na_3PO_4(c)$ in Na_2CO_3 was measured using the exact procedures described by Felmy et al. (3) for studying the solubility of $SrCO_3$ in Na_2CO_3 except that 0.3 grams of $Na_3PO_4 \cdot 12H_2O$ (Aldrich) were added to each 30-ml centrifuge tube.

$Fe(OH)_3(am)$

The solubility of the hydrous ferric oxide (HFO) at 1m Na_2CO_3 was studied in the presence of 0.01 m organic chelates (EDTA, HEDTA and gluconate) with NaOH ranging from 0.01 to 5.0 M. All the solutions were prepared from ACS reagent-grade chemicals, inside a N_2-filled chamber, to avoid CO_2 contamination. The Fe content in the HFO slurry, prepared according to Ainsworth et al. (4), was 0.0895 mol/L.

Na_2CO_3 salt was added to 50-mL Oak Ridge centrifuge tubes, and transferred afterwards to the N_2- controlled chamber. Then, deionized-degassed water was added to every centrifuge tube, followed by known volumes of NaOH, and chelate solutions to reach the desired concentrations. Finally, a known volume of HFO slurry was added to obtain 0.024 mol/L Fe in each tube. The tubes were agitated at a rotating shaker during the length of the experiment.

After 7-day equilibration periods, the samples were centrifuged, and filtered, through previously prewashed Amicon-type F-25 Centriflo membrane cones, with an approximate pore size of 0.0018 μm. The pH of the samples was measured using an Orion-glass electrode calibrated at pH 7, and 10. The sample filtrates were acidified with 2% HNO_3, and Fe concentration was determined by inductively coupled plasma mass spectrometer (ICP-MS) Hewlett Packard 4500.

$Ni(OH)_2$ and $Ca(OH)_2$ in NaOH

The solubility of $Ni(OH)_2$ in the presence and absence of $Ca(OH)_2$ was studied in the presence of 0.008 m EDTA, at various NaOH concentrations (extending to 10 M). This study was conducted in a controlled atmosphere chamber, filled with N_2 gas to prevent contamination of solutions from atmospheric CO_2, at room temperature.

In the studies where both solids were present, an equal amount (0.1 g) of $Ca(OH)_2$, and (0.1 g) of $Ni(OH)_2$ (ACS reagent-grade) solids was added to Oak Ridge centrifuge tubes (50 mL). The tubes were transferred into N_2-filled chamber. Known volumes of low-carbonate concentrated NaOH (J.T. Baker Dilut-it), and concentrated EDTA (1 m) solutions were added to the centrifuge tubes to reach the desired base, and chelate concentration. Final volume in each tube was 30 mL. All solutions were prepared with deionized-degassed water (18.3 $M\Omega cm^{-1}$). The tubes were agitated using a rotating shaker during the course of the experiment.

Several times, subsamples were taken for chemical analyses, to ensure that equilibrium or steady-state conditions were reached. The subsampling procedure consisted of centrifuging the suspensions at 2000 rpm for 10 min, and filtering 4 mL of supernatant through previously pre-washed (30 mL deionized-degassed water and 1 mL sample) centrifugal filter devices (Centricon-plus 20 – approximate 0.0036µ pore size). The filtered samples were diluted and acidified afterwards with 2% HNO_3. They were analyzed for Ni and Ca by inductively coupled plasma-optical unit (ICP-OES) Perkin Elmer Optima 3000, with an error less than 5%.

The XRD analysis of selected samples revealed only $Ni(OH)_2$ and $Ca(OH)_2$ crystalline patterns.

Results and Discussion

In order to develop an accurate thermodynamic model to predict the chemical behavior of Sr, we systematically studied the reactions of Sr with the most important inorganic (beginning with hydroxide and carbonate) ligands and organic (EDTA, HEDTA) chelators in the tanks. These systems were studies over a broad range of ionic strengths, ligand concentrations, and temperatures.

Inorganic Chemical Systems

Na-Sr-OH-H₂O System

Owing to the nearly ubiquitously high concentrations of hydroxide in the tank wastes, the first chemical system studied was the Na-Sr-OH-H$_2$O system. The results showed that the hydrolysis of Sr^{2+} was fairly weak and that the entire range of NaOH concentrations extending to 6 M NaOH could be described using Pitzer ion-interaction parameters for Sr^{2+}-OH$^-$ interactions along with a Na$^+$-Sr^{2+} mixing term. No clear evidence was found for the formation of SrOH$^+$ or other hydrolysis species. In fact, our experimental solubilities of Sr(OH)$_2$.8H$_2$O could not be fit with any ion pairing or hydrolysis species formulation since the solubilities predicted with increasing NaOH concentration considering only the species Sr^{2+} and utilizing Pitzer's form of the extended Debye Hückel equation exceeded the experimental values (Figure 1). Therefore any attempt to include a hydrolysis species in such a model would only act to increase the predicted solubilities (see Felmy et al. [3]).

This thermodynamic analysis was at least partially supported by density functional theory (DFT) calculations of gas phase clusters. The DFT calculations showed changes in the attachment of the hydroxyl ions to the solvated Sr^{2+} ion as a function of added water molecules. When the hydration number for the Sr^{2+} cation exceeded seven water molecules, the hydroxyl detached from the central metal and formed hydrogen bonds with the solvated water molecules in an outer sphere like complex (Figure 2) again indicating the relatively weak attachment of hydroxyl with the large Sr^{2+} cation.

These studies also identified the thermodynamically stable Sr hydroxide solid under high base conditions (Sr(OH)$_2$ 8H$_2$O) and developed the first solubility product for this phase.

Na-Sr-CO₃-H₂O System

With an accurate model for the Na-Sr-OH-H$_2$O system complete, our studies then emphasized the development of an aqueous thermodynamic model for the Na-Sr-CO$_3$-H$_2$O system owing to the high concentration of carbonate in the waste tank supernatants (Table II). Figure 3 shows examples of the solubility data for SrCO$_3$(c) and CaCO$_3$(c) in Na$_2$CO$_3$ solutions that indicated the presence of previously unidentified Sr(CO$_3$)$_2$$^{2-}$ and Ca(CO$_3$)$_2$$^{2-}$ complexes. The calculated stability constants for these complexes from the data shown in Figure 3 were then shown to correlate well with the known stability constants for other Ca and Sr complexes based upon their DFT binding energies.

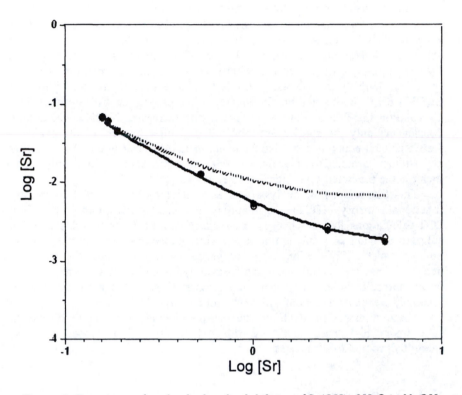

Figure 1. Experimental and calculated solubilities of Sr(OH)$_2$·8H$_2$O in NaOH. Patterned line represents calculations with Sr^{2+}-OH interactions described solely with the use of Pitzer's form of the extended Debye-Hückel equation. Solid line represents the calculations of our final thermodynamic model, which includes values for the Pitzer ion interaction parameters. Total concentrations in units of molarity. From (3).

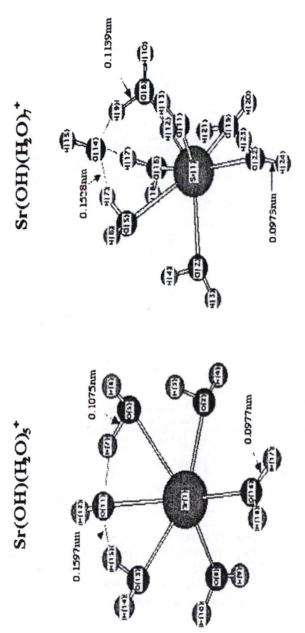

Figure 2. Structures of (a) gaseous $Sr(H_2O)_7\text{-}OH^+$, and (b) gaseous $Sr(H_2O)_5OH^+$ calculated by DFT. (Reproduced with permission from reference 3. Copyright 1998 Elsevier.)

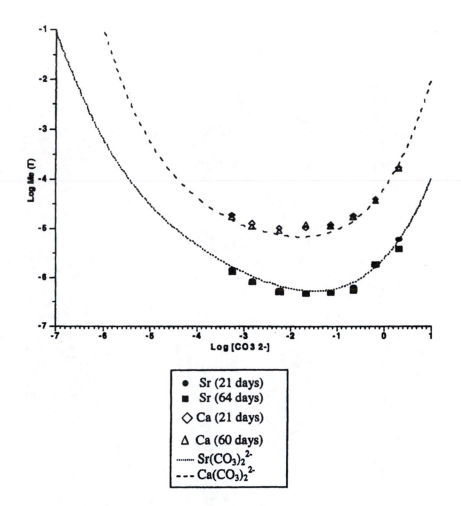

Figure 3. Experimental and calculated Sr and Ca concentrations in equilibrium with SrCO₃ and CaCO₃ respectively. Model calculations assume the formation of dicarbonate Ca and Sr complexes. The notation Me represents either Ca or Sr. (T) represents total concentration.

(Reproduced with permission from reference 3. Copyright 1998.)

The formation of strong Sr-CO$_3$ aqueous species was a clear indication that the predominate inorganic aqueous complex in solution was likely to be Sr(CO$_3$)$_2$$^{2-}$. If true, it will be the formation and ion-interactions of this species with the bulk electrolyte ions (principally Na$^+$) that will determine the overall solubility behavior of the tank solutions in the absence of the organic chelates. This would mean that interactions of minor species, such with the uncomplexed Sr^{2+} ion, would be unimportant unless they were sufficiently strong to perturb or displace the Sr(CO$_3$)$_2$$^{2-}$ complex.

However, the tank solutions also contain significant concentrations of other potentially important inorganic ligands (i.e., NO$_3$, NO$_2$, Cl, F, and PO$_4$). Of these, most either interact weakly with Sr^{2+} or the thermodynamic data for these are well known (see the following section). However, the situation involving trivalent phosphate was unclear. The currently recommended stability constants for Sr-PO$_4$ complexes all involve only protonated forms of PO$_4$ (i.e., SrH$_2$PO$_4$$^+$ and SrHPO$_4$(aq)), indicating that such interactions at high base may be unimportant (see Martell and Smith [5]) However, there are also several Ca-PO$_4$ solids that are known to be insoluble under high base conditions and in fact have been identified in tank wastes (e.g., hydroxyapatite). Since Ca^{2+} is an analog for Sr^{2+} it was deemed wise not to ignore PO$_4$ interactions. However, before we invested large amounts of research efforts on studying PO$_4$ interactions with Sr^{2+}, we elected to conduct competition studies on the effects of PO$_4$ in the presence of CO$_3$.

Na-Sr-CO$_3$-PO$_4$-H$_2$O

Our first objective was simply to determine if studies of Sr-PO$_4$ complexation/precipitation would be important in determining the overall chemical behavior of the tank waste system. To do this we set up solubility studies of SrCO$_3$(c) and Sr$_3$(PO$_4$)$_2$(c) as a function of added Na$_2$CO$_3$. Sr$_3$(PO$_4$)$_2$(c) was the phase that readily precipitated from solution under basic conditions in the presence of phosphate. The initial dissolved phosphate concentration in these scoping studies was set at a reasonable value for the tank solutions (i.e., 0.03 M). These previously unpublished results, Figure 4, showed at lower added Na$_2$CO$_3$ the solubilities in solutions in contact with SrCO$_3$(c) were always lower than in solutions in contact with Sr$_3$(PO$_4$)$_2$(c). As the Na$_2$CO$_3$ concentration increased, a clear phase transition occurred, and the observed solubilities in the presence and absence of phosphate became identical. XRD analysis of the precipitates showed that when the total number of moles of carbonate in solution exceeded the total number of moles of Sr tied up as the added Sr$_3$(PO$_4$)$_2$(c) solid phase, that complete conversion of the added Sr$_3$(PO$_4$)$_2$(c) occurred with formation of SrCO$_3$(c). The solubilities of the systems in the presence and absence of phosphate then became identical. These

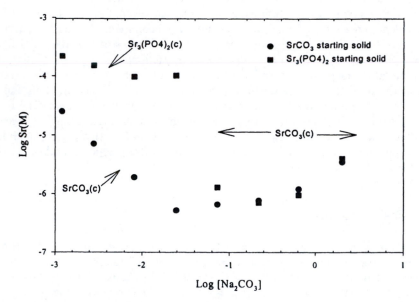

Figure 4. *Stability of SrCO₃(c) and Sr₃(PO₄)₂(c) starting materials as a function of added Na₂CO₃. Initial PO₄ concentration in the Sr₃(PO₄)₂(c) experiments was 0.03 M. Samples equilibrated for 357 days.*

scoping studies established two facts. First, the presence of phosphate at the concentrations expected in the tank wastes is unlikely to displace the carbonate from the stable $Sr(CO_3)_2^{2-}$ complex and that $SrCO_3(c)$ appears to be the most stable precipitate. These studies not only completed our studies of the inorganic reactions for Sr^{2+} but also set the stage for the studies of the organic chelates.

Other Inorganic Ligands

Examining the remaining inorganic ligands in the tank wastes (Table II) shows that all of the Sr^{2+} interactions are either known (i.e., Sr^{2+}-NO_3^- and Sr^{2+}-Cl^-, Pitzer [6]; Sr^{2+}-SO_4^{2-}, Felmy et al. [7, 8]) or expected to be quite weak Sr-NO_2^-, and Sr^{2+}-F^-. In the case of Sr^{2+}-F^-, the aqueous complexes are known to be very weak (Log K for SrF^+ formation equals only 0.14), the solubility products of possible precipitates, $SrF_2(c)$, are well known (Log K_{sp} = -8.58) (5), and the reported analytical data for F^- are greatly inflated owing to the inclusion of low molecular weight organics (formate and acetate) in the IC analysis of F^-. These facts combined with the high concentration of carbonate meant that it was extremely unlikely that either F^- or NO_2^- could form an aqueous complex stronger than the expected dominant $Sr(CO_3)_2^{2-}$ complex. Specific studies with F^- and NO_2^- were therefore not conducted.

Organic Chelate Studies

The inorganic speciation studies revealed that in the absence of organic chelates, the dominant aqueous species is likely to be the previously unidentified $Sr(CO_3)_2^{2-}$ complex and the likely stable precipitate would be $SrCO_3(c)$. These findings made it possible to now identify the most important organic chelate complexes of Sr^{2+} in a more systematic manner. Specifically, in order for the organic-chelate complex to be important enough to require thermodynamic study, the complex must be strong enough to outcompete the dominant inorganic species $Sr(CO_3)_2^{2-}$. Second, the most likely precipitate is likely to be $SrCO_3(c)$, so solubility studies designed to evaluate the impact of organic complexes on the precipitation/dissolution behavior of Sr^{2+} in tank waste should focus on use of this phase. Fortunately, the solubility products of $SrCO_3(c)$ are well known as a function of temperature (9) making this an excellent solid for such studies.

With these factors in mind, we initiated a series of organic ligand competition studies in which the carbonate concentration was varied to evaluate the range of conditions over which the chelates were able to displace the carbonate from the $Sr(CO_3)_2^{2-}$ complex. The objective was to reduce the list of possible organic chelates requiring study to a select few for which a highly accurate model could then be developed. Although an initial screening of the potential organic chelates could be accomplished by comparing standard state equilibrium constants for the Sr-chelate complexes, this does not represent a definitive test since multiple ligand complexes can form and/or the ionic strength effects on certain of the complexes could be quite strong.

As an example of the results, Figure 5 shows the solubility of $SrCO_3(c)$ in the presence of four organic chelates (EDTA, HEDTA, NTA, and IDA), representing a wide range of expected binding energies, at different concentrations of Na_2CO_3. The results show that at 0.1 M added Na_2CO_3 (Figure 5a), Sr-EDTA, and Sr-HEDTA complexes are the predominant species in solution across the entire range of added organic ligand. The $Sr(CO_3)_2^{2-}$ complex is the dominant species in solutions with added NTA until the added NTA concentration reaches approximately 0.001 M, whereupon the Sr-NTA complexation reactions dominate *(10)*. The results in 1.0 M added Na_2CO_3 (Figure 5b) show that Sr-EDTA complexes are the predominant species in solution across the entire range of added organic ligand; Sr-HEDTA complexes predominate at a ligand concentration greater than 10^{-4} M; and Sr-NTA complexes predominate only above about 10^{-2} M.

With these studies complete it became obvious that any organic chelate with a stability constant less than that for NTA (i.e., Log K for formation of SrNTA⁻ is 4.99 [5]) is unlikely to be important in the tank waste system in the presence of high concentrations of carbonate. As a result, it did not appear that thermodynamic studies were needed for glycolate, gluconate, IDA, or citrate complexation with Sr since their stability constants for Sr complexation were too low (log K's range from 0.80 (glycolate) to 2.99 (citrate)). The two most important ligands requiring study were clearly EDTA and HEDTA. Studies with NTA were also considered, but in the complex waste matrix the concentration of NTA would need to be more than an order of magnitude greater to outcompete HEDTA and more than two orders of magnitude higher to outcompete EDTA. Hence, all further studies of Sr complexation focused on these two chelators, EDTA and HEDTA.

With the key organic ligands now defined (EDTA and HEDTA), we initiated a comprehensive study of the thermodynamic of these complexes over the temperature range 25 to 75 °C, from 0.01 m to 2.0 m Na_2CO_3, and extending to an ionic strength of 9. Examples of our final thermodynamic model are shown in Figure 6. From a scientific perspective, these studies provided the first definitive values for the standard state equilibrium constants for $SrEDTA^{2-}$ and $SrHEDTA^{-}$ complexes, as opposed to the commonly used 0.1 M reference state values. The necessary Pitzer ion-interaction parameters were developed for both the $SrEDTA^{2-}$ and $SrHEDTA^{-}$ species as a function of temperature. In addition, during the modeling effort it became apparent that the interactions of the Na^+ ion with the uncomplexed $EDTA^{4-}$ and $HEDTA^{3-}$ ligands were an important factor in determining the final standard state equilibrium constants and ion-interaction parameters for the $SrEDTA^{2-}$ and $SrHEDTA^{-}$ complexes. This realization necessitated the development of improved models for describing the Na^+-$EDTA^{4-}$ and Na^+-$HEDTA^{3-}$ interactions. A comprehensive analysis of all of the available data on Na^+-$EDTA^{4-}$ and Na^+-$HEDTA^{3-}$ complexation including analysis of apparent equilibrium constants for formation of $NaEDTA^{3-}$, the effects of Na^+ concentration on the protonation of EDTA and HEDTA species, and the solubility of Na_3HEDTA in $NaNO_3$ was conducted $(11, 12)$. Not unexpectedly, the final model describing Na^+-$EDTA^{4-}$ interactions required the inclusion of a $NaEDTA^{3-}$ ion pair into the model, whereas the Na^+-$HEDTA^{3-}$ interactions were described solely by the use of Pitzer ion-interaction parameters. These differences in Na^+ complexation can be seen in Figure 6, where the increasing $NaNO_3$ concentration has a much greater effect on the solubilities in the presence of EDTA (Figure 6a) than on the presence of HEDTA (Figure 6b).

These comprehensive studies completed our research on the Sr complexation with organic ligands.

Studies of Competing Metal Ions

The other key factor in predicting the aqueous chemistry of Sr in the tank waste solutions is to unravel the effects of competing metal ions. The importance of competing metal ions is clear from the metals concentration data shown in Table II. The four metals present at highest concentration are Al, Fe, Ca, and Ni. Of these only the dissolved concentration of Al appears to be easily explainable in terms of inorganic speciation reactions. The dissolved Al concentration computes to be in equilibrium with respect to gibbsite and the dominant aqueous species is the aluminate ion $(Al(OH)_4^-)$ (2). This is not the case for the other metal ions that all compute to be over saturated by orders of magnitude with respect to likely precipitates, $(Fe(OH)_3(am), Ca(OH)_2/CaCO_3,$ and $Ni(OH)_2)$, if organic complexes in solution are ignored (2). In the case of Fe, ferric iron is known to form extremely strong complexes with gluconate (13) as well as with EDTA (5). In addition, the analytical data in Table II show near identical iron, gluconate, and EDTA concentrations. These results indicate that the Fe is present as a ferric gluconate complex or a ferric EDTA complex. If present as a ferric gluconate complex, Fe would not be a competing metal ion for the EDTA or HEDTA chelates of concern in the case of Sr. To at least partially confirm these assumptions we conducted solubility studies of amorphous $Fe(OH)_3(s)$ in the presence and absence of the strongest binder chelates EDTA, HEDTA, and gluconate. These results, Figure 7, show near identical solubilities for Fe in the presence and absence of EDTA and HEDTA, but near stoichiometric saturation of the added gluconate with Fe. Clearly, EDTA and HEDTA are not capable of solubilizing significant amounts of iron at high base concentration, and the high dissolved Fe concentration in the tank

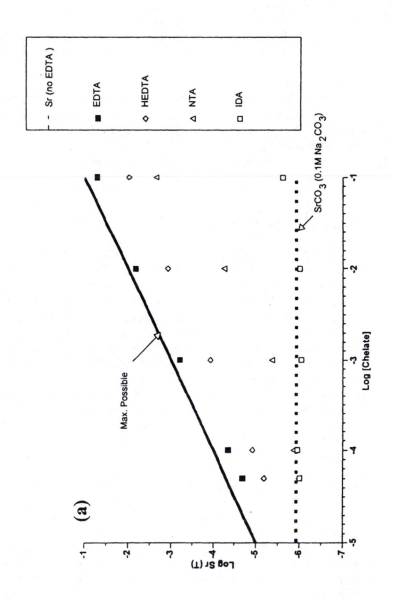

(a)

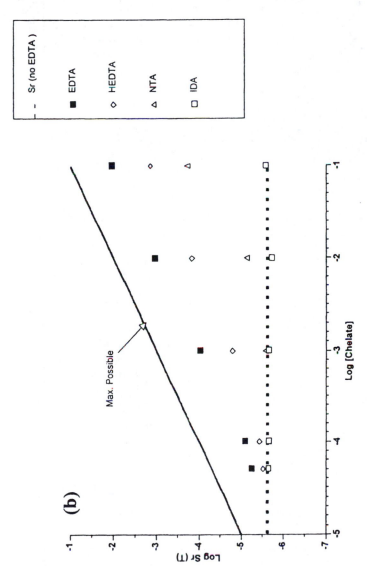

Figure 5. The solubility of SrCO₃(c) in the presence of organic chelates at different concentrations of Na₂CO₃: (a) 0.1 M Na₂CO₃, and (b) 1.0 M Na₂CO₃.

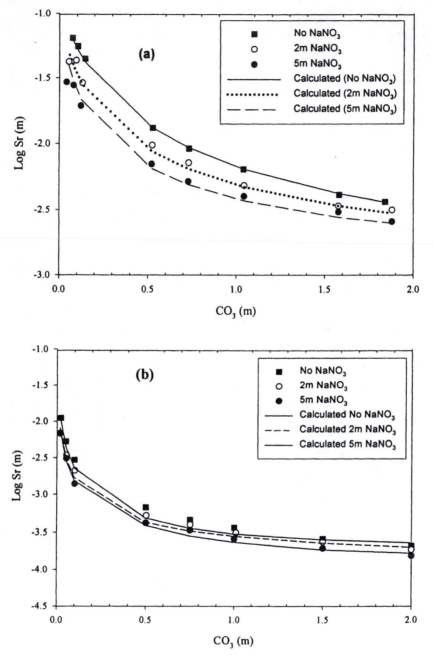

Figure 6. Experimental and calculated solubilities of SrCO₃ as a function of Na₂CO₃ concentration and in the presence of (a) EDTA, and (b) HEDTA with varying concentrations of NaNO₃ (Reproduced with permission from references 11 and 12. Copyright 2003 Springer.)

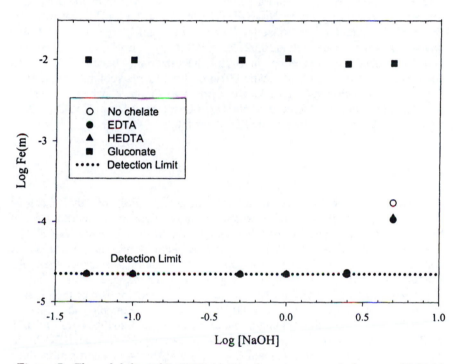

Figure 7. The solubility of Fe(OH)₃(am) in the presence and absence of 0.01 M EDTA, HEDTA, or gluconate as a function of NaOH concentration. Equilibration time, 7 days.

solutions is most likely present as a ferric gluconate complex. Therefore, Fe, though present in the tank supernatants at high concentration, is unlikely to be complexed by the key ligands of concern in modeling the behavior of Sr (e.g., EDTA and HEDTA).

However, this is certainly not the case for the cationic components Ca and Ni, which are known to form extremely strong complexes with the organic chelates, especially EDTA. In fact the 0.1 M reference state values for these complexes are considerably higher for $CaEDTA^{2-}$ (Log K = 10.65) and $NiEDTA^{2-}$ (Log K = 18.4) than for $SrEDTA^{2-}$ (Log K = 8.74). In addition, Ni is known to form a very strong $NiOHEDTA^{3-}$ species at high pH. Interestingly, the combined total concentrations for Ca and Ni (Table II) are very close to the analytical EDTA concentration (0.021 m vs 0.024 m). Clearly, these metal ions could be effectively competing for Sr for the important chelator EDTA, and such reactions could have a dramatic effect on the observed solubilities of Sr in the tank waste solutions. This fact is evident in Figure 8, which shows the dramatically reduced solubilities of $SrCO_3(c)$ in the presence of $CaCO_3(c)$ as a result of the Ca complexation of the EDTA chelate.

Ca Studies

With these factors in mind we began studies of the aqueous thermodynamics of the Ca-EDTA complexation reactions at high carbonate and base concentrations. The aqueous thermodynamics of the inorganic Na-Ca-OH-H_2O and Na-Ca-CO_3-H_2O chemical systems were previously examined as companion studies with the corresponding Sr containing systems (see, for example, the data for $CaCO_3(c)$ in Figure 3). The organic chelate studies examined the solubility of $CaCO_3(c)$ in the presence and absence of the organic chelators, principally EDTA and HEDTA (see, for example, the data in Figure 7). One interesting finding was that the observed molality ratio of Ca/Sr could be accurately predicted for solutions containing EDTA and HEDTA in equilibrium with both $SrCO_3(c)$ and $CaCO_3(c)$ using only standard state values for the solubility products and aqueous complexation constants *if* the values of the activity coefficients for analogous Sr and Ca species (i.e., $CaEDTA^{2-}$, $SrEDTA^{2-}$, $CaHEDTA^{-}$, $SrHEDTA^{-}$) were assumed to be identical (*10*). This analogy proved to be quite effective in predicting the observed solubilites of $CaCO_3(c)$ in EDTA (see Figure 9) and HEDTA solutions as well as predicting the solubility of $SrCO_3(c)$ in the presence of $CaCO_3(c)$ (see the "predicted" line in Figure 8 for EDTA and Figure 10 for HEDTA). The use of the Pitzer ion-interaction parameters for $SrEDTA^{2-}$ and $SrHEDTA^{-}$ in place of the corresponding values for $CaEDTA^{2-}$ and $SrHEDTA^{-}$ also allowed the extrapolation of the well-known 0.1 M reference state values for these complexes to the standard state. This extrapolation is quite difficult owing to

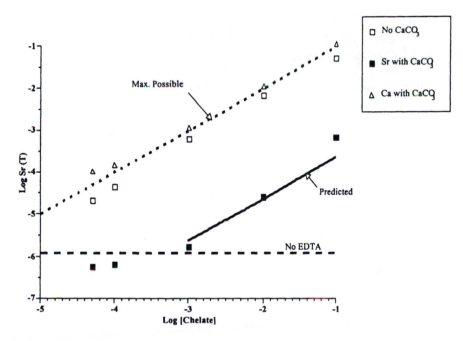

Figure 8. Solubilities of SrCO₃(c) in the absence (open squares) and presence (solid squares) of CaCO₃(c). The observed Ca concentrations (open triangles) essentially complex all of the added EDTA and reduce the Sr concentrations in equilibrium with SrCO₃(c). (Reproduced with permission from reference 10. Copyright 1998 Springer.)

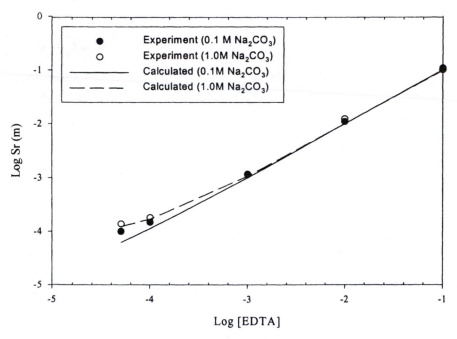

Figure 9. Experimental and calculated CaCO₃(c) solubilities as a function of EDTA concentration. Experimental data from Felmy and Mason

(Reproduced with permission from reference 11. Copyright 2003 Springer.)

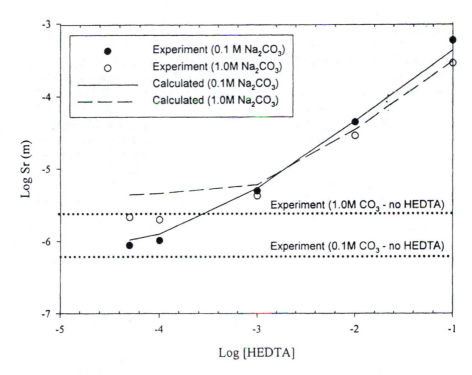

Figure 10. Experimental and calculated SrCO₃(c) solubilities as a function of HEDTA concentration in the presence of CaCO₃(c).
(Reproduced with permission from reference 12. Copyright 2003 Springer.)

the charges on the bare ions (Ca^{2+}, $EDTA^{4-}$) and equals 1.7 log units (e.g., 10.65 at I = 0.1 and 12.36 in the standard state). These estimates were also in good agreement with previous extrapolations (5), even though parameters for $SrEDTA^{2-}$ were used in the actual extrapolation. This close analogy then allowed the prediction of the activity coefficients and standard state properties for the competing metal Ca without the need for the extensive studies required to evaluate the parameters for the Sr containing systems.

Ni Studies

As described above, Ni is known to form very strong complexes with EDTA, and these complexes have the potential to be important even at high base concentration due to the expected formation of mixed Ni-OH-EDTA complexes (i.e., $NiOHEDTA^{3-}$). In our research program prior to that funded by the DOE Environmental Management Science Program (EMSP) (see Mattigod et al. [14]), we had examined the solubility of $Ni(OH)_2$(c) in NaOH and evaluated the solubility and hydrolysis behavior of Ni(II) to high base concentration. We therefore initiated studies on the solubility of $Ni(OH)_2$(c) in EDTA solutions as a function of added NaOH. Although still on-going, the initial results, open and filled circles in Figure 11, showed that as expected the addition of EDTA resulted in the formation of very strong Ni-EDTA aqueous complexes that were effective in solubilizing $Ni(OH)_2$ to very high base concentration (i.e., 2 to 3 M NaOH). Unfortunately, the complexation reactions were so strong that essentially total saturation of the EDTA chelate with Ni occurred. In such circumstances it is difficult to extract unambiguous thermodynamic data for the Ni-EDTA complexation reactions since only "greater than" values can be obtained. To obtain more precise value required introducing a strong competing metal ion for which the aqueous thermodynamics are well known. The aqueous thermodynamics of the ion of interest (e.g., Ni) can then be determined by comparing the changes in solubility that occur in response to the introduction of the competing metal. In this case the strongest competing metal ion for which we have developed an accurate thermodynamic model in the presence of EDTA at high base concentration is Ca^{2+}. Studies were therefore conducted on the solubility of $Ni(OH)_2$ in EDTA solutions with added Ca. The greatest Ca concentration that can be maintained in these solutions is when the solutions are saturated with respect to $Ca(OH)_2$. These dual solubility (i.e., saturation with respect to both $Ni(OH)_2$ and $Ca(OH)_2$) data are also shown in Figure 11, squares. The addition of $Ca(OH)_2$ significantly reduces the observed Ni concentrations as a result of displacement of Ni from the EDTA chelate by the added Ca concentration (Ca data not shown). Equilibration of the solutions with $Ni(OH)_2$ in these experiments was also tested by spiking the solutions with $NiEDTA^{2-}$ and thus approaching equilibrium from the oversaturation direction.

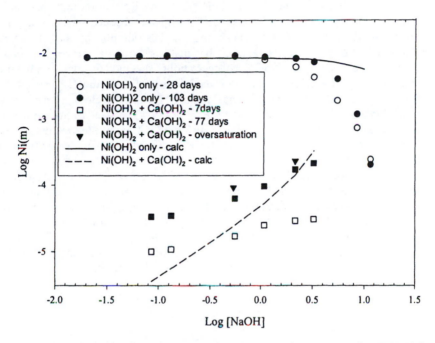

Figure 11. The solubility of Ni(OH)₂(c) in EDTA solutions as a function of the concentration of NaOH. EDTA concentration 0.01 M. Data indicated by circles are for solutions in contact with Ni(OH)₂(c) only. Data in squares and triangles are for solutions in contact with both Ni(OH)₂ and Ca(OH)₂. The term oversaturation indicates solutions, which have been spiked with Ni to approach equilibrium from the oversaturation direction. Calculations are described in the text.

Although these studies are still a work in progress, we do show initial thermodynamic modeling calculations of the Ni-EDTA system. In these calculations, the previously described thermodynamic model for the Na-Ca-OH-EDTA-H$_2$O system was used in conjunction with equilibrium constants for NiEDTA^{2-} and NiOHEDTA^{3-} from Martell and Smith (5) and the solubility product of Ni(OH)$_2$(c) from Mattigod et al. (14). Ion-interaction parameters for the NiEDTA^{2-} species were set equal to the analogous values for SrEDTA^{2-} as was shown to be effective for the CaEDTA^{2-} species. Although this initial model appears to work well for the solutions in the absence of Ca, this good agreement is actually fortuitous as the models simply predict complete saturation with the chelate across the majority of the range in NaOH concentration. At higher NaOH concentrations, the models over predict the solubility of Ni(OH)$_2$(c). Also, this model does not accurately predict the Ni(OH)$_2$(c) solubilities in the presence of Ca(OH)$_2$(c). Developing an accurate thermodynamic model for the Na-Ni-OH-EDTA-H$_2$O system is the subject of our ongoing research program. Clearly, such data will be needed to finalize our thermodynamic model for the solubility of Sr compounds in the waste tank chemical system (Table II).

Finally a note should be made about Al. The dissolved Al concentrations appear to be easily explained by simple inorganic complexation reactions (i.e., Al(OH)$_4^-$). However, the dissolved Al concentration is quite high, and it is still possible that small concentrations of Al-organic complexes could form that could compete with Sr for complexation with the chelates. Therefore, competition studies are planned to examine the influence of dissolved Al on the solubility of SrCO$_3$(c) at high base and carbonate concentration.

Summary

This paper describes the approach taken to reduce the number of chemical systems to a subset, the Na-Ca-Sr-OH-CO$_3$-NO$_3$-EDTA-HEDTA-H$_2$O system, which incorporates all of the strongest complexation/ion-interactions for Sr in the more complex actual Hanford tank wastes. A compilation of the previously developed thermodynamic data is presented along with previously unpublished experimental data for certain chemical systems used to define the key systems needing detailed research. It is hoped that combining all of these results into one manuscript will both facilitate the use of the data and provide additional essential detail on the choice of the chemical systems and the validity of the overall model to the complex tank wastes.

Finally, besides the obvious implications of these results to processing high-level tank wastes, they also have a dramatic impact on the potential migration of ^{90}Sr in the subsurface environment. For example, current EMSP studies (15) have focused on the adsorption or attenuation reactions of Sr under low carbonate concentrations found far from the vicinity of a tank leak. Under such conditions, the Sr is present as the divalent Sr^{2+} ion and the concentration in solution is controlled by cation exchange reactions. This situation is expected to be dramatically different in the vicinity of a tank leak where the high carbonate concentration will result in the ^{90}Sr being present as anionic $Sr(CO_3)_2^{2-}$ complexes or as $SrEDTA^-$ if the waste tank contained chelates. The migration of these species through the subsurface will be dramatically different than the studied Sr^{2+} cation.

Acknowledgments

Preparation of this manuscript was supported by the EMSP under a high level waste project entitled "Chemical Speciation of Strontium, Americium, and Curium in High Level Waste," Project No. 26753, and an EMSP Subsurface Science Project "The Aqueous Thermodynamics and Complexation Reactions of Aqueous Silica Species to High Concentration," Project No. 30944.

References

1. Urie, M. W.; Bredt, P. R.; Campbell, J. A.; Farmer, O. T.; Greenwood, L. R.; Jagoda, L. K.; Schele, R. D.; Soderquist, C. Z.; Swoboda, R. G.; Thomas, M. P.; Wagner, J. J. *Chemical Analysis and Physical Properties Testing of 241-AN-102 Tank Waste-Supernatant and Centrifuged Solids*; WTP-RPT-020; Pacific Northwest National Laboratory: Richland, WA.
2. Felmy A. R. *Thermodynamic Modeling of Sr/TRU Removal*; PNWD-3044, BNFL-RPT-037 Rev 0; British Nuclear Fuels Laboratory: Richland, WA; 2000.
3. Felmy, A. R.; Dixon, D. A.; Rustad, J. R.; Mason, M. J.; Onishi, L. M. *J. Chem. Thermodyn.* **1998**, *30*, 1103–1120.
4. Ainsworth, C. C.; Pilon, J. L.; Gassman, P. L.; Van Der Sluys, W. G. *Soil Sci. Soc. Am. J.* **1994**, *58*, 1615–1623.
5. Martell, A. E.; Smith, R. M. Critically Selected Stability Constants of Metal Complexes Database, Version 2.0; NIST Standard Reference Data Program: Gaithersburg, MD, 1995.
6. Pitzer, K. S. *Activity Coefficients in Electrolyte Solutions*; CRC Press: Boca Raton, FL; 1991.
7. Felmy, A. R.; Rai, D.; Amonette, J. E. *J. Solution Chem.* **1990**, *19*, 175–185.
8. Felmy, A. R.; Rai, D.; Moore, D. A. *Geochim. Cosmochim. Acta* **1993**, *57*, 4345–4363.

9. Busenberg, E.; Plummer, L. N. *Geochim. Cosmochim. Acta* **1984**, *48*, 2021–2035.

10. Felmy, A. R.; Mason, M. J. *J. Solution Chem.* **1998**, *27*, 435–454.

11. Felmy, A. R.; Mason, M. J. *J. Solution Chem.* **2003**, *32* (4), 283–300.

12. Felmy, A. R.; Mason, M. J.; Qafoku, O. *J. Solution Chem.* **2003**, 301–318.

13. Sawyer, D. T. *Chem. Rev.* **1964**, 633–643.

14. Mattigod, S. V.; Rai, D.; Felmy, A. R.; Rao L. *J. Solution Chem.* **1997**, *26*, 391–403.

15. Jardine, P. M.; Mayes, M. A.; Fendorf, S.; Pace, M. N.; Yin, X.; Zachara, J. M.; Mehlhorn, T. L. *Abstracts of Papers*, 226th National Meeting of the American Chemical Society, New York, Sep 7–11, 2003; American Chemical Society: Washington, DC, 2003; ENVR 238.

16. Felmy, A. R.; Rustad, J. R.; Mason, M. J.; de la Bretonne, R. *A Chemical Model for the Major Electrolyte Components of the Hanford Waste Tanks;* TWRS-PP-94-090; Westinghouse Hanford Co.: Richland, WA; 1994.

17. Mizera, J.; Bond, A. H.; Choppin, G. R.; Moore, R. C. In *Actinide Speciation in High Ionic Strength Media*; Reed et al., Eds.; Kluwer Academic/Plenum Publishers: New York, 1999; pp 113–124.

18. Harvie, C. E.; Moller, N.; Weare, J. H. *Geochim. Cosmochim. Acta* **1984**, *48*, 723–751.

19. Felmy, A. R.; MacLean, G. T., *Development of an Enhanced Thermodynamic Database for the Pitzer Model in ESP: The Fluoride and Phosphate Components;* WTP-RPT-018 Rev. 0; Bechtel National, Inc.: Richland, WA; 2001.

Appendix

This section presents, in Tables A-I and A-II, a complete summary of the thermodynamic data for the $Na-Sr-Ca-OH-CO_3-NO_3-EDTA-HEDTA-H_2O$ system. The temperature dependent parameters for the Pitzer ion-interaction parameters were fit to an equation of the form,

$$P(T) = a_1 + a_2T + a_3/T + a_4\ln T + a_5/(T-263) + a_6T^2 + a_7/(680-T) + a_8/(T-227) \tag{A1}$$

where $P(T)$ is a temperature-dependent ion-interaction parameter and T is in degrees Kelvin.

Table A-I. Parameters for the Temperature Dependent Expression (Eq A1) for the Pitzer Ion-Interaction Parameters for the Na-Sr-Ca-OH-CO$_3$-NO$_3$- EDTA-HEDTA-H$_2$O System *Continued on next page.*

Species	Parameter	a1	a2	a6	Reference
Na$^+$- CO$_3$$^{2-}$	β^0	-2.37265980e+00	1.42978840e-02	-2.08566020e-05	(16)
	β^1	-6.56843130e+00	5.18062300e-02	-8.20256870e-05	(16)
	C$^\phi$	5.20600000e-03	0.0	0.0	(16)
Na$^+$- NO$_3$$^-$	β^0	-2.64744648e+00	1.52565224e-02	-2.13146667e-05	(16)
	β^1	-7.77643468e+00	4.58638400e-02	-6.34666667e-05	(16)
Na$^+$- OH$^-$	β^0	-6.60527020e-01	4.55226810e-03	-6.86137740e-06	(11)
	β^1	0.0	0.0	2.83791820e-06	(11)
Na$^+$- EDTA^{4-}	C$^\phi$	1.04625300e-01	-5.6347570e-04	7.59648980e-07	(11)
	β^0	1.1			(11)
	β^1	15.6			(11)
Na$^+$ - NaEDTA^{3-}	C$^\phi$	0.001			(17)
	β^0	0.59			(11)
	β^1	5.39			(11)
Na$^+$HEDTA^{3-}	β^0	0.23			(12)
	β^1	5.22			(12)
Na$^+$- SrEDTA^{2-}	C$^\phi$	-0.479114	.00156		(12)
	β^0	0.3			(11)
Na$^+$ - SrHEDTA$^-$	C$^\phi$	-4.57312920e+00	2.89647688e-02	-4.55760000e-05	(11)
	β^0	0.62482	-.0028		(12)
	β^1	0.29			(12)

Table A-I. Continued.

Species	Parameter			Ref.
Na+ - Sr(CO3)2^2-	C^ϕ	-0.32778	0.0012	(12)
Na+ - Ca(CO3)2^2-	β^0	0.15		(3)
Na+ - Sr^2+	β^0	.095		(3)
Sr^2+ - OH-	θ	.07		(3)
	β^0	-.061		(3)
	β^1	1.655		(3)
Sr^2+ - Cl-	β^0	.286		(6)
	β^1	1.67		(6)
	C^ϕ	-.00065		(6)
Sr^2+ - NO3-	β^0	.135		(6)
	β^1	1.38		(6)
	C^ϕ	-.01		(6)
Sr^2+ - SO4^2-	β^0	0.20		(7)
	β^1	3.1973		(7)
	C^ϕ	-54.24		(7)
Ca^2+ - OH-	β^0	-.1747		(18)
	β^1	-.2303		(18)
	β^2	-5.72		(18)
OH- - CO3^2-	θ	0.1		(18)
OH- - CO3^2- - Na+	ψ	-.017		(18)
OH- - NO3-	θ	-.00005		(16)
NO3- - CO3^2-	θ	0.14		(19)
NO3- - NaEDTA^3-	θ	0.12		(11)
NO3- - HEDTA^3-	θ	0.17		(12)

Table A-II. Logarithms (base 10) of the Thermodynamic Equilibrium Constants of Aqueous Phase Association Reactions and Solid Phase Dissolution Reactions (K_{sp}) Used in this Study

Species	Temperature (°C)	Log K^0	Reference
SrEDTA^{2-}	22-23	10.45	(11)
	50	10.36	(11)
	75	10.25	(11)
SrHEDTA$^-$	22-23	7.75	(12)
	50	7.67	(12)
	75	7.60	(12)
NaEDTA^{3-}	22-23	2.70	(11)
	50	2.60	(11)
	75	2.50	(11)
CaEDTA^{2-}	22-23	12.36	(11)
CaHEDTA$^-$	22-23	9.20	(12)
SrCO$_3$(c)	25	-9.27	(9)
	50	-9.37	(9)
	75	-9.59	(9)
SrSO$_4$(c)	25	-6.62	(7)
SrF$_2$(c)	25	-8.58	(5)
CaCO$_3$(c)	25	-8.41	(18)
CaCO$_3$(aq)	25	3.15	(18)
SrCO$_3$(aq)	25	2.81	(9)
Sr(CO$_3$)$_2^{2-}$	25	3.31	(3)
Ca(CO$_3$)$_2^{2-}$	25	3.88	(3)
NaNO$_3$(aq)	25	-1.04	(16)
	50	-0.53	(16)
	75	-0.58	(16)

Environmental Sensing
and Monitoring

Chapter 12

Environmental Monitoring Using Microcantilever Sensors

Thomas G. Thundat[1], Hai-Feng Ji[2], and Gilbert M. Brown[3]

[1]Life Sciences Division, Oak Ridge National Laboratory,
Oak Ridge, TN 37831
[2]Department of Chemistry and Institute for Micromanufacturing,
Louisiana Tech University, Ruston, LA 71272
[3]Chemical Sciences Division, Oak Ridge National Laboratory,
Oak Ridge, TN 37831

The environmental remediation program of the U.S. Department of Energy needs rugged, low-cost sensing systems for real-time, in situ chemical, physical, and radiological sensors in the characterization and monitoring of mixed waste, groundwater, contaminated soil, and process streams. At the present time an array of sophisticated instrumentation is frequently required to simultaneously achieve the desired selectivity and sensitivity for the complicated samples and matrices encountered in the environment. These instruments may use a preliminary separation step followed by a sensitive but not necessarily selective detector. These limits may soon be alleviated with the advent of new micromechanical sensors that are under development. These miniature sensors are extremely sensitive and highly selective. In addition, these sensors work with ease under solution. An array of microcantilever sensors will have applications ranging from monitoring plume containment and remediation to determination of location, chemical composition, and levels of contaminants.

Introduction

The unprecedented sensitivity of microcantilever physical, chemical, and biological sensors, as demonstrated by our group at Oak Ridge National Laboratory (ORNL) and many other laboratories around the world, suggests that in the years to come, microcantilever sensors will be an integral part of many sensor devices (1–5). The microcantilever concept will serve as a general platform for a myriad of extremely sensitive, highly selective real-time microsensors that can be mass-produced using conventional techniques.

The resonant frequency of microcantilevers, such as those used in scanning force microscopy, varies as a function of molecular adsorption. In addition, cantilevers undergo bending due to mechanical forces involved in molecular adsorption—one of the most overlooked, yet fascinating aspects of adsorption. A literature survey has shown that such forces were observed as early as 1858. In 1909, Stoney derived a relation between surface stress and the radius of curvature of the substrate (6). Surprisingly, until recently adsorption-induced stresses did not receive much attention and the concept remained obscure for over one hundred years. In fact, almost all the scientific interest in adsorption is in the context of adsorption energy rather than adsorption-induced force. These adsorption-induced forces are so large that on a clean surface they can rearrange the lattice locations of surface and sub-surface atoms, producing surface reconstructions and relaxations.

Although the mechanics of adsorption-induced surface stress have been understood for almost a century, no attempt has been made to exploit this concept for practical sensor applications—until now. We have observed that adsorption-induced forces can be easily detected on so called "real surfaces" such as the surface of a microcantilever. Using this concept, we have demonstrated the feasibility of chemical detection of a number of vapor phase analytes with sub parts-per-billion sensitivity. The primary advantages of the microcantilever method are sensitivity based on the ability to detect cantilever motion with subnanometer precision and the ability to micromachine into a multi-element array sensor. Another advantage is that, since cantilevers have an extremely small thermal mass, they can be heated to hundreds of degrees in less than a millisecond for regeneration purposes. No other technology offers such versatility.

Advantage has been taken of the nanomechanical effects of molecular adsorption, confined to one surface of a microcantilever, to utilize these devices for sensors. Surface adsorption can be characterized as either physisorption or chemisorption. Physisorption is a result of, among other effects, London-type dispersion forces resulting from an induced dipole and induction forces caused by surface electric fields. On the other hand, chemisorption is accompanied by charge transfer between adsorbed molecules and the surface or the presence of covalent bonds between the adsorbate and the surface atoms. The mechanical strengths of bonds are such that hydrogen bonds are of the order of magnitude 10^{-11} N, whereas covalent bonds have strength of the order of magnitude 10^{-9} N. The nanomechanical manifestation of these adsorption forces is a tangential force component. If the spring constant of the microcantilever beam is of the same order of magnitude as the free energy change accompanying adsorption, the result is bending of the microcantilever.

The deflections of a microcantilever can be measured with sub-Angstrom resolution using current techniques perfected for the AFM, such as optical reflection, piezoresistive, capacitance, and piezoelectric detection methods. One great advantage of the cantilever technique is that five resonance response parameters (resonance frequency, phase, amplitude, Q-factor, and deflection) can be simultaneously detected.

Although microcantilevers offer extremely high sensitivity, they are inherently unselective to molecular absorption. Selective sensors can be prepared, however, by coating or covalently binding a molecular recognition agent (a molecule or a polymer that has a strong and specific noncovalent interaction with a guest molecule) to one surface of a microcantilever. Adsorption or intercalation of the analyte markedly changes the differential surface stress of the microcantilever and can be detected as changes in the cantilever deflection. Because adsorption can be reversible or irreversible depending on the coating chemistry, both instantaneous and dosimetric responses are possible, and both types have been experimentally realized in the work to be discussed in this paper.

Advances in surface chemistry and molecular recognition, especially as developed for other sensors (e.g., quartz crystal microbalance, surface acoustic wave devices, and chemical field effect transistors), serve as guides in developing chemically selective coatings for microcantilever-based ones (*7, 8*). The challenge lies in adapting this chemistry to the microcantilever platform while maintaining chemical selectivity, sensitivity, reproducibility, and stability. Chemically selective coating strategies applicable to microcantilevers include self-assembly (e.g., alkanethiol or organosilane films), dip coating (e.g., polymeric resins), adsorption/evaporation (e.g., sol-gel matrices), or direct covalent attachment of molecular receptors.

Another compelling feature of microcantilever sensors is that they can be operated in air, vacuum, or liquid (*9*). The damping effect in a liquid medium, however, reduces the resonance response of a microcantilever. In most liquids, the observed resonance response is approximately an order of magnitude smaller than that in air. The bending response, however, remains unaffected by the presence of a liquid medium (*10–14*). Therefore, the feasibility of operating a microcantilever in a solution with high sensitivity makes the microcantilever an ideal choice for biosensors. In addition, highly selective biochemical interactions can be used to regulate molecular adsorption and tune microcantilever response.

Microcantilever sensors offer improved dynamic response, greatly reduced size, high precision, and increased reliability compared with conventional sensors. They are the simplest micromechanical systems that can be mass-produced with conventional micromachining techniques. They can be fabricated into multi-element sensor arrays and fully integrated with on-chip electronic circuitry. Because the thermal mass of microcantilevers is extremely small, they can be heated and cooled with a thermal time constant of less than a millisecond. This is advantageous for rapid reversal of molecular absorption processes and regeneration purposes. Therefore, the micromechanical platform offers an unparalleled opportunity for the development and mass production of extremely sensitive, low-cost sensors for real-time, in situ sensing of many chemical and biological species.

Adsorption-Induced Bending of Microcantilevers

As pointed out in the last section, microcantilever deflection changes as a function of adsorbate coverage. The exact mechanism of adsorbate-induced bending still remains to be solved. By manipulating the surface stress directly or factors controlling the surface stress, one would be able to increase the sensitivity and selectivity of the sensor performance.

Microcantilevers, such as those used in our experiments, have an intrinsic deflection due to unbalanced stresses on the opposing surfaces. For example, cantilevers with thin films (many atomic layers) of another material show bending. Although bending can be expected for films of many atomic layers due to differences in physical parameters such as elastic and lattice constants, bending due to submonolayer coverages as small as 10^{-3} monolayers (ML) is not intuitive. One monolayer of gold surface has 1.5×10^{15} atoms/cm^2 while on Si(111) surface the density is 7.4×10^{14} atoms/cm^2. Therefore, 10^{-3} ML corresponds to approximately 10^8 atoms on the surface of the cantilever.

One way to explain the cantilever bending is by using free energy changes involved in adsorption. Change in free energy is equivalent to change in the surface stress. Surface stress is one of the most significant properties of a solid surface. However, measuring surface stress of a solid is extremely difficult. Measuring variation in surface stress of a solid due to adsorption is even more challenging. Due to these technical challenges, accurately measuring the variation in surface stress still remains unexplored.

Microcantilever deflection changes as a function of adsorbate coverage when adsorption is confined to a single side of a cantilever (or when there is differential adsorption on opposite sides of the cantilever). Using Stoney's formula (6), we can express the radius of curvature of cantilever bending due to adsorption as

$$\frac{1}{R} = \frac{6(1-\upsilon)}{Et^2}\delta\sigma \qquad (1)$$

where R is the cantilever's radius of curvature; υ and E are Poisson's ratio and Young's modulus for the substrate, respectively; t is the thickness of the cantilever; and $\delta\sigma$ is the differential surface stress. Since we do not know the absolute value of the initial surface stress, we can only measure the variation in surface stress. The differential surface stress is the difference between the surface stress of the top and bottom surfaces of the cantilever beam in units of N/m or J/m^2. Typically, the microcantilevers have spring constants in the range of 0.1 N/m. Using geometry, a relationship between the cantilever displacement and the differential surface stress can be expressed as:

$$h = \frac{3L^2(1-\upsilon)}{Et^2}\sigma \qquad (2)$$

where L is the length and h is the deflection. Eq 2 shows a linear relation between cantilever bending and differential surface stress. Therefore, any variation in the differential surface stress can result in cantilever bending. When confined to one surface, molecular adsorption on a thin cantilever can cause large changes in surface stress.

Operation under Liquid

One of the challenges of most mass sensors is operation under liquid. Vibrations damp very fast in liquid due to viscosity of the liquid. Therefore, the sensitivity of a resonating sensor is significantly decreased under solution (the higher the frequency of operation, the higher the loss). Since the microcantilevers are low frequency devices, the loss is not very significant. The resonance frequency shifts by a factor of ten when operated under water. For example, a cantilever with a resonance frequency of 24 kHz in air shows a frequency of 5.5 kHz under water. The quality factor and the amplitude of vibration also decrease by a factor of 10 in solution. However, in these experiments the variation in resonance frequency was within the noise limit. Therefore, we utilized the frequency only as a characterization tool for integrity of the cantilever. The resonance frequency of the cantilever due to Brownian motion shows that the cantilever was undamaged during surface modification.

One of the most significant aspects of the cantilever sensor is that the adsorption-induced cantilever bending is not affected by operation in solution. We were able to detect the Cs^+ ion at concentrations as small as 10^{-12} M in solution using the cantilever deflection method. This corresponds to a coverage of 4×10^{-4} monolayers!

Experimental

To demonstrate groundwater monitoring using cantilever sensors, we have detected a number of ions such as Cs^+ (15), CrO_4^{-2} (16), Cu^{2+} (17), and Hg^{2+} (18). Chemical selectivity is achieved by using self-assembled monolayers (SAM) coupled with molecular recognition agents. Differential stress was realized by making the cantilevers bimaterials with a gold layer on one side. The gold side of the cantilever was modified with the chemical recognition SAM.

Cantilevers

Silicon or silicon nitride cantilevers, such as those used in atomic force microscopy, are typically 100–200 μm long, 20–40 μm wide, and 0.6-μm thick (available from Digital Instruments, CA, and Park Scientific, CA). We have used commercially available silicon microcantilevers (Park Instrument, CA) in these experiments. The dimensions of the V-shaped microcantilevers were 200-μm length, 20-μm width, and 0.7-μm thickness. One side of the cantilever had a thin film of chromium (3 nm) followed by a 40-nm layer of gold deposited by e-beam evaporation.

Cantilever Excitation

The cantilevers are set into vibration by a piezoelectric cantilever holder driven by an external alternating current (ac) voltage. The cantilevers, however, also resonate in response to ambient conditions such as room temperature or acoustic noise without requiring any external power. Since our detection concept was based on cantilever bending mode, we did not excite the cantilever. However, the Brownian motion frequency was determined using a spectrum analyzer as a diagnostic tool for cantilever integrity.

Cantilever Readout

The most common readout technique for cantilever motion is the optical beam deflection technique, which can detect cantilever motion with sub-Angstrom resolution. In the optical beam deflection technique, a light beam from a diode laser is focused at the end of the cantilever. The reflected beam is then allowed to fall on a position sensitive detector (PSD). The cantilever motion can be recorded by laser deflection with the output of the photodetector sent to a recording device. The bending of the cantilever changes the radius of curvature of the cantilever, resulting in a large change in the direction of the reflected beam. A single detection technique can be used for measuring the resonance frequency, resonance amplitude, and bending of the cantilever simultaneously. The ac signal in the PSD corresponds to the frequency of vibration of the cantilever, while the direct current signal is proportional to cantilever bending. However, since the PSD is measuring the curvature rather than the displacement, higher modes of cantilever resonance will have higher amplitudes.

Flow-Through System

All the experiments were done under fluid flow control. The modified cantilevers were mounted in the liquid cell (~250 μL volume) and allowed to equilibrate in solution until a stable baseline was achieved for flow-through experiments (all at 25 °C; 2 mL/h to 10 mL/h flow rate for flow experiments). Flow was controlled using a syringe pump (IITC, Inc., Woodland Hills, CA) equipped with a low pressure liquid chromatography injector valve and injection loop (Upchurch Scientific, Oak Harbor, WA). For flow experiments, the analyte solution was injected into the flow stream using a 1-ml injection loop. The PSD signal voltage change due to cantilever deflection (analyte adsorption on the cantilever) was monitored in situ throughout each experiment (1 V = 100 nm deflection).

Cantilever Modification

For achieving chemical selectivity, the cantilevers are modified with appropriate chemically selective coatings. The appropriate self-assembled monolayers were immobilized on the gold side of the cantilever as discussed below. In all applications, the Si cantilevers (with Au on one side) were cleaned in piranha solution (7:3 H_2SO_4 (96%):H_2O_2 (30%)) for 1 min and rinsed with H_2O (3 times) and EtOH (2 times) before measurement or SAM preparation.

(Caution: Piranha solution reacts violently with many organic materials and should be handled with great care.)

Surface Modification for CrO₄²⁻ Ion detection

For CrO_4^{2-} detection we have used tri-N-ethyl-12-mercaptododecylammonium bromide. The formation of tri-N-ethyl-12-mercaptododecylammonium bromide SAM on the gold-coated cantilever was achieved by immersing the cantilever into 10^{-3} M solution of tri-N-ethyl-12-mercaptododecylammonium bromide in EtOH for seven days. The cantilevers were rinsed in EtOH three times and air dried. The predicted surface structure of the modified microcantilever used in this experiment is shown in Figure 1.

Surface Modification for Cs+ Ion Detection

The cesium recognition agent used in this work was 1,3-alternate 25,27-bis(11-mercapto-1-undecanoxy)-26,28-calix[4]benzocrown-6 (see Figure 2), bound to a gold-coated microcantilever. The crown cavity of the 1,3-alternate conformation of calix[4]benzocrown-6 has been shown to be very suitable for accommodating cesium ions, with Cs/Na and Cs/K selectivity ratios in excess of 10^4 and 10^2, respectively, determined using a solvent extraction technique. Binding constant values of 10, 2×10^4, and 2.5×10^6 have also been determined by a fluorescence technique for Na^+, K^+, and Cs^+ in 1:1 methanol/methylene chloride solution. Based on the high selectivity exhibited by these compounds, the receptor molecule shown in Figure 2 was designed and anchored onto the gold surface of the microcantilever by standard techniques. 1-Decanethiol was co-absorbed onto the gold surface in a 2:1 ratio to fill the gaps present between the two alkyl thiol arms shown in Figure 2 and the adjacent molecules and to enhance the aqueous stability of the SAM.

Surface Modification for Cu²⁺ Ion Detection

A SAM of L-cysteine on a gold-coated cantilever was demonstrated to be selective for Cu(II) in a pH 5 phosphate buffer. The SAM forms upon immersion of the gold-coated cantilever in a 10^{-3} M solution of L-cysteine with the pH adjusted to the isoelectric point, pH 5, using a 10^{-2} M phosphate buffer for 12 hrs. This was followed by rinsing three times with the pH 5 phosphate buffer. The surface structure, shown in Figure 8, is thought to involve a Cu(II) bound by two L-cysteine molecules, each L-cysteine having bidentate coordination to the Cu(II).

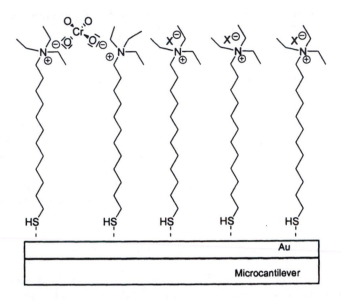

Figure 1. *Schematic representation of the molecular structure of chromate sorbed to a SAM of the receptor molecule triethyl-12-mercaptododecyl-ammonium bromide on the gold surface of microcantilever.*

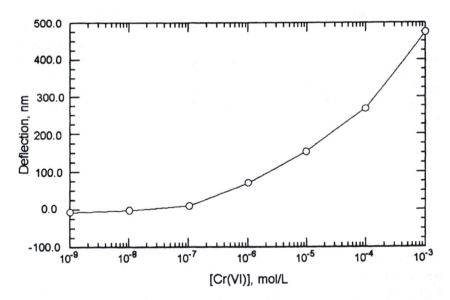

Figure 2. *Bending response for a silicon cantilever coated with self-assembled monolayer of triethyl-12-mercaptododecyl-ammonium bromide on the gold surface of a microcantilever as a function of the change in concentration of* CrO_4^{2-}.

Results and Discussions

Cr Detection

Surface modified microcantilevers were tested (*16*) for CrO_4^{2-} by initially exposing the cantilever to a constant flow (10mL/h) of distilled water, and its deflection was set as the baseline. The CrO_4^{2-} containing solution was then directed through the cell also at 10mL/h flow rate. The cantilever bent up (toward the gold-coated surface) in response to the CrO_4^{2-}, and it reached an equilibrium position within 100 s. It takes approximately 100 s to flush the cell volume once at 10 mL/h, thus the flow rate and not the binding kinetics currently limit the sensor response time. Although the response improved at higher flow rate, the current setup is not designed for flow dynamics, and higher flow rates resulted in increased noise. During adsorption of the CrO_4^{2-} ions on the surface, they displace the halide ions, combining with the cationic ammonium group to form a stable ionic association, altering the surface stress, resulting in a cantilever deflection.

The magnitude of the cantilever deflection varies with the concentrations of CrO_4^{2-} ions, the solution concentration being varied in the range 10^{-10} M to 10^{-3} M. The response time to CrO_4^{2-} ion also depends on the CrO_4^{2-} ion concentration. When the solution was switched from CrO_4^{2-} to water, the deflection of the cantilever bending decreased back to the original position. Similar results were observed when using 10^{-3} M NaCl as the background electrolyte with both the CrO_4^{2-} solution and the flushing solutions. In the absence of added electrolyte, it is not clear what anion remains associated with the surface bound quarternary ammonium cation, but the important point is that the deflection response of the cantilever to changes in CrO_4^{2-} concentration is reversible. The deflection response depends on the mechanical properties of the cantilever as well as the concentration of CrO_4^{2-} in solution. Two closely matched cantilevers (as characterized by resonance frequency) deflected 270 and 285 nm in response to 1×10^{-4} M CrO_4^{2-}, indicating the measurements are reproducible between cantilevers, if the mechanical properties and surface coatings are optimized.

During CrO_4^{2-} ion adsorption the SAM undergoes a tensile stress. Adsorption of CrO_4^{2-} decreased the surface stress with respect to the opposite side creating a bending towards the SAM side. An isotherm of equilibrium bending as a function of CrO_4^{2-} ion concentration is shown in Figure 3. The plot

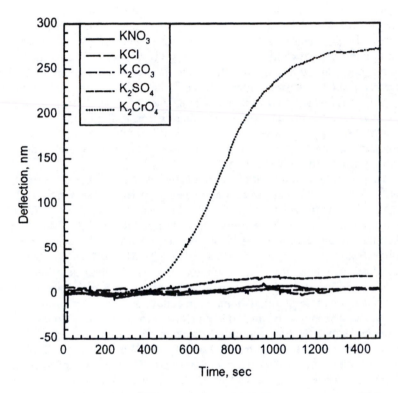

Figure 3. Bending response of the SAM coated microcantilever to CrO_4^{2-} and other anions at the same concentration ($1x10^{-4}M$) in water.

in Figure 3 shows that the deflection of a microcantilever can be observed at a CrO_4^{2-} concentration as low as 10^{-9} M, which indicates a higher sensitivity than ion-selective electrode (ISE) or extraction methods. The data in Figure 3 indicates the adsorption isotherm for CrO_4^{2-} on the SAM modified microcantilever is nonlinear, indicating a cooperative effect at the highest concentration investigated.

During CrO_4^{2-} ion adsorption the SAM undergoes a tensile stress. The triethyl-12-mercaptododecylammonium bromide modified cantilever bends up toward the gold side of cantilever (tensile stress on the SAM) upon sorption of CrO_4^{2-}. As illustrated in scheme 1, the sorption of CrO_4^{2-}, dissolved in solution, at the modified surface requires two quaternary ammonium molecules, sorbed to the cantilever surface, to maintain electrical neutrality.

$$CrO_4^{2-}(soln) + 2\ HS\text{-}R\text{-}NEt_3^+\text{-}Br^-\ (surface) \rightarrow$$
$$2\ Br^-(soln) + 2\ HS\text{-}R\text{-}NEt_3^+\text{-}CrO_4^{2-}\ (surface) \qquad scheme\ (1)$$

This ionic association changes the stress on the surface. Two singly charged anions are being replaced with a divalent anion. A compact layer of adsorbed anions (Helmholtz layer) will exist at the cantilever surface, analogous to the double layer at the surface of a charged electrode, and a more diffuse Stern layer will be present in the adjacent solution.

The high sensitivity is due in part to the extraordinarily large deflection response due to Cr(VI) sorption. At the highest concentration of CrO_4^{2-} investigated (10^{-3} M), the bending response of the microcantilever was 0.25% of the length of the cantilever. The experiment could not be continued with higher concentrations of CrO_4^{2-} ions because the bending signal reached the maximum limit for the position sensitive detector. Although monolayer adsorption was expected to result in a saturation (flattening) of cantilever bending, the lack of a saturation plateau in Figure 3 cannot be interpreted as due to multilayer formation.

The selectivity of this sensor for CrO_4^{2-} ions over other anions was investigated. Figure 4 shows the deflections of a modified cantilever to the same concentrations (1×10^{-4} M) of CrO_4^{2-}, NO_3^-, Cl^-, CO_3^{2-} (and HCO_3^-), and SO_4^{2-}, and the bending response of most of the anions was insignificant. The SO_4^{2-} anion causes approximately 20-nm deflection of the cantilever, but it is smaller than the effect of CrO_4^{2-}.

The CrO_4^{2-} selectivity of this cantilever can be explained by an electrostatic model of anion transfer from the aqueous phase to the surface of microcantilever, where the anions associated with the SAM go into the aqueous solution and the CrO_4^{2-} anions becomes associated with quaternary ammonium cations. This situation is analogous to sorption in an ion exchange resin. The influence of hydration on resin anion exchange was first discussed by Diamond and Whitley (19). The hydration energy of CrO_4^{2-} is smaller than other anions,

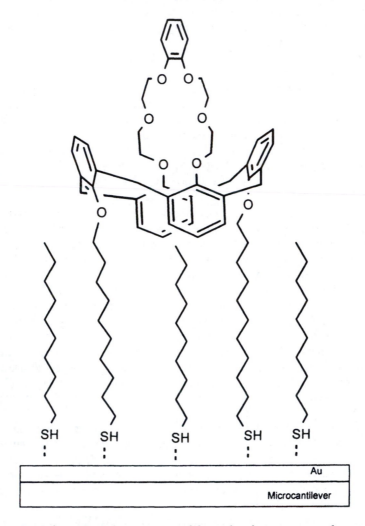

Figure 4. Schematic representation of the molecular structure of receptor molecule 1,3-alternate 25,27-bis(11-mercapto-1-undecanoxy)-26,28-calix[4]-benzocrown-6 co-absorbed with 1-decanethiol on the gold surface of a microcantilever.

such as Cl⁻, Br⁻, SO_4^{2-}, HCO_3^-, and CO_3^{2-} that are commonly found in groundwater. Low hydration energy (which generally decreases as the size of the anion increases) favors the exchange of the anion associated with the SAM to the higher dielectric medium of the aqueous solution. These anions with higher hydration energies appear to have a weaker association with the quaternary ammonium cations of the SAM (16).

Cs⁺ Ions

Cesium ions have been shown to have a high affinity for derivatives of 1,3-alternate-calix[4]arene-benzocrown-6. This chemistry was utilized in designing a microcantilever sensor to detect Cs⁺ ions in a flowing solution (15). The high affinity of this ion-selective SAM-coated microcantilever for Cs⁺ allows it to be capable of detecting cesium ions in the presence of high concentrations of potassium or sodium ions. Our data show that the sensitivity of this cantilever-based sensor for in situ measurements is several orders of magnitude better than the sensitivities of currently available ISE.

Figure 5 shows the time dependence of the bending response of the SAM-coated cantilevers to Cs⁺ concentration. The most dramatic change in bending response occurs when the concentration of cesium is in the range of 10^{-7} to 10^{-11} M. In this same general concentration range, the cantilever response to K⁺ (the most prevalent interfering ion) is relatively small at short response times (e.g., 20-nm deflection for 1×10^{-8} M solution of K⁺ compared to 330-nm deflection for Cs⁺ at the same concentration). The equilibrium bending response of the cantilever for Cs⁺ concentrations > 10^{-6} M reaches its maximum value at around 330 nm (Figure 5). A blank test performed on a gold-coated silicon nitride cantilever without the SAM revealed that even at high Cs⁺ concentration (e.g., 10^{-3} M), the cantilever showed negligible bending response, indicating that the molecular recognition agent was required for bending.

A direct comparison of the bending response of the SAM-coated microcantilever for Cs⁺, K⁺, and Na⁺ complexation was also conducted for the same concentration of each ion (10^{-5} M) (Figure 6). The results indicated that the SAM-coated microcantilever was much more selective towards Cs⁺ ions than towards K⁺ and Na⁺ ions. A relationship between the cantilever displacement, h, and the molar concentration of Cs⁺ can be obtained by manipulating Stoney's equation (eq 1). Since surface stress is directly proportional to ion absorption by the microcantilever, eq 2 can be rearranged to the form shown in eq 3,

$$h = B(\frac{3(1-v)L^2}{t^2 E})(\frac{K[M]R_0}{1+K[M]}) \tag{3}$$

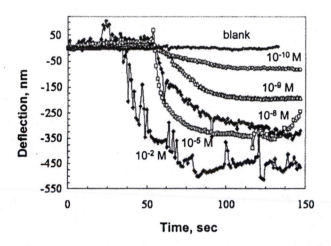

Figure 5. Bending response of three SAM coated microcantilevers as a function of time after exposure to different concentrations of Cs^+ ions in solution.

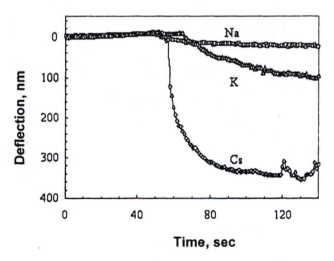

Figure 6. Microcantilever deflection response for Cs^+, K^+, and Na^+ ions (10^{-5} M).

where B is a constant, K is the complexation constant between the ion receptor and the ions present in solution (1:1 stoichiometery), $[M]$ is the concentration of ions in solution, and R_0 is the moles of ion receptor present on the cantilever before complexation. The concentration of ions in solution is essentially unchanged by adsorption on the cantilever, and the concentration of the metal ion $[M]$ will be the same as the initial concentration of ions in solution. The complexation constants determined from the plot of $1/h$ vs $1/[M]$ for cesium (K_{Cs}) and potassium (K_K) using eq 3 are 2×10^9 M^{-1} and 1.6×10^7 M^{-1}, respectively. These values are much higher (by three orders of magnitude) than the corresponding values observed for the derivatives of similar compounds in 1:1 methanol/methylene chloride solution, but the ratio K_{Cs}/K_K is essentially the same. The value for R_0 was chosen from geometric considerations and the assumption of a monolayer, and it is possible that this may be inappropriate. Similar enhancements in the association constant values for other SAM systems (relative to the free molecule) have also been reported. The large value of association constant observed for Cs^+ ions indicates that binding is essentially irreversible. Cycling the loaded cantilever with pure water for several hours failed to regenerate the sensor. Based on solvent extraction experiments, it is expected that a dilute nitric acid solution would have removed (or stripped) the cesium from the SAM. Na^+ ions have a minimal effect (if any) on the bending response, while K^+ ions exhibit enough sensitivity to interfere (as a perturbing ion when present) in the detection of cesium ions.

Detection of Cs^+ In Tank Waste Simulant

We have carried out detection of Cs ions in a tank waste simulant using silicon nitride cantilevers. The cantilevers were vacuum-deposited with a 40-nm gold on one side. The 3-nm Cr layer was served as the adhesion layer between the silicon nitride cantilever and gold film. The tank waste simulant had a pH of 14.19. The simulant consists of NaOH (3.3 M), $Al(NO_3)3.9H_2O$ (0.4 M), sodium dichromate (7×10^{-3} M), Na_2CO_3 (1.3 M), $NaNO_3$ (1.3 M), KNO_3 (1.7 M), Na_2SO_4 (0.2 M), NaCl (0.1 M), NaF (0.05 M), and $NaNO_2$ (0.8 M). The concentration of Cs^+ ions in the solution was 10^{-5} M. The concentration of total and free OH^- was 3.2 M and 1.5 M respectively. Figure 7 shows the cantilever response in tank waste simulant. The cantilever response in tank waste shows an increased noise probably due to high pH value of the solution.

Cu²⁺ Ions

Electrochemical sensors for Cu^{2+}, based on the complexation to an L-cysteine self-assembled monolayer sorbed on a gold electrode, have been reported (20—23). For these sensors the current for the chemically reversible reduction of Cu(II) to Cu(I) is used to quantify this ion. We have taken advantage of these developments and have observed that a microcantilever sensor based on an L-cysteine monolayer, sorbed to the gold surface, is effective for the in situ detection of Cu^{2+}.

The anticipated surface structure of the L-cysteine SAM on gold and its complex with Cu^{2+} is shown in Figure 8. Electrochemical and XPS data suggest one Cu per two cysteine ligands (23). Literature reports indicate that the best complexation performance for the electrochemical sensors was at pH = 5, the isoelectric point of L-cysteine. Similar optimal performance at pH = 5 was also observed for microcantilever deflection experiments. The deflection amplitude caused by Cu^{2+} complexation with the L-cysteine monolayer was much less for experiments performed at pH = 3 or pH = 7, than at pH = 5.

L-cysteine coated cantilevers were initially exposed to 10^{-2} M phosphate buffer (pH = 5.0) and equilibrated to obtain a stable baseline (i.e., 0 nm deflection). Once equilibrated, solutions having a range of Cu^{2+} concentrations (10^{-10} to 10^{-3} M, pH = 5.0) were injected into the fluid cell, each time using a new cantilever, to study the interaction of Cu^{2+} with the SAM. The cantilever rapidly bent downwards for approximately the first 5 min when a 10^{-5} M solution of Cu^{2+} was injected into the fluid cell. At this concentration the bending rate leveled off after approximately 15 min, as shown in Figure 9, appearing to reach equilibrium after about 50 min. A period of approximately 15 min was required for the cantilever to return to its original position after replacing the copper solution with phosphate buffer, suggesting that the interaction and binding between L-cysteine and Cu^{2+} was very strong.

Figure 10 shows that the cantilever deflections varied at different concentrations of Cu^{2+} in solution (10^{-10} to 10^{-3} M). With increasing concentration, the cantilevers bent down with systematically steeper bending rates. As shown in Figure 10, bending rates and amplitudes upon complexation with Cu^{2+} at 10^{-3} M and 10^{-4} M suggested that the Cu^{2+} sorbance on the L-cysteine monolayer had reached its maximum. Bending rates and amplitudes indicated that the cantilever was most sensitive under these conditions for concentrations ranging from 10^{-7} to 10^{-3} M.

Better sensitivity was achieved with increased time of exposure of the cantilever to the Cu^{2+} containing solution. Figure 11 shows the deflection recorded at specific times after injection (5 min, 10 min, 15 min, and 30 min) versus different concentrations of Cu^{2+}. The plot shows that the deflection of a modified microcantilever can be used as a Cu^{2+} sensor with concentrations as low as 10^{-10} M, if the user is willing to wait at least 30 min. Sensitivity to

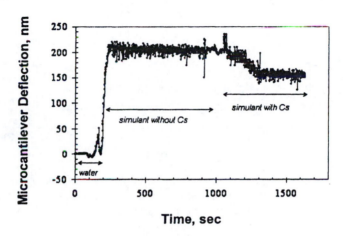

Figure 7. Detection of Cs⁺ in a tank waste simulant (pH 14) with and without Cs⁺ ions. The noise in the response can be reduced by using cantilevers of inert materials such as SiC.

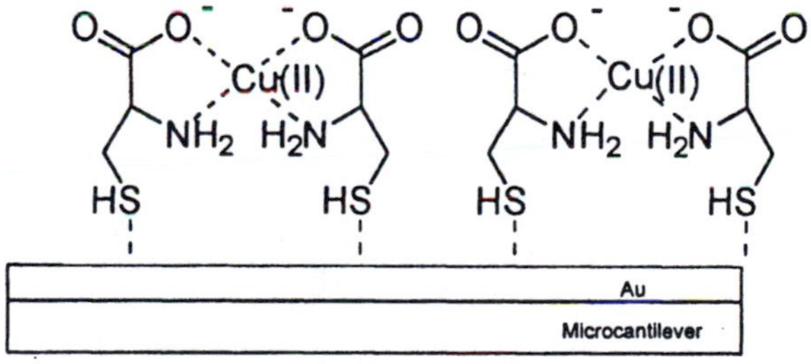

Figure 8. Schematic representation of the molecular structure of the L-cysteine self-assembled monolayer on the gold surface of the microcantilever and its complexation with Cu^{2+}.

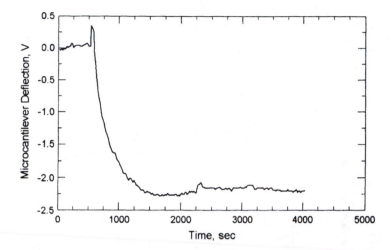

Figure 9. *The bending deflection response as a function of time, t, for a silicon cantilever coated with a self-assembled monolayer of L-cysteine on the gold surface, after injection of 10^{-5} M concentration of Cu^{2+} solution in 10^{-2} M phosphate buffer at pH = 5.0.*

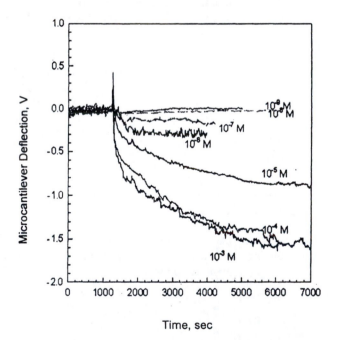

Figure 10. *The bending deflection response as a function of time, t, for a silicon cantilever coated with a self-assembled monolayer of L-cysteine on the gold surface before and after exposure to the different concentrations of Cu^{2+} solutions in 10^{-2} M phosphate buffer at pH = 5.0.*

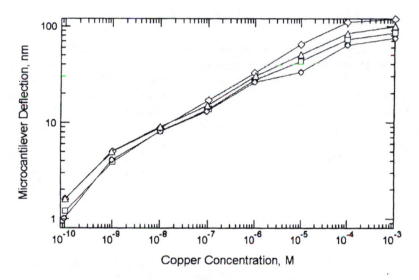

Figure 11. Bending response of a silicon cantilever coated with self-assembled monolayer of L-cysteine on the gold surface of a microcantilever as a function of the change in concentration of Cu^{2+} in 10^{-2} M phosphate buffer at pH = 5.0 at different times. Circles, five minutes; squares, ten minutes; triangles, fifteen minutes; diamonds, thirty minutes.

10^{-10} M is comparable to the detection limits that have been reported elsewhere for L-cysteine-modified surface-based copper sensors (20–23). The effect of exposure of the L-cysteine modified cantilever to the ions Na^+, K^+, Pb^{2+}, Zn^{2+}, Cd^{2+}, Ni^{2+}, and Ca^{2+}, at a concentration of 10^{-5} M, on cantilever deflection was studied. None of these cations caused much deflection of the cantilever, while the response to Cu^{2+} was significant. This suggested that this L-cysteine modified cantilever was a good Cu^{2+} sensor with both high sensitivity and selectivity.

Conclusions

We have demonstrated an extremely sensitive sensor platform for groundwater monitoring. The sensitivity of the cantilever sensor depends on cantilever dimension, while the selectivity depends on the selectivity of the surface coating for chemical interactions. Research is underway to develop cantilever arrays for simultaneous multi-analyte detection. The primary advantages of the microcantilever method are (1) the sensitivity of microcantilevers based on their ability to detect cantilever motion with subnanometer precision; (2) their ability to be fabricated into multi-element sensor arrays; and (3) their ability to work in a liquid environment.

Acknowledgments

We would like to thank Drs. E. Finot, Z. Hu, and A. Mehta for help with experiments. Support from the DOE Office of Biological and Environmental Research and the Environmental Management Science Program is gratefully acknowledged. Oak Ridge National Laboratory is managed by UT-Battelle, LLC, for the U.S. Department of Energy under contract DE-AC05-00OR22725.

References

1. Thundat, T; Warmack, R. J.; Chen, G. Y.; Allison, D. P. *Appl. Phys. Lett.* **1994**, *64*, 2894–2896.
2. Gimzewski, J. K.; Gerber, Ch.; Mayer, E.; Schlitter, R. R. *Chem. Phys. Lett.* **1994**, *217*, 589.
3. Chen, G. Y; Thundat, T.; Wachter, E. A.; Warmack, R. J. *J. Appl. Phys.* **1995**, *77*, 3618–3622.
4. Thundat, T.; Chen, G. Y.; Warmack, R. J.; Allison, D. P.; Wachter, E. A. *Anal. Chem.* **1995**, *67*, 519–521.

5. Thundat, T.; Wachter, E. A.; Sharp, S. L.; Warmack, R. J. *Appl. Phys. Lett.* **1995**, *66*, 1695–1697.
6. Stoney, G. G. The Tension of Metallic Films Deposited by Electrolysis. *Proc. R. Soc. London, Ser. A* **1909**, *82*, 172–175.
7. Grate, J. W.; Snow, A.; Ballantine, D. S.; Wohltjen, H.; Abraham, M. H.; McGill, R. A.; Sasson, P. *Anal. Chem.* **1998**, *60*, 869–875.
8. McGill, R. A.; Abraham, M. H.; Grate, J. W. *CHEMTECH* **1994**, *24*, 27–37.
9. Oden, P. I.; Chen, G. Y.; Steele, R. A.; Warmack, R. J.; Thundat, T. *Appl. Phys. Lett.* **1996**, *68*, 3814–3816.
10. Butt, H. J. *J. Colloid Interface Sci.* **1996**, *180*, 251–260.
11. Raiteri, R.; Butt, H. J. *J. Phys. Chem.* **1995**, *99*, 15728–15732.
12. Fritz, J.; Baller, M. K.; Lang, H. P.; Rothuizen, H.; Vettiger, P.; Mayer, E.; Guntherodt, H. J.; Gerber, Ch.; Gimzewski, J. K. *Science* **2000**, *288*, 316–318.
13. Wu, G.; Datar, R.; Hansen, K. M.; Thundat, T.; Cote, R. J.; Majumdar, A. *Nature Biotechnology* **2001**, *19*, 856–860.
14. Hansen, K. M.; Ji, H. F.; Wu, G.; Datar, R.; Cote, R.; Majumdar, A.; Thundat, T. *Anal. Chem.* **2001**, *73*, 1567–1571.
15. Ji, H. F.; Finot, E.; Dabestani, R.; Thundat, T.; Britt, P. F.; Brown, G. M. *Chem. Commun.* **2000**, *6*, 457–458 .
16. Ji, H. F.; Thundat, T.; Dabestani, R.; Brown, G. M.; Britt, P. F.; Bonnesen, P. V. *Anal. Chem.* **2001**, *73*, 1572–1576.
17. Xu, X.; Zhang, N.; Thundat, T.; Brown, G. M.; Ji, H.-F. *Anal Chem*, in press.
18. Xu, X.; Thundat, T. G.; Brown, G. M.; Ji, H.-F. *Anal. Chem.* **2002**, *74*, 3611–3615.
19. Diamond, R. M.; Whitley, D. C. In *Ion Exchange, A Series of Advances;* Marinsky, J., Ed.; Marcell Dekker: New York, 1966; Vol. 1, pp 207–302.
20. Liu, A-C.; Chen, D-C.; Lin, C-C.; Chou, H-H.; Chen, C-H. *Anal. Chem.* **1999**, *71*, 1549–1552.
21. Arrigan, Damien W. M.; Bihan, Loic Le. *Analyst* **1999**, *124*, 1645–1649.
22. Yang, Wenrong; Gooding, J. Justin; Hibbert, D. Brynn. *J. Electroanal. Chem.* **2001**, *516*, 10–16.
23. Bai, Yan; Ruan, Xiangyuan; Mo, Jinyuan; Xie, Youqin. *Anal. Chim. Acta* **1998**, *373*, 39–46.

Chapter 13

Spectroelectrochemical Sensor for Technetium: Preconcentration and Quantification of Pertechnetate in Polymer-Modified Electrodes

David J. Monk[1], Michael L. Stegemiller[1], Sean Conklin[1], Jean R. Paddock[1], William R. Heineman[1], Carl J. Seliskar[1], Thomas H. Ridgway[1], Samuel A. Bryan[2], and Timothy L. Hubler[3]

[1]Department of Chemistry, University of Cincinnati, Cincinnati, OH 45221
[2]Radiochemical Processing Laboratory and [3]Environmental Molecular Sciences Laboratory, Pacific Northwest National Laboratory, Richland, WA 99352

A remote spectroelectrochemical sensor and instrumentation package is being developed for the detection of technetium as aqueous pertechnetate (TcO_4^-) in the vadose zone and associated groundwater. This sensor would be used to monitor the integrity of low-level and high-level nuclear waste containment at U.S. Department of Energy sites. Electrochemical studies of TcO_4^- reduction at bare indium doped tin oxide (ITO) optically transparent electrodes (OTEs) show a poorly formed reduction wave for cyclic voltammetry and precipitation of technetium oxide (TcO_2) on the electrode surface. Similar experiments at ITO OTEs coated with thin films containing cationic polymers show partitioning of TcO_4^- into the films. Three films were investigated: poly(dimethyldiallylammonium chloride) (PDMDAAC) and quaternized poly(4-vinylpyridine) (QPVP), both immobilized in porous glass by the sol-gel process, and poly(vinylbenzyltrimethylammonium chloride) (PVTAC) copolymerized with poly(vinylalcohol). The largest enhancement in cyclic voltammetry reduction wave for TcO_4^- was for QPVP. The electrochemical mechanism changes to favor the formation of a relatively long-lived soluble species that ultimately converts to TcO_2. The electrodeposition of technetium oxide in these films was shown to be a method for the quantitative spectroelectrochemical determination of TcO_4^- and has been verified using radiochemistry dose measurements and scanning electron microscopy.

Introduction

One of the goals of the U.S. Department of Energy (DOE) Environmental Management Science Program (EMSP) is to support basic research addressing fundamental issues that may be critical to advancing the remediation of Department of Energy (DOE) sites nationwide. One critical area of this remediation is the need to monitor radiochemical constituents in various areas, ranging from the containment of low- and high-level radioactive waste to monitoring of contaminant plumes in subsurface water. Present methods of analysis are hazardous and time consuming, usually requiring lengthy sampling, preparation, analysis, and data interpretation. A better approach would be to use sensors to perform rapid, sensitive, and economic in situ analyses for various constituents of interest.

Technetium is one such constituent of radioactive waste where the need for a chemical means of detection exists, but a sensor does not. Technetium is not found in appreciable quantities in nature; however, the isotope ^{99}Tc is a byproduct of the thermal nuclear fission of ^{235}U, ^{233}U, and ^{239}Pu at 6.1%, 4.8%, and 5.9% yields, respectively ([1]), and significant quantities of ^{99}Tc exist at many DOE sites. ^{99}Tc exhibits rather weak β^- decay ($E_{max} = 0.292$ keV), but it is of particular concern for two reasons: its long half life (2.13×10^5 y) and the high solubility of its most common form in aqueous environmental media, pertechnetate (TcO_4^-) ([2]). Pertechnetate does not readily adsorb to most minerals, and therefore in aqueous form and under suitable conditions, it may rapidly present itself to subsurface waters ([3], [4]).

The goal of our current EMSP funded research has been to develop a sensor for Tc, capable of exceeding the U.S. Environmental Protection Agency (EPA) 900 pCi/L ($\sim 5 \times 10^{-10}$ M) standard for drinking water. Our previous research focused on a novel sensing technology combining three modes of selectivity (electrochemistry, spectrophotometry, and selective partitioning) into a single sensing technique, spectroelectrochemistry. We demonstrated that spectroelectrochemistry performed using ion-exchange-polymer-doped films coated on waveguides that also function as optically transparent electrodes (OTEs) provides a unique strategy to detect analytes in the presence of interferences (5). The ability of the ion-exchange-polymer-doped film to perform charge exclusion while increasing the concentration of the target analyte near the electrode surface combined with electrochemical modulation of an optical signal for quantification provides an effective strategy for the detection of many analytes.

One such analyte that has been successfully detected and quantified by the spectroelectrochemical sensor is ferrocyanide. Ferrocyanide is a good example of an analyte that exhibits a large optical change when subjected to electrochemical modulation between ferrocyanide and ferricyanide. The concentration of ferrocyanide in some high-level radioactive waste storage tanks at the DOE Hanford Site is of importance. In the previous EMSP symposium and subsequent publication, we described the concept of sensing ferrocyanide using the spectroelectrochemistry approach (6). The sensor was shown to be effective for the rapid detection of ferrocyanide in simulated tank waste solutions (7). An actual tank waste sample was then analyzed for ferrocyanide using a portable spectroelectrochemical cell (8). The successful detection of ferrocyanide in samples with such high levels of potentially interfering species was a good test of the spectroelectrochemical sensor's high degree of selectivity.

The strategy for the detection and quantification of TcO_4^- is similar to that of ferrocyanide in that it is based on the measurement of an electrochemically modulated optical signal associated with technetium. However, TcO_4^- must first be converted into another chemical form with the required electrochemical and optical properties to achieve the selectivity and low detection limit needed for many DOE applications. This is necessary because neither TcO_4^- nor its electrochemical reduction products in aqueous media are strong chromophores or fluorophores. Development of the sensor is being accomplished in two major steps:

- Development of a charge selective film into which TcO_4^- partitions. This film increases the concentration of TcO_4^- near the detection interface while providing the first level of selectivity by rejecting cations that might interfere.
- Conversion of the TcO_4^- that accumulates in the film into a coordination compound with electrochemical and spectroscopic properties that render it detectable by electrochemical modulation of an optical signal. The coordination compound is electrogenerated by electrochemically reducing TcO_4^- in the presence of an appropriate ligand that is immobilized in the film. To achieve this goal, we have investigated the electrochemistry of

TcO$_4^-$ and have synthesized ligands that will form technetium coordination compounds that either absorb or fluoresce strongly to give high sensitivity.

Here we describe the partitioning of TcO$_4^-$ into thin films doped with cationic polymers that are coated onto OTEs and the electrochemical behavior of TcO$_4^-$ in these films. The OTE consists of a thin film of indium tin oxide (ITO), which is a semiconductor, coated on a glass substrate in which multiple internal reflectance spectroscopy can be conducted. The thin films containing polymer consist of poly(dimethyldiallylammonium chloride) (PDMDAAC) and quaternized poly(4-vinylpyridine) (QPVP), both doped in porous glass, and poly(vinylbenzyltrimethylammonium chloride) (PVTAC) copolymerized with poly(vinyl alcohol) (PVA). We also demonstrate the ability to electrochemically reduce TcO$_4^-$ to give a change in optical absorbance due to electrochemical deposition of technetium oxide (TcO$_2$) in the anion-exchange film. This formation of a precipitate is demonstrated as a simple method to produce a spectroelectrochemical determination of TcO$_4^-$ concentration.

Materials and Methods

Caution

^{99}Tc is a known beta emitter (E$_{max}$ = 0.292 keV) with a half life of 2.13 x 10^5 y. Amounts of ^{99}Tc greater than 20 mg may cause secondary X-rays produced by the action of the β$^-$ particles on glass (Brehmstralung). All experiments described herein were performed at Pacific Northwest National Laboratory in the Radiochemical Processing Laboratory, a DOE Hazard Category II Non-reactor Nuclear Facility, by a properly equipped analyst with Radiological Worker II Certification.

Chemicals and Reagents

A pertechnetate stock solution was prepared by dissolving 1.5703 g of ^{99}Tc metal in 100 mL of 2 M HNO$_3$. The activity was measured to be 5.95 x 10^8 dpm/mL. PDMDAAC solution containing 20 wt % solution in water, poly(4-vinylpyridine) (PVP, MW 160,000), PVA (MW 85,000–146,000, 98-99% hydrolyzed), tetraethyl orthosilicate (99.999%), and 3-aminopropyltriethoxysilane were purchased from Aldrich. PVTAC (MW 400,000) solution containing 30 wt % in water was obtained from Scientific Polymer Products. Glutaraldehyde solution (50 wt % in water) was obtained from Fisher Scientific.   Stock solutions for polymer blends were prepared either by dilution of commercial solutions or by dissolution of solid polymer in deionized water from a purification system.   All the other compounds (ACS-certified Reagent Grade) were used without further purification.

Apparatus

Electrochemistry on pertechnetate was performed with a remotely operable potentiostat, designed in our laboratory (9). All potential values were measured vs Ag/AgCl, 3 M NaCl reference electrode (Cypress Systems).   ITO coated glass (Corning 1737F and 7059, 11-50 Ω/sq, 150-nm thick film on 1.14-mm glass) was purchased from Thin Film Devices and was cut into 1-cm x 4.5-cm slides and used as the working electrode. For thin film electrochemistry, a special cell that employed a large-area platinum coated titanium mesh as the auxiliary electrode was designed.   Spectrophotometry for technetium experiments was performed using a Hewlett-Packard diode array spectrophotometer.   Radiochemical dose rate measurements were performed using a vacuum-tube ionization chamber, commonly called a cutie-pie (CP).

Thin Film Preparation

Silica sols were prepared according to a previously reported protocol (10). Incorporation of PDMDAAC and QPVP into the silica sol was performed by mechanically blending the polyelectrolyte solution with the silica sol just prior to spin-coating.   PVTAC-PVA films were prepared following a procedure previously described (11). In all cases, the mixture of polymer and matrix was spin-coated to the desired thickness on the surface of ITO coated slides with a Headway photo resist spinner. QPVP was prepared from PVP by dissolving 0.5 g PVP in 50 mL methanol in a round bottom flask, to which is added 1 mL CH_3I (Fisher, 99.8%) and some glass beads. The solution is refluxed for 6 h (heating mantle, condenser). Several portions (~25 mL each) of diethyl ether are added to complete precipitation and decanted after the product has settled for a couple of hours, to remove some of the solvent and also remove unwanted organics (byproducts, contaminants, etc.). The remaining solvent is evaporated on a rotovap. The QPVP-iodide salt is weighed, and dissolved in DI water sufficient to make 20% solution (weight/volume). 0.1 g $AgNO_3$ is added for each milliliter of QPVP solution to remove iodide counter ions by precipitation of AgI and replace them with nitrate counter ions. The solution is centrifuged (15,000 rpm for 15 min, several times) to remove solid AgI. KCl (5-10 grains/crystals) is then added to precipitate excess silver, and the solution is centrifuged again to remove AgCl. This process is repeated until the addition of KCl does not cause precipitation of AgI. Failure to remove all excess silver resulted in poor film quality.

Results and Discussion

Materials for Selective Films

A critical component of the sensor is the selective film into which the TcO_4^- partitions for detection by spectroelectrochemical modulation. To be effective, the film must possess a number of characteristics:

- Preconcentrate TcO_4^- into the film
- Provide a level of selectivity by excluding at least some potentially interfering substances—cations that might have similar electrochemical and optical properties as the electrogenerated technetium coordination compound
- Exhibit optical transparency at the wavelength used for detection
- Be electrochemically inactive at the potentials used for reduction of TcO_4^- and for electrochemical modulation of the formed technetium coordination compound
- Not interfere with electrochemistry at the underlying electrode (i.e., passivate the electrode)
- Form thin, uniform films free of pin holes and cracks
- Be sufficiently robust for the intended application
- Serve as host for ligands to form coordination compounds with technetium that enhance sensitivity
- Be reasonably priced

Consequently, coincident with our work on sensor design, we have developed chemically selective optical materials that satisfy these criteria (*10–14*). We have developed two classes of optical quality chemically selective films based on two different host materials, namely, sol-gel processed silica (SiO_2) and cross-linked PVA. These hosts can be bonded to oxide surfaces, for example, glass or ITO yielding sufficiently robust films on sensor surfaces. We have made composites of these host materials with three ionomers, the structures of which are shown below.

PDMDAAC   PVTAC   QPVP

These cationic ionomers have been shown to accumulate anions and so were chosen for investigation with TcO_4^-. These polymer films are spin coated onto the ITO OTEs, and the resulting films are amazingly uniform as shown in the scanning electron micrographs in Figure 1. Our experience with these films has shown that they preconcentrate analytes from very dilute solutions up to film concentrations approaching one molar, which leads to an enormous enhancement in sensitivity of the sensor. Adapting these films to fluorescence spectroelectrochemical sensing will involve combining selective fluorophores into the films.

Voltammetry of TcO_4^- at Bare ITO

The electrochemistry of TcO_4^- was investigated at ITO because it is the OTE that has worked so well for our spectroelectrochemical sensors. Voltammograms at bare ITO gave a poorly defined reduction wave for TcO_4^- and complicated electrochemistry analogous to that reported by Mazzocchin et al. on platinum (15). A representative voltammogram is show in Figure 2 with the electrode reactions assigned by Mazzocchin et al. (15). The cyclic scan starts at 500 mV and proceeds negative. As the scan approaches -800 mV and through -1000 mV, the reduction of TcO_4^- occurs as a cathodic shoulder on the large wave for the evolution of hydrogen. The current increases with the evolution of hydrogen, and on the return scan, a cross over the initial scan may be noted around -900 mV, indicative of a changing electrode surface due to the electrodeposition of TcO_2. As the scan proceeds positive, a slight anodic wave occurs at -500 mV. The larger anodic wave at approximately 150 mV corresponds to the oxidation of some of the TcO_2 back to TcO_4^-. The poorly defined reduction wave for TcO_4^- caused by its catalytic nature makes it difficult to establish a quantitative relationship between TcO_4^- reduction and TcO_2 oxidation back to TcO_4^- by comparing the charges for the two processes (i.e., comparing the areas under the reduction and oxidation waves). It is not a

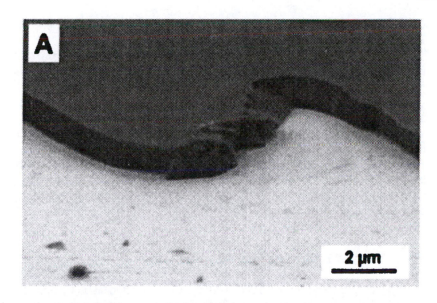

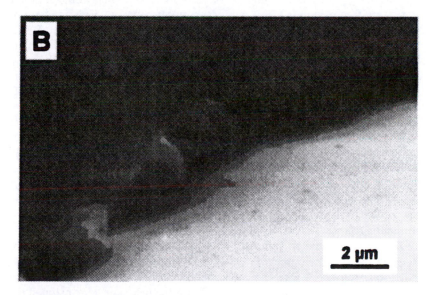

Figure 1. Scanning electron micrographs of two types of films on ITO: (A) QPVP-SiO₂ and (B) PVTAC-PVA. Thicknesses are about 500 nm.

chemically reversible redox system, as TcO_2 eventually coats the bare electrode in visible quantities with continuous cycling.

Two key observations can be made relating to the development of a spectroelectrochemical sensor for TcO_4^-. First, TcO_4^- is reducible to a lower oxidation state on ITO, which is a critical step in developing the spectroelectrochemical sensor for TcO_4^-. Second, the reduction product is a precipitate on the electrode surface (TcO_2), not a soluble lower oxidation state species of technetium with a strong spectrum in the visible range. Thus, under these conditions, a spectroelectrochemical sensor would have to be based on the electrochemical modulation of the TcO_4^-/TcO_2 couple. Since TcO_4^- is nonabsorbing in the visible range and TcO_2 scatters light over a wide range of wavelength, this means of optical detection could be made at any wavelength in the visible range.

Voltammetry of TcO_4^- at Polymer-Modified ITO

The electrochemistry of TcO_4^- was then investigated at ITO coated with anion exchange films to determine if TcO_4^- is preconcentrated from solution into the sensing film, which is a key step in the sensor concept. Voltammograms of TcO_4^- were run on the three anion exchange films discussed above: PVTAC-PVA, PDMDAAC-SiO$_2$, and QPVP-SiO$_2$. Voltammograms of TcO_4^- (Figure 3) showed a much more sharply defined reduction wave at all three films tested. Three features of these voltammograms are important. First, the current for the reduction wave of TcO_4^- (at about -1100 mV) is considerably larger at all of the film-coated ITOs compared to bare ITO, which is indicative of a preconcentration of TcO_4^- into all of the films. Second, there are differences in the enhancement of the TcO_4^- reduction wave by the three films, even though they are all nominally attracting TcO_4^- electrostatically by interaction with a quaternary ammonium group. These differences indicate the importance of film choice. Third, the voltammogram has now taken on the look of a chemically reversible redox couple for the reduction wave of TcO_4^- by the appearance of a new anodic wave on the reverse scan. The appearance of this new anodic wave (at about -400 mV) is accompanied by a proportionate disappearance of the anodic wave attributed to TcO_2 (at 150 mV). This behavior is clearly seen from the repetitive voltammograms at QPVP-SiO$_2$ in Figure 4. The first scan of the cyclic voltammogram, also shown as the QPVP-SiO$_2$ scan in Figure 3, shows some oxidation of TcO_2 back to TcO_4^-. This is the behavior demonstrated in the bare electrode scan of Figure 1, and would be expected to be the behavior noted for multiple scans with the coated electrode. This is not the case—in subsequent scans the anodic wave at 150 mV disappears almost entirely. With this disappearance, the cathodic wave at -1050 mV in the first scan begins to shift more positive. This shift becomes more pronounced on subsequent scans, and

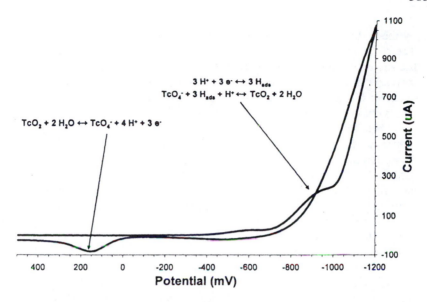

Figure 2. Cyclic voltammogram of TcO₄⁻ on bare ITO electrode. 5.0 x 10⁻⁴ M TcO₄⁻ in pH 7 phosphate buffer as supporting electrolyte, 25 mV/s scan rate, Ag/AgCl reference electrode, solution deoxygenated for 30 min prior to scan.

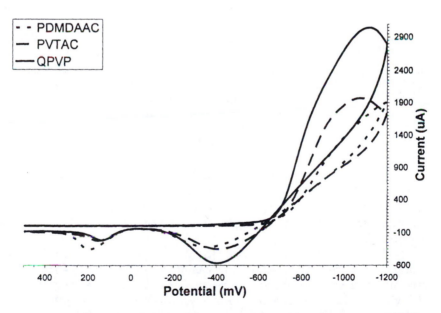

Figure 3. Cyclic voltammogram of TcO₄⁻ on ITO coated with films with QPVP-SiO₂, PVTAC-PVA, and PDMDAAC-SiO₂ after 30 min of immersion in sample. 1.6 x 10⁻³ M TcO₄⁻ in pH 7 phosphate buffer as supporting electrolyte, 25 mV/s scan rate, Ag/AgCl reference electrode, Pt auxiliary electrode, solution deoxygenated for 30 min prior to scan.

by the sixth scan, a reasonably well defined peak is present at about -950 mV. The change in the cathodic reduction wave of TcO_4^- is paralleled by a change in the anodic wave that appeared at -400 mV for the coated electrodes. This wave increases in size until it becomes nearly proportional to the cathodic wave. This behavior further supports the formation of a chemically reversible redox couple that is only evident with a film coated electrode.

Thus, the presence of the film has caused a change in mechanism of the reduction of TcO_4^- that is consistent with the formation of a soluble, relatively stable lower oxidation state of Tc, rather than the insoluble TcO_2 observed on bare ITO. This lower oxidation state is apparently stabilized by the presence of the polymer film, perhaps by electrostatic interaction with the positively charged quaternary ammonium groups (i.e., ion pairing). It is clearly stable on the time scale of the voltammetric scans shown in Figure 4 (i.e., tens of minutes). This change in mechanism for the electrochemical process is important to our strategy of converting TcO_4^- into a coordination compound for spectroelectrochemical detection.

Additionally, the peak heights for both the TcO_4^- reduction wave and this anodic peak are proportional to TcO_4^- concentration, indicating a quantitative response. This is shown in Figure 5 for three concentrations of TcO_4^- at a QPVP-SiO_2 coated electrode. Similar behavior was observed with films containing PDMDAAC and PVTAC.

Interestingly, the reduced Tc species slowly forms TcO_2 as a precipitate in the film. The precipitated TcO_2 in turn scatters light, and at high concentration this can be easily visualized. The deposited TcO_2 in the films of the coated electrodes is shown in Figure 6. The three electrodes to the left are QPVP-SiO_2 coated electrodes. The one on the far left was cycled in a solution of 1.6×10^{-3} M TcO_4^-, the second from the left in 5×10^{-4} M TcO_4^-, and the second from the right 1×10^{-4} M TcO_4^-; the PVTAC-PVA coated electrode on the right side was cycled in 5×10^{-4} M TcO_4^-. Measurement of the absorbance (scattering) of the precipitate by passing light perpendicularly through the film and the substrate ITO showed a linear relationship between concentration of TcO_4^- in the solution to which the electrode was exposed and absorbance and radioactivity from ^{99}Tc by beta counting (Figure 7), which further confirms that TcO_4^- uptake into the film is quantitative, as would be required for sensing. However, the sensitivity of such an absorbance based system is far from ideal because neither TcO_4^- nor the products of its reduction in aqueous solution are sufficiently strong chromophores or fluorophores to give sensitive detection with electrochemical modulation.

Scanning electron microscopy (SEM) X-ray analysis confirms the appearance of precipitated TcO_2. This can be seen in Figure 8, where the left hand side of the photo is the glass substrate, the thin line down the center is the ITO electrode, and the pale white line to the right is the ion exchange film. Particles of TcO_2 show up as bright white spots in the film. It is interesting to

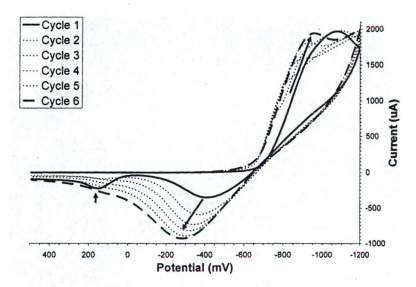

Figure 4. Sequential cyclic voltammograms of TcO₄⁻ at ITO coated with PVTAC-PVA film after 30 min of immersion in sample. 1.6 x 10⁻³ M TcO₄⁻ in pH 7 phosphate buffer as supporting electrolyte, 25 mV/s scan rate, Ag/AgCl reference electrode, solution deoxygenated for 30 min prior to scan.

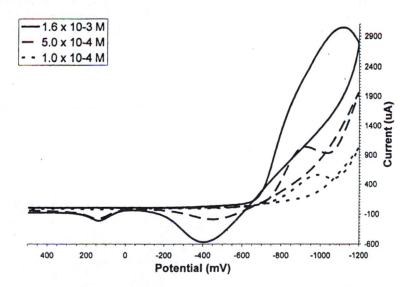

Figure 5. Cyclic voltammograms of three concentrations of TcO₄⁻ on ITO coated with film of QPVP-SiO₂ after 30 min of immersion in sample. 1.6 x 10⁻³, 5.0 x 10⁻⁴, 1.0 x 10⁻⁴ M TcO₄⁻ in pH 7 phosphate buffer as supporting electrolyte, 25 mV/s scan rate, Ag/AgCl reference electrode, solution deoxygenated for 30 min prior to scan.

318

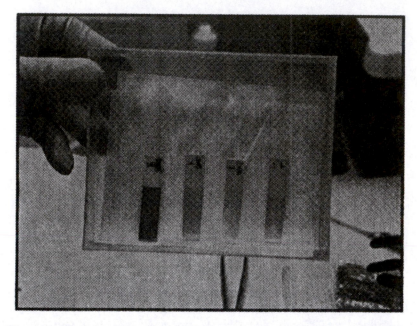

Figure 6. The visual deposition of TcO_2 in the films of the coated electrodes. The three electrodes to the left are $QPVP$-SiO_2 coated electrodes. The one on the far left was cycled in a solution of 1.6×10^{-3} M NH_4TcO_4, the second from the left in 5×10^{-4} M NH_4TcO_4, and the second from the right 1×10^{-4} M NH_4TcO_4; the PVTAC-PVA coated electrode on the right side was cycled in 5×10^{-4} M NH_4TcO_4.

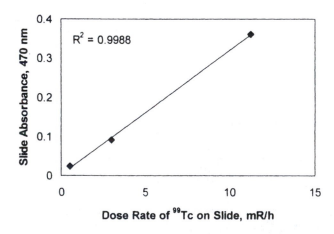

Figure 7. Plot of absorbance versus ^{99}Tc dose rate for three concentrations of TcO_4^-.

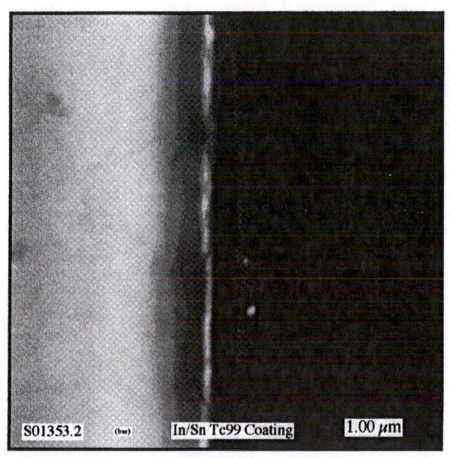

Figure 8. An SEM image of an electrode used in an electrochemical experiment shows from left to right; glass substrate (white area), thin layer ITO electrode (bright white strip near center), and anion exchange film (faint white strip). Several inclusions in the ion exchange film are evident. X-ray analysis has determined these particles to be technetium.

note that the TcO$_2$ is found predominantly within the polymer film rather than directly on the ITO electrode surface, which would be expected if TcO$_2$ were formed immediately by the electrode process as is the case with reduction of TcO$_4^-$ at bare ITO. This finding is consistent with the cyclic voltammetry of TcO$_4^-$ at ITO coated with polymer film that points to reduction of TcO$_4^-$ to a soluble technetium species that can diffuse some distance from the electrode before it ultimately converts to TcO$_2$. This electrogenerated species is available to react with an appropriate ligand immobilized in the film to form a strongly absorbing or fluorescing coordination compound that would be electromodulated for sensitive detection.

Conclusions

The ability to preconcentrate TcO$_4^-$ into an anion exchange film on an optically transparent electrode has been demonstrated. All three films evaluated showed a concentration dependent uptake of pertechnetate. Thus, a critical step toward the building of a spectroelectrochemical sensor for pertechnetate has been achieved. A second critical step for sensor development is the ability to reduce pertechnetate electrochemically. This has been demonstrated for both non-modified and polymer-modified ITO electrodes. The presence of a film is shown to change the nature of Tc electrochemistry forming a chemically reversible redox couple. The ultimate formation of TcO$_2$ due to this electrochemistry has been shown to be one basis for spectroelectrochemical determination of technetium concentration and has been verified via beta counting and SEM. This method for determination of TcO$_4^-$ concentration would prove inadequate for the detection limit of 900 pCi/L required by the sensor. Future works will focus on methodologies such as selective coordination compound formation and metal templating to increase selectivity and lower the detection limit.

Acknowledgments

Financial support provided by the DOE EMSP (Grant DE-FG07-96ER62311) is gratefully acknowledged. DJM acknowledges support provided by a Lange Fellowship from the Department of Chemistry, University of Cincinnati.

References

1. Siegel, J. M. *Rev. Mod. Phys.* **1946**, *18*, 513–544.
2. Schwochau, K. *Technetium Chemistry and Radiopharmaceutical Applications*, Wiley-VCH, Weinheim, 2000.
3. Strickert, R.; Friedman, A. M.; Fried, S. *Nucl. Technol.* **1980**, *49*, 253–266.
4. El-Wear, S.; German, K. E.; Peretrukhin, V. F. *J. Radioanal. Nucl. Chem.* **1992**, *157*, 3–14.
5. Shi, Y.; Seliskar, C. J.; Heineman, W. R. *Anal. Chem.* **1997**, *69*, 4819–4827.
6. Maizels, M.; Stegemiller, M; Ross, S.; Slaterbeck, A.; Shi, Y.; Ridgway, T. H.; Heineman, W. R.; Seliskar, C. J.; Bryan, S. A. In *Nuclear Site Remediation; First Accomplishments of the Environmental Management Sciences Program;* Eller, P. G., Heineman, W. R., Eds.; ACS Symposium Series 778; Oxford University Press: Washington, DC, 2001; pp 364–378.
7. Maizels, M.; Seliskar, C. J.; Heineman, W. R. *Electroanalysis* **2002**, *14*, 1345–1352.
8. Stegemiller, M. L.; Heineman, W. R.; Seliskar, C. J.; Ridgway, T. H.; Bryan, S. A.; Hubler, T. L.; Sell, R. L. *Environ. Sci. Technol.* **2003**, *37*, 123–130.
9. Monk, D. J.; Ridgway, T. H.; Heineman, W. R.; Seliskar, C. J.; *Electroanalysis* **2003**, *15*, 1198–1203.
10. Shi, Y.; Seliskar; C. J. *Chem. Mater.* **1997**, *9*, 821–829.
11. Gao, L.; Seliskar, C. J. *Chem. Mater.* **1998**, *10*, 2481–2489.
12. Gao, L.; Shi, Y.; Slaterbeck, A. F.; Seliskar, C. J.; Heineman, W. R. *Proc. SPIE* **1998**, *3258*, 66–74.
13. Hu, Z.; Slaterbeck, A. F.; Seliskar, C. J.; Ridgway, T. H.; Heineman, W. R. *Langmuir* **1999**, *15*, 767–773.
14. Heineman, W. R.; Seliskar, C. J.; Richardson, J. N. *Aust. J. Chem.* **2003**, *56*, 93–102.
15. Mazzocchin, G. A.; Mangno, F.; Mazzi, U.; Portanova, R. *Inorg. Chim. Acta* **1974**, *9*, 263–268.

Chapter 14

Sensors and Automated Analyzers for Radionuclides

Jay W. Grate, Oleg B. Egorov, and Matthew J. O'Hara

Pacific Northwest National Laboratory, Richland, WA 99352

Novel chemical and instrumental approaches have been developed that enable application of classical radiochemical analysis principles to in situ sensing and automated monitoring of beta and alpha emitting radionuclides. The overall approach is based on the integration of sample treatment, separation, and radioactivity detection steps within a single functional sensor or analyzer device. New radionuclide sensors and analyzers can provide timely information on radiological composition in environmental and process monitoring applications and represent significant advancement over manual analytical techniques performed in centralized laboratories. The concept of preconcentrating minicolumn sensors using dual function sorbent/scintillating materials is described for low-level environmental sensing applications. Equilibrium-based sensing approach is described that enables reagentless operation for long-term measurements. Integration of automated microwave-assisted sample treatment, column separations, and flow-through radiometric detection is used to develop radionuclide analyzer instrument for rapid radiochemical measurements in nuclear waste process monitoring. Sensors and analyzers for ^{99}Tc have been engineered into fully functional practical devices for operation in the field or plant settings.

Introduction

The production of nuclear weapons materials has generated a legacy of nuclear waste and contaminated environmental sites (*1*). The remediation of radiochemically contaminated sites and processing of stored wastes into stable waste forms will require characterization procedures throughout all phases of these activities (*2*). Consequently the demands for radiochemical analysis will increase rapidly in the future. Methods are required to do these analyses rapidly and precisely. To meet these characterization needs, new automated techniques are required in order to provide improved precision, consistent analytical protocols, reduced worker exposure to toxic and/or radioactive samples, greater sample throughput, reduced costs, and reduced secondary waste generation. Furthermore, methods are required that provide automatic analyses in settings other than analytical laboratories.

Analyses of radionuclide content are needed at-site or in situ for monitoring surface waters, groundwater, and process waters. The greatest concern is radionuclides that are mobile in the environment and cannot be practically detected and identified using gamma spectrosopy. Radionuclide measurements are required to support baseline monitoring of surface and subsurface waters; to support characterization of contaminant migration; to assess efficacy of remediation processes in the field (e.g., pump-and-treat and barriers); to support operation of remediation processes (e.g., column breakthrough); and in long-term stewardship. In current practice, water samples are collected and transported to the analytical laboratory where costly radiochemical analyses are performed. Sample turnaround times are several days to several weeks.

In addition, new automated radiochemical analysis techniques will be required to enable rapid characterization of the nuclear processing streams, for example, during radionuclide removal steps. These analyzers must operate continuously within the nuclear waste processing plant. Such new monitoring techniques would facilitate more precise and reliable control of waste processing operations, and offer considerable cost savings. For example, replacement of separation resins in radionuclide removal steps is very costly due to the expense of the selective resins. Monitoring of column performance will allow replacement only when needed, rather than scheduled periodic replacement at arbitrary intervals. In addition, treated waste that does not meet specifications will have to be run through the treatment process again, at considerable cost. Using automated monitors to ensure effective process operations will avoid such costs and schedule delays.

Although there are many radioactivity detectors available, there has been very little development of analytical instruments or radiochemical sensors suitable for rapid and selective quantification of beta- and alpha-emitting radionuclides such as ^{99}Tc, ^{90}Sr, and TRU actinides in water or process streams. Current baseline analytical methods for these analytes are based on tedious manual radiochemical analysis methods performed in centralized laboratories.

For example, in 1994, Koglin, Poziomek, and Kram reviewed "Emerging Technologies for Detecting and Measuring Contaminants in the Vadose Zone" in the *Handbook of Vadose Zone Characterization and Monitoring* (*3*). The section of this article on radiochemical sensors is notably lacking in examples. Instead, the section discusses various ways of detecting and quantifying radionuclides and heavy metals, including general radioactivity detection techniques, inductively coupled plasma mass spectrometry (ICP-MS), inductively coupled plasma atomic emission spectroscopy (ICP AES), neutron activation analysis (NAA), and x-ray fluorescence (XRF) spectrometry. Drawbacks and limitations of these techniques for field analysis were noted. Radioactivity detection "usually requires some form of sample preparation to concentrate the radionuclides prior to counting to achieve a reasonable degree of sensitivity." The article states that "much of current analytical work is still done in fixed chemical laboratories using conventional radiochemical analysis" and that "conventional detection techniques ... are confined to fixed or mobile laboratories." This represents a significant technical gap that has been identified in several technology needs throughout the U.S. Department of Energy (DOE) complex. Moreover, significant scientific challenges exist in order to meet these technical needs.

Considerable progress has been made in developing automated methods for radiochemical analysis using sequential injection fluidics and extraction chromatography separations (*4-12*). These separation processes capture radionuclides of interest while washing away matrix components and interferences. The captured radionuclides are then released selectively by sudden changes in the solution composition passing through the column. Automated analytical separations for ^{90}Sr, ^{99}Tc, and actinides have been described, using flow through scintillation counting or ICP-MS for detection and quantification. The present paper turns its attention to new approaches for sensing and monitoring.

Methods and Results

Preconcentrating Minicolumn Sensors for Radionuclides

Design Principles

The fundamental challenges in radionuclide sensing of alpha- and beta-emitters are the short ranges of these particles in condensed media (e.g., water) and the ultra-trace analytical detection limit requirements. Chemical sensing methods may be applicable only to those radionuclides where the required detection limits are not too stringent, such as uranium. For other radionuclides of interest, where the required detection limits are far beyond the capabilities of any known or conceivable chemical sensing method, radioactivity detection is necessary. Radioactivity detection is also required when radioactive isotopes of concern must be distinguished from stable natural isotopes.

For detection on the basis of alpha or beta emissions, some method of separating the radionuclide of interest from interfering species and localization in the detection volume of a detector must be incorporated into the sensing mechanism. A preconcentrating minicolumn concept has been developed for a radionuclide sensor that is based on dual functionality minicolumns (13-17). The materials in these columns incorporate selective separation chemistry to capture and retain the radionuclides of interest, and scintillating fluors so that radioactivity of the captured species results in a measurable light output. The minicolumn is located between two photomultiplier tubes of a scintillation detection system to collect the signal. The selectivity is determined mainly by the separation chemistry used to preferentially capture and preconcentrate the analyte of interest. This design is shown schematically in Figure 1.

This configuration meets all the functional requirements for effective radionuclide sensors. The packed column format provides for efficient fluidic processing of the sample for preconcentration. The detection method is radiometric via the process of scintillation. Selective chemistry is in very close proximity to the scintillation material and retains the radionuclides for counting. These new column sensors represent a novel and advantageous approach for detection of radionuclides when selective preconcentration is required.

A number of approaches have been developed for obtaining selective uptake and a light output signal within the same column. In addition, two different operating methods have been developed, one based on quantitative uptake, and the other based on equilibration of the uptake material with ambient concentrations. These approaches and methods are described below.

Selective Scintillating Microspheres

The initial development of preconcentrating minicolumn sensors began with the development of dual functionality extraction chromatographic materials containing organic scintillator fluors (13-17). These have been called extractive scintillating resins or selective scintillating microspheres (SSMs). The scintillating fluors can be incorporated by various methods, including (1) co-depositing scintillating fluors with the extractant into the bead pores; (2) diffusing the scintillator fluors into the bead structure prior to impregnating with extractant; or (3) introducing scintillating fluors into the bead structure during bead polymerization. Examples of fluors used are 2,5-diphenyloxazole,

(PPO) as the primary fluor, and secondary scintillator fluors such as 1,4-bis ˙ -methylstyryl)-benzene, (bis-MSB), or 1,4-bis-(4-methyl-5-phenyl-2-oxazolyl)benzene (DM-POPOP).

For example, SSMs for ^{99}Tc-sensing were prepared by impregnating porous acrylic ester beads with a mixture of Aliquat 336 (a liquid anion exchanger based on long chain quaternary ammonium ions) and a pair of scintillating fluors (13). These SSM materials were characterized for their uptake of pertechnetate from neutral to acidic conditions. High analyte uptake was found at low-acid to neutral pH range, as expected for pertechnetate uptake by Aliquat 336. The uptake in neutral conditions is important for enabling detection in groundwater. The instrumental pulse height spectra were also obtained to evaluate the luminosity of the sensor material.

The sensing properties of this material are illustrated in Figure 2. Injection of an aliquot of ^{99}Tc standard in dilute acid results in analyte capture and measurable scintillation light output. In this example the analyte capture was quantitative and the signal persists as the sensor column is washed with dilute acid, which leaves the Tc retained on the column. On the other hand, injection of a radioactive species that does not exhibit affinity to the sensor material results in only a transient peak signal as shown for ^{137}Cs in Figure 2. Interfering species are promptly removed from the system using a small volume of wash solution. In this manner, the sensing method is selective towards the target analyte.

Because the sensor material exhibits high binding affinity towards Tc(VII), large sample volumes can be preconcentrated using a small sensor column. Analytes can be quantified from either the slope of the uptake signal or from the steady state signal after washing to remove interferences. The absolute detection efficiency for ^{99}Tc detection was 45%.

Using the pertechnetate-selective scintillating beads prepared by co-immobilization of selective extractants and scintillating fluors, we have demonstrated the detection of ^{99}Tc in acidified Hanford groundwater (13, 14). The resin was used in conjunction with a scintillation detector flow-cell and an automated fluid handling system. The results of these tests clearly indicated that radioactive pertechnetate can be readily detected in acidified groundwater samples below the U.S. Safe Drinking Water Act maximum concentration level of 900 pCi/L.

Composite Bed Sensors

Nevertheless, sensor materials based on impregnated fluors and extractants exhibited limited long-term stability in their scintillating properties. This prompted us to explore an alternative approach for developing preconcentrating minicolumn sensors. Scintillation and chemical selectivity were combined in a composite sensor column by creating an intimate mixture of separate solid phase

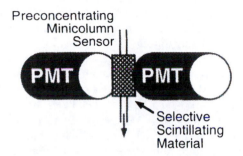

Figure 1. Preconcentrating minicolumn sensor concept for selective capture of radionuclides with transduction to a light output signal by the process of scintillation.

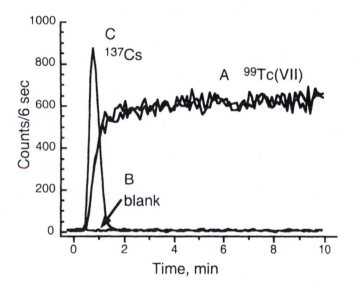

Figure 2. SSM-based sensor response to ^{99}Tc(VII) analyte (A, traces from duplicate runs) and potentially interfering species (^{137}Cs, trace C) unretained by the sensor material. A blank run is indicated by trace B. Flow rate 1 mL/min, injected sample volume 0.1 mL. Following the injection the sensor bed is washed with 10 mL of 0.02 M nitric acid. Reproduced from Analytical Chemistry, 71 (1999) 5420-5429. Copyright 1999 American Chemical Society.

extraction beads and nonporous scintillating beads. The nonporous scintillating beads were found to have much higher chemical stability in sensing and regeneration solutions. In addition, this approach facilitates the use of solid phase extraction materials for radionuclide uptake that are not readily impregnated with scintillating fluors.

Because of the relatively long range of the beta particles, this approach appears to be particularly suitable for detection of beta-emitters such as ^{99}Tc and [90]Sr. The ranges in water for beta particles emitted by [90]Y (E_{max} = 2282 keV), [90]Sr (E_{max} = 546keV), and [99]Tc (E_{max} = 294 keV) are 1.1 cm, 1.8 mm, and 750 µm respectively. As long as the range of the particle is comparable or greater than that of the sorbent bead diameter, scintillating beads mixed with the sorbent beads can transduce radiation energy to a light signal, while the fluidic advantages of the packed column geometry are maintained.

Composite bed sensing columns for ^{99}Tc sensing in unacidified groundwater were made using a strongly basic anion exchanger (AGMP1) for pertechnetate uptake. AGMP1 sorbent has very high uptake affinity and good selectivity towards Tc(VII) ions in basic to weakly acidic media. In addition, low uptake values in strong acid indicate that material can be regenerated by eluting retained analyte with the nitric acid. The composite sensor bed was prepared by mixing AGMP1 material (particle size 20–50 µm) with BC400 plastic scintillator beads (particle size 100-250 µm) at a 1:20 weight ratio. The total bed volume was just 50 µL. Because of the very high uptake affinity (K_d = 2.5 x 10^5 for the [99]Tc (VII) in unacidified Hanford groundwater), small volumes of the sorbent material can be used to preconcentrate analyte from large volumes of groundwater.

The absolute detection efficiency of the composite sensor column was 34% and analyte recovery (loading efficiency) was 97%. The [99]Tc(VII) selective composite bed sensor can be regenerated using a small volume of 2 M nitric acid solution, resulting in rapid elution of the retained analyte without loss of the scintillation properties.

Thus, using mixed-bed sensor columns, we demonstrated the feasibility of ^{99}Tc sensing in unacidified groundwater with excellent long-term sensor column stability. This approach appears especially promising for the development of a ^{99}Tc sensor probe suitable for downwell groundwater monitoring.

Equilibrium-Based Reagentless Sensors

The radionuclide selective sensing approaches described thus far were based on quantitative capture of radionuclides from sub-breakthrough volumes on a mini- or microcolumn sensor. In this scenario, no analyte losses due to the column breakthrough occur during sample load steps. Sensor regeneration or renewal procedures are required to facilitate subsequent measurements of new samples. Sensor regeneration was typically accomplished using eluent solutions that strip the captured analyte from the sensor column. Alternatively, we have demonstrated the feasibility of sensor column renewal using automated fluidic procedures to repack the sensor column for each measurement (7).

Quantitative capture sensing approaches are well suited for analytical characterization in the field, where sensor regeneration or renewal procedures are acceptable. Although renewable reagent or renewable surface sensors could be engineered for in situ monitoring, provided that the consumable volumes per measurement are very small, reagentless sensors would be preferable for extended periods of in situ operation.

Recent emphasis on reagentless sensing for long-term stewardship prompted us to explore an alternative sensing approach based on equilibrium sensing rather than quantitative capture. In this scenario, the sample is delivered to the sensor column until complete breakthrough has occurred and the entire column reaches dynamic equilibrium with the sample solution. Chromatographic theory indicates that under these dynamic equilibrium conditions, analyte concentration on the sensor column is proportional to the analyte concentration in the sample used for equilibration. Consequently, when using the sensor equilibration approach, there is no need for sensor regeneration between measurements. If reversible analyte capture can occur from chemically untreated groundwater, the equilibration approach forms the basis for reagentless radionuclide selective sensing.

The degree of analyte capture or retention within the sorbent layer is determined by the uptake affinity or analyte distribution coefficient, D:

$$D = \frac{C_s}{C_a}$$

where C_s is the equilibrium analyte concentration in the sorbent layer (stationary phase) and C_a is the equilibrium analyte concentration in the aqueous (mobile) phase. Under complete column breakthrough conditions, the equilibrium concentration is established throughout the entire length of the sorbent column. The sensor column is in equilibrium with the flowing sample. Under these conditions no further analyte preconcentration by the sensor column is possible, and the analyte concentrations in the flowing sample before and after the sorbent column are equal. The maximum degree of analyte preconcentration has been attained. The amount of analyte captured by the sensor column under equilibrium conditions can be described by the maximum effective sample volume, V_{max}, which corresponds to the equivalent sample volume from which the analyte has been quantitatively captured by the sorbent layer.

In a continuous flow chromatographic system, V_{max} is equal to the analyte retention volume, V_r. The analyte retention volume is related to the chromatographic system capacity factor, k', which is defined by the following formula:

$$k' = D\frac{V_s}{V_m}$$

where D is the distribution coefficient for the analyte uptake, V_s is the volume of the stationary phase in the sensor column and V_m is the volume of the mobile phase or free column volume. The effective maximum preconcentrated volume, V_{max} can be determined according to the following formula:

$$V_{max} = V_r = k' V_m = DV_s$$

The amount of analyte, A, captured by the sensor column after equilibration can be determined as follows:

$$A = V_{max}C_a = DV_sC_a$$

where C_a is the analyte concentration in the sample stream.

This last equation indicates that the amount of analyte present on the sorbent column is proportional to the analyte concentration in the sample. For a given analyte concentration in the sample solution, the amount of analyte captured by the sensor column is determined by the uptake affinity (distribution coefficient D) and the size of the sensor column (volume of the stationary phase in the sensor column, V_c).

The sample volume, V, that must be delivered to the sensor column in order to achieve equilibrium conditions can be determined from the following formula:

$$V \geq k' V_m \left[1 + 3\sqrt{\frac{H}{L}} \right]$$

where H is the sensor column plate height and L is the length of the sensor column. This formula indicates that equilibration volume is determined by the degree of analyte uptake (k'), physical dimensions of the sensor column (V_m and L), and its chromatographic efficiency (H).

The feasibility of equilibrium sensing was investigated using the AGMP1 composite bed sensor for Tc described previously (50 µL column volume / AGMP1 anion exchange material / BC400 plastic scintillator beads / 1:20 weight ratio). The detector traces in Figure 3 correspond to sensor responses obtained by delivering successive 100-mL samples of ^{99}Tc calibration standards in 0.02 M nitric acid (^{99}Tc distribution coefficient in 0.02 M nitric acid is ~10000). Equilibrium conditions and plateau responses are obtained after approximately

50 mL of each sample. The calibration line in Figure 4 indicates that the equilibrium sensor response is indeed dependent on the sample concentration; background subtracted equilibrium count rates are higher for samples with greater ^{99}Tc activity. Furthermore, the delivery of Tc-free 0.02 M nitric acid at the end of the run (Figure 3) shows that the sensor is reversible. It equilibrates with the unspiked sample, the ^{99}Tc is released from the sensor column, and the sensor signal returns to the baseline level. These results illustrate the concept of "reagentless" equilibrium radionuclide sensing.

Further experiments were conducted using chemically untreated Hanford groundwater spiked with ^{99}Tc. The sensor column was equilibrated with 1.9 liters of unspiked groundwater, followed by 1.9 liters of groundwater containing 2 dpm/mL of ^{99}Tc. Note that the ^{99}Tc level of 2 dpm/mL corresponds to the regulatory drinking water standard. This level was detected with a signal that was easily distinguished from the background.

Because of the high distribution coefficient for ^{99}Tc uptake by the sensor material in unacidified groundwater ($K_d = 2.5 \times 10^5$), high degrees of analyte preconcentration are achieved using a microscale sensor column (total bed volume 50 μL). Approximately 1 L of the sample solution was required to achieve sensor equilibration.

Having demonstrated that reagentless equilibration sensing is feasible, the remaining challenges have focused on potential interferences in real groundwater. In the initial experiments, it was found that pumping large volumes of untreated Hanford groundwater through the sensor leads to decreases in sensitivity over time. We propose that these are due to retention of traces of humic acid on the anion exchange material. If present, organic acids are expected to be retained on anion exchange materials from the slightly basic Hanford groundwater.

Visual observation of the sensor column revealed the presence of a yellow color on the sensing material. The color was partially removed from the sensor column using 2 M nitric acid. Analysis of this fraction using ICP AES did not reveal the presence of metal ion species that may contribute to the yellow color on the sorbent material.

Recently, we have found two strategies that are effective in removing this interference. One is the selection of a different anion exchange material for pertechnetate capture. The second involves the use of a guard column in front of the sensor that captures the interference while allowing pertechnetate to proceed through to the sensor minicolumn.

Engineered Reagentless Sensor Probe

The laboratory-based experiments for sensor development placed the sensor flow cell within the detection zone of a commercial flow through liquid scintillation counter, using the photomultiplier tubes of this instrument to detect

332

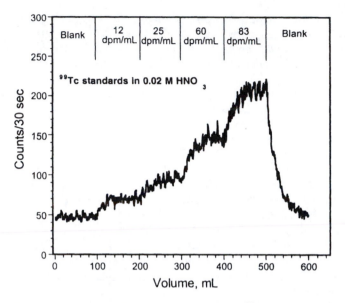

Figure 3. *Detector traces illustrating equilibrium ^{99}Tc sensing using ^{99}Tc standards in 0.02M nitric acid. The sample volume is 100 mL; flow rate is 1 mL/min. Plateau response region corresponds to attainment of equilibrium conditions between the sensor column and sample solution.*

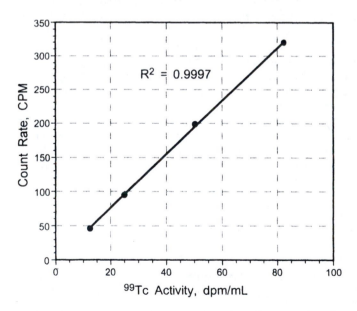

Figure 4. *Equilibrium sensor response is directly proportional to the analyte concentration in the sample.*

the sensor flow cell light output. We have recently developed an engineered sensor probe in a tubular configuration with a diameter of less than four inches. This probe incorporated the sensor flow cell, light measurement electronics to collect the light from the flow cell, and an anticoincidence shield around the sensor. The flow cell contained the composite dual functionality column material. Images of the sensor during assembly are shown in Figure 5. The anticoincidence shield intercepts and measures radiation entering the sensor from the outside, and is used to reduce the background counts. In fact, the use of the shield in this engineered probe reduced the background counts and the detection limits by a factor of ten compared to the laboratory measurements. Thus, the engineered probe provided superior performance compared to the laboratory measurements.

This probe was used to measure the activity of groundwater samples from the Hanford site. The measurements were in agreement with those from standard laboratory-based measurements.

Automated Analyzer for Radionuclides in Nuclear Waste Process Streams

Determination of non-gamma emitting radionuclides in nuclear waste process streams is a significant challenge. The combination of the sample matrix, the characteristics of the species to be detected, and the detection limit requirements preclude most existing analytical methods.

In the United States the chemical processing of low activity waste (LAW) prior to vitrification at the Hanford site requires removal of ^{90}Sr, transuranics, ^{137}Cs, and ^{99}Tc. The process streams are caustic brine matrixes with complex and varying composition. The basicity, nitrite content, organics, complexants, and aluminum ion concentrations vary with the source of the waste. The radiological characteristics are also quite variable. Radionuclides such as ^{99}Tc may not be present as a single chemical species, and in some cases the actual speciation is not known.

Our specific emphasis has been on the process monitoring needs associated with the operation of the chemical pretreatment or LAW processing operations. In particular, we have been developing an approach for monitoring the Tc removal process, which is based on a semi-continuously operated ion-exchange column. The current preferred method of Tc measurement is off-line analysis of a liquid sample using ICP-MS or classical wet radiochemical analysis, which requires taking a process sample and sending it to a centralized laboratory. The process of sampling and laboratory analysis is too time-consuming to provide timely feedback for continuous process control.

Instead, an automated radiochemical process monitor is required that takes a sample aliquot from the process, adjusts the sample matrix and the analyte speciation, separates the analyte from interfering species, and finally quantifies

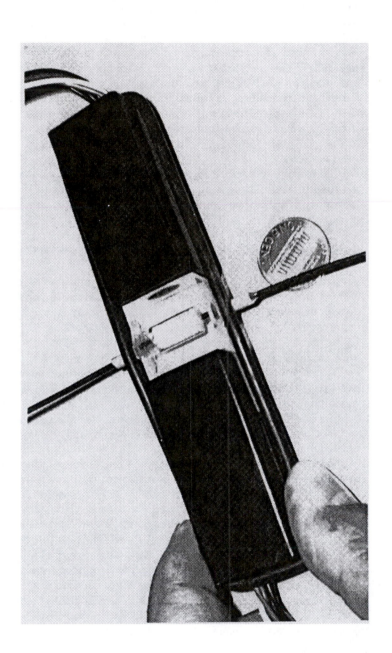

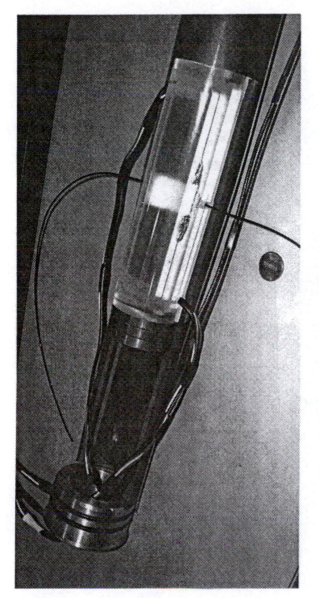

Figure 5. Photographic mages of the engineered sensor probe for Tc monitoring using the reagentless equilibrium sensing approach. (Top) The sensor flow cell with two photomultiplier tubes. (Middle) The sensor within the anticoincidence shield in the sensor probe housing. (Bottom) The complete probe prior to assembling the outer housing. Reproduced with permission from the "Handbook of Radioactivity Analysis, Second Edition" Chapter 14, page 1158. Copyright 2003 Elsevier USA.
Continued on next page.

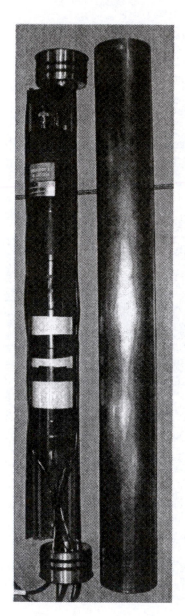

Figure 5. *Continued.*

the purified analyte. This overall concept for a radiochemical process monitor is shown in Figure 6.

The wet analytical separation chemistries for ^{99}Tc are generally effective only for Tc in the pertechnetate form. Therefore, for the analysis of total Tc in nuclear waste using a radiochemical measurement, it is necessary to fully oxidize the Tc present so that it is all in the pertechnetate state. We have developed a fluidic analyzer system that takes a complex nuclear waste stream sample, acidifies it, and oxidizes Tc to pertechnetate in an automated microwave digestion system using peroxidisulfate. It then performs a chemical separation and detects the purified ^{99}Tc by radioactivity detection. The monitor determines total Tc to the required detection limits and the total analysis time is less than 15 minutes. Furthermore, it is fully automated for continuous operation (18-23). A schematic diagram of the automated Tc measurement process in shown in Figure 7. A photograph of the prototypical analyzer unit is shown in Figure 8.

The sample treatment protocol starts with sample acidification followed by heating. This initial treatment ensures removal of the nitrites, which are expected to interfere with the subsequent oxidation, and promotes rapid dissolution of the $Al(OH)_3$ precipitate that forms during acidification. The nitric acid concentration and volume are selected to ensure complete dissolution of the $Al(OH)_3$ species upon heating, while maintaining relatively high pertechnetate uptake values during subsequent sample loading on the anion exchange separation column. The initial sample acidification is followed by digestion using sodium peroxidisulfate oxidizing reagent to convert the reduced Tc species to pertechnetate.

Pertechnetate separation is accomplished using a strongly basic anion exchange resin. The separation selectivity using anion exchange is adequate for the analysis of aged LAW sample matrixes and provides reliable separation of pertechnetate from the major radioactive constituents (^{90}Sr/^{90}Y, ^{137}Cs) and the minor constituents (e.g., isotopes of Sn, Sb, and Ru). A combination of column washes using dilute nitric acid, nitric-oxalic acid, sodium hydroxide, and moderately concentrated nitric acid has been developed for reliable separation of pertechnetate from anionic species.

The purified pertechnetate eluted from the column was detected and quantified with a flow through scintillation detector using a lithium glass solid scintillator. This scintillator material exhibited excellent stability in the strong nitric acid solutions used for pertechnetate elution.

An automated standard addition technique was implemented as a part of the analytical protocol. This approach provides a reliable method for remote, matrix matched Tc monitor instrument calibration. The monitor instrument was fully automated for continuous operation with software performing asynchronous device control and detector data acquisition. The data were automatically processed including a peak search and integration, as well as the instrument calibration via standard addition. Total analysis time was 12.5 minutes per

338

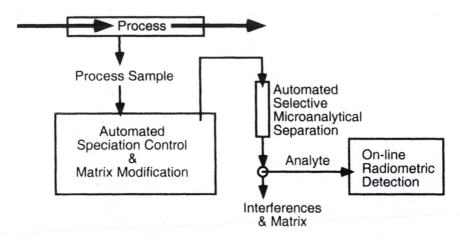

Figure 6. *Overall instrumental approach for process monitoring. The specific steps required will depend on the analyte.*

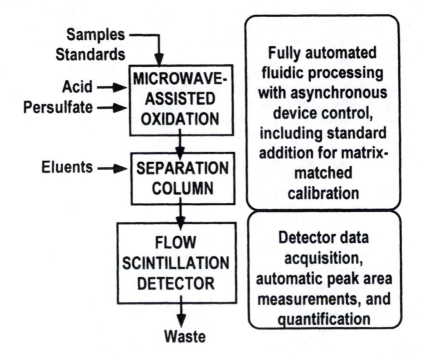

Figure 7. *Schematic diagram of the functions of the fully automated Tc analyzer for process monitoring applications.*

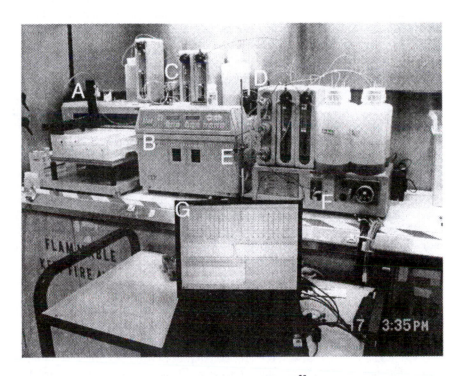

Figure 8. Photograph of the fully automated total ^{99}Tc analyzer instrument in the laboratory. The labeled components are: (A) robotic autosampler; (B) microwave digestion unit; (C) fluid handling components for sample injection, automated standard addition, sample acidification/digestion; (D) separation fluidics including syringe pumps, flow reversal, and diversion valves; (E) separation column; (F) flow scintillation detector; and (G) control computer with automation software. Reproduced with permission from the "Handbook of Radioactivity Analysis, Second Edition" Chapter 14, page 1152. Copyright 2003 Elsevier USA.

sample. The total analysis time for the standard addition measurement was 22 minutes, including analysis of both sample and spiked sample.

Tests were conducted on various LAW samples from the Hanford site, including high organic content AN107 waste in which the majority of Tc is present in reduced oxidation states. The fully automated instrument correctly quantified the total Tc in these challenging samples, as demonstrated by comparisons with independent analysis by ICP-MS.

Discussion

Our initial work in automating laboratory analysis of radionuclides in nuclear waste has now advanced to the development of sensors and instruments for nonlaboratory settings. Our sensor for ^{99}Tc in groundwater was selected for prototype engineering development, and a field test was conducted at the Hanford site in September 2002. This unit successfully quantified the ^{99}Tc activity in Hanford groundwater samples. The automated Tc analyzer has been under development for use at the Hanford site, and was selected in a competitive process for incorporation into the Hanford Waste Treatment Plant.

These developments represent significant success in the development of radiochemical analysis methods for at site or in situ radiochemical measurements that can provide timely information on radiological composition in environmental or process monitoring applications.

Acknowledgments

The authors thank coworkers and collaborators Mike Knopf and Timothy DeVol for their contributions and enthusiasm. This research was supported by the Environmental Management Science Program of the Office of Biological and Environmental Research, U.S. Department of Energy (DOE). Sensor probe engineering was funded by the Advanced Monitoring Systems Initiative of the DOE. The Pacific Northwest National Laboratory is a multiprogram national laboratory operated for the DOE by Battelle Memorial Institute.

References

1. Murray, R. L. *Understanding Radioactive Waste*, 4th ed.; Battelle Press: Richland, WA, 1994; p 212.
2. Erickson, M. D.; Aldstadt, J. H.; Alvarado, J. S.; Crain, J. S.; Orlandini, K. A.; Smith, L. L. *J. Hazard. Mater.* **1995**, *41*, 351–358.

3. Koglin, E. N.; Poziomek, E. J.; Kram, M. L. In *Handbook of Vadose Zone Characterization & Monitoring*; Wilson, L. G., Everett, L. G., Cullen, S. J., Eds.; Lewis Publishers: Ann Arbor, MI, 1994; pp 675–676.
4. Grate, J. W.; Egorov, O. B. *Anal. Chem.* **1998**, *70*, 779A–788A.
5. Grate, J. W.; Strebin, R. S.; Janata, J.; Egorov, O.; Ruzicka, J. *Anal. Chem.* **1996**, *68*, 333–340.
6. Egorov, O.; Grate, J. W.; Ruzicka, J. *J. Radioanal. Nucl. Chem.* **1998**, *234*, 231–235.
7. Egorov, O.; O'Hara, M. J.; Grate, J. W.; Ruzicka, J. *Anal. Chem.* **1999**, *71*, 345–352.
8. Egorov, O. B.; O'Hara, M. J.; Ruzicka, J.; Grate, J. W. *Anal. Chem.* **1998**, *70*, 977–984.
9. Grate, J. W.; Egorov, O. *Anal. Chem.* **1998**, *70*, 3920–3929.
10. Grate, J. W.; Fadeff, S. K.; Egorov, O. *Analyst* **1999**, *124*, 203–210.
11. Grate, J. W.; Egorov, O. B.; Fiskum, S. K. *Analyst* **1999**, *124*, 1143–1150.
12. Egorov, O. B.; O'Hara, M. J.; Farmer, O. T., III; Grate, J. W. *Analyst* **2001**, *126*, 1594–1601.
13. Egorov, O. B.; Fiskum, S. K.; O'Hara, M. J.; Grate, J. W. *Anal. Chem.* **1999**, *71*, 5420–5429.
14. DeVol, T. A.; Egorov, O. B.; Roane, J. E.; Paulenova, A.; Grate, J. W. *J. Radioanal. Nucl. Chem.* **2001**, *249*, 181–189.
15. DeVol, T. A.; Duffey, J. M.; Paulenova, A. *J. Radioanal. Nucl. Chem.* **2001**, *249*, 295–301.
16. Roane, J. E.; DeVol, T. A. *Anal. Chem.* **2002**, *74*, 5629–5634.
17. DeVol, T. A.; Roane, J. E.; Williamson, J. M.; Duffey, J. M.; Harvey, J. T. *Radioact. Radiochem.* **2000**, *11*, 34–46.
18. Egorov, O.; O'Hara, M.; Grate, J. *Abstracts of Papers,* 223rd National Meeting of the American Chemical Society, Orlando, FL, April 7–11, 2002; American Chemical Society: Washington, DC, 2002; NUCL-006.
19. Egorov, O.; O'Hara, M.; Grate, J. W. *Abstracts of Papers,* 223rd National Meeting of the American Chemical Society, Orlando, FL, April 7–11, 2002; American Chemical Society: Washington, DC, 2002; NUCL-062.
20. Grate, J. W.; Egorov, O. B. *Abstracts of Papers,* 222nd National Meeting of the American Chemical Society, Chicago, IL, Aug 26–30, 2001; American Chemical Society: Washington, DC, 2001; IEC-010.
21. Grate, J. W.; Egorov, O. B. *Abstracts of Papers,* 222nd National Meeting of the American Chemical Society, Chicago, IL, Aug 26–30, 2001; American Chemical Society: Washington, DC, 2001; NUCL-059.
22. Egorov, O.; DeVol, T.; Grate, J. *Abstracts of Papers,* 222nd National Meeting of the American Chemical Society, Chicago, IL, Aug 26–30, 2001; American Chemical Society: Washington, DC, 2001; NUCL-183.
23. Unpublished results.

Chapter 15

Airborne Particle Size Distribution Measurements at DOE Fernald

N. H. Harley[1], P. Chittaporn[1], M. S. A. Heikkinen[1], R. Medora[2], and R. Merrill[3]

[1]New York University School of Medicine, Nelson Institute of Environmental Medicine, New York, NY 10016
[2]Oak Ridge Associated Universities, Oak Ridge, TN 37831
[3]Fluor Fernald Radiation Control Section, 7400 Willey Road, Hamilton, OH 45013–9402

The size of inhaled aerosol particles is the major determinant of lung dose. For this reason an integrating miniature particle size sampler was developed for the U.S. DOE Environmental Management Science Program for use at the former uranium processing facility in Fernald, OH during remediation. The sampler can be worn with a belt pump or used as an area detector. The tracer for the aerosol particles in this study is ^{210}Po, the naturally occurring radioactive decay product of ^{222}Rn (radon), but the size sampler can be adapted to any environment with selection of the measurement method. About two years of data have been measured approximately quarterly at Fernald and at a suburban home in New Jersey used as a quality control site. The particle size distribution at these two locations is shown for comparison for the winter season 2003 to 2004 and indicate a median size of 150–200 nanometers.

Introduction

No particle size sampler presently available commercially can operate continuously for up to three months in severe outdoor conditions to provide integrated, averaged size distribution information. The particle size sampler developed under the U.S. Department of Energy (DOE) Environmental Management Science Program (EMSP) provides this capability. In the present study, environmental concentrations of the naturally occurring long lived radon decay products ^{210}Pb, ^{210}Po were used as tracers for the short lived radon decay products of dosimetric interest. The miniature particle size sampler (4-cm diameter x 2-cm height; weight 55 g) used in this study was a commercial unit modified to include an inlet impactor stage that could be alpha counted directly and with 4 internal fine mesh screens and a back up filter. A new lightweight version in electrically conducting plastic is now available with an impactor stage, up to 6 fine mesh screen holders, and a back up filter. The radioactivity on all filter stages was measured in a low background alpha counter.

Fernald Feed Materials Production Center (FMPC)

The Fernald Feed Materials Production Center (FMPC), 16 miles from downtown Cincinnati, processed uranium ore and thorium from 1951 through 1988. FMPC processed raw ore (feed material) using nitrate digestion, followed by oxide reduction, tetrafluoride production, and then reduction to the metal. The uranium was separated from the nitrate digestion by solvent extraction, the radium was separated by barium sulfate coprecipitation, and the resulting salt cake was stored initially in drums, and later transferred to two silos. The remaining liquid residue (raffinate) containing the other radionuclides in the uranium natural activity series (^{230}Th, ^{210}Pb) was ponded in raffinate pits. Scrap from the metal fabrication was also recovered. Two large silos each contain about 7.4 x 10^{13} Bq (2000 Ci) of ^{226}Ra. Fernald is now undergoing restoration for return of the land to the State for future public use. The purpose of our measurements is to provide accurate personal exposure assessment of inhaled radioactivity, primarily radon and decay products, during the removal of relatively large quantities of radium- and thorium-bearing ore residues. Also, the measurements are to provide detailed data on the airborne transport of material within or outside the boundaries of the site, and assess current dispersion models.

Particle Size Distribution Measurements

The present sampler configuration uses a miniature commercial sampler head modified to hold a ZnS alpha phosphor impaction stage at inlet, followed by four, very fine mesh stainless steel screen filters 200 to 500 mesh), and a 0.8-μm Millipore membrane exit filter to collect all residual particles (1). A low flow pump (4 to 6 lpm) draws the atmosphere sampled through the sampler for periods of up to three months to yield an integrated particle size sample. Six units have sampled outdoors at the DOE Fernald site for about two years. The site locations are as follows: (1) on top of and between the two radium storage silos, (2) near the raffinate waste pits, (3) near the soil dryer that dehydrates material removed from the raffinate pits, and (4) several movable locations selected during the ongoing restoration process.

Radon and thoron measurements are always made concurrently with the particle size measurements using the miniature 4-chamber radon, thoron alpha track detector developed in conjunction with the particle size sampler for complete exposure assessment at Fernald (2). Radon gas is present in all atmospheres, and most of its short lived decay products attach immediately to the ambient aerosol particles, ultimately decaying to long lived ^{210}Pb/[210]Po. The alpha emitting [210]Po is measured directly and used as the aerosol particle tracer. The [210]Po on all 6 particle filtration stages is measured in our ultra low background scintillation counters (5 counts per day) (3), and a deconvolution program is used to estimate the particle size distribution from the measurements (4). Extensive calibrations were performed using monodisperse NaCl particles, and freshly formed (unattached) [222]Rn decay products, to determine the deposition characteristics of the sampler over the size range from nanometer size (defined here as < 20 nm) to ultrafine sized particles. Two size modes are generally present in outdoor field samples, a nanometer size (measured 1 to 10 nm) and an ultrafine/accumulation (100 to 500 nm) mode. There is generally a small contribution from a micron sized mode.

The screens and backup filter are counted repetitively over several months to measure the build up of [210]Po from the ^{210}Pb deposited directly on these filtration elements, along with the [210]Po arising from the filtered short-lived [222]Rn decay product chain. The build up measurements ensure that only [210]Po is being counted. If there is any discrepancy in the 138 day build up half time of [210]Po, alpha spectrometry is also performed to identify any other radionuclides present. To show that ^{210}Pb could be measured accurately by direct counting of the filters, data were accumulated for four months and measured both by alpha counting and radiochemical analysis of the filters. The alpha counting of the backup filters with low background counters was shown to provide excellent agreement with dissolution of the filter material and radiochemical measurement of the [210]Pb (5).

The collection efficiency of particulates on fine mesh screens has been studied theoretically and experimentally for many years (*6–10*). Two publications provide data on intercomparison exercises using various screen diffusion batteries (*11, 12*). A deconvolution program is necessary to generate the particle size distributions from the alpha counting data on the 6 collection stages. This is accomplished using an Extreme Value Estimation program (EVE) that provides a best estimate of the particle size distribution (*4*).

Two background locations in New Jersey are monitored during approximately the same time intervals to obtain a typical, suburban particle size distribution. These background data are used for quality control and comparative purposes.

Results

Figure 1 shows the particle size sampler and pump. Figure 2 shows the particle size distribution measurements made on top of the ^{226}Ra silos at Fernald from December 2003 to March 2004. Figure 3 shows the measurements made at one New Jersey suburban home from January to February 2004. Figures 2 and 3 also indicate the cumulative particle frequency curve, i.e., a summation of the particle size distribution data, for the two locations. The cumulative spectrum facilitates the calculation of bronchial dose because it yields the median (50%) particle diameter directly so that a dose factor can be applied accordingly. Figures 2 and 3 show that the air concentrations of ^{210}Pb are similar at Fernald and suburban New Jersey, 500 and 750 micro Bq m^{-3} respectively.

The outdoor measurements at Fernald have a median particle diameter similar to that of the New Jersey home (about 150 to 200 nm). It was expected that the Fernald silos would show a somewhat larger median particle diameter because of the ongoing remediation activities. However, the location on top of the silos does show a wider dispersion than this suburban residential New Jersey area. The Fernald site is located in a very remote area, and the New Jersey site is in a town of about 15,000 with mainly single family homes. The nanometer mode diameter does differ by a factor of two in size, with Fernald having the larger size, about 3 nm compared with less than 2 nm for the New Jersey location. The growth of nanometer sized particles is related to the concentration of other airborne constituents such as water vapor. Also, fewer airborne particulates may allow less growth for the nanometer size mode. However, the total mass on all of the filters was measured over the intervals studied, and was 12 µg m^{-3} at both locations.

The dosimetric consequence of particle size determination is that a realistic bronchial dose may be calculated (*13*). In this case, essentially the same dose conversion factor applies at both Fernald and at the residential home in New Jersey because the median diameter is identical. UNSCEAR 2000 (*14*) gives the

Figure 1. The Integrating Miniature Particle Size Sampler.

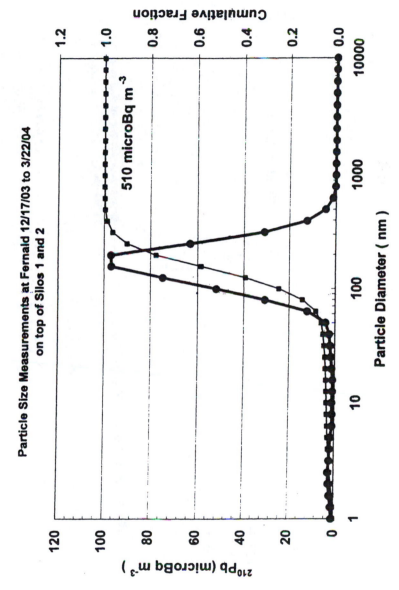

Figure 2. Particle Size Measurements and cumulative fraction on top of and between the Radium Silos 1 and 2 at Fernald, 12/17/03 to 3/22/04.

348

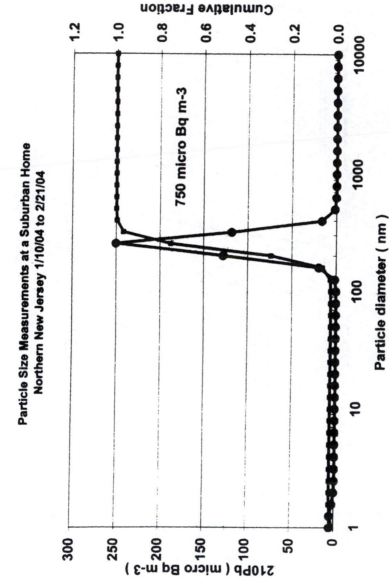

Figure 3. Particle size distribution measurements and cumulative fraction outdoors at a suburban New Jersey home, 1/10/04 to 2/21/04.

graph of dose conversion factors for radon decay products in units of equilibrium equivalent radon concentration (EEC) and as a function of both the median particle size and Fp, the fraction of potential energy of the unattached or nanometer sized mode. UNSCEAR (*14*) selects an average equilibrium factor of 60% for outdoor air. Therefore, the ^{222}Rn Bq m^{-3} EEC for outdoor radon will be (0.6) (outdoor concentration). UNSCEAR (*14*) uses a dose conversion factor of 6 nGy per (Bq hr m^{-3} EEC) for a median diameter of 200 nm.

Summary

The personal aerosol particle size sampler developed for EMSP is capable of measuring most integrated airborne particulate substance of interest depending upon the measurement technique applied to the filtration stages. In this study ^{210}Pb was measured as a tracer for radon decay product aerosol size and was accomplished at normal outdoor concentrations at the DOE Fernald site and a home in northern New Jersey.

Very little has been done for personal exposure assessment with regard to the inhaled particle size distribution. This can be attributed to practical difficulties in assessing particle size under field conditions. The measurements with alternative instruments can be very labor intensive, cover only short time periods, mainly require real time instrumentation, are not suitable for personal use, and are expensive.

The inhaled aerosol particle size is the major determinant of the radiation dose to the lung, because this determines the fraction that will be deposited on the airways and in the lower lung (*13*). As yet, there are no long-term particle size distribution data for airborne radionuclides at any site during DOE remediation activities with the exception of Fernald. In most cases when occupational lung dose is calculated, the particle size is assumed to be of nominal value, namely 1000 nm. At Fernald and in New Jersey, the median diameter was measured for about a 3-month period and found to be about 150 to 200 nm. This new integrating particle size sampler can fill an important gap in exposure assessment.

The sampler is now available in a lightweight electrically conducting plastic. It can accommodate up to six screen filters along with the impactor and back up filter stages. It is available from the molding company on request.

Acknowledgments

Support for the project under EMSP Contract DE-FG07-97ER62522 is gratefully acknowledged. Dr. Edith Robbins, New York University School of Medicine, Department of Cell Biology, performed the scanning electron microscope studies of the screen filter elements, thus providing the necessary data for the screen collection efficiency.

References

1. Harley, N. H.; Chittaporn, P.; Fisenne, I. M.; Perry, P. ^{222}Rn Decay Products as Tracers of Indoor and Outdoor Aerosol Particle Size. *J. Environ. Radioactivity* **2000**, *51*, 27–35.

2. Harley, N. H.; Chittaporn, P.; Medora, R.; Merrill, R. Measurements of Outdoor Radon and Thoron at Fernald, OH, New York City and New Jersey. *Health Physics* **2001**, *80* (Suppl. 6), S171.

3. Hallden, N. A.; Harley, J. H. An Improved Alpha-Counting Technique. *Anal. Chem.* **1960**, *32*, 1861.

4. Paatero, P. *The Extreme Value Estimation Deconvolution Method with Applications in Aerosol Research*; University of Helsinki Report Series in Physics, HU-P-250; University of Helsinki: Helsinki, Finland, 1990.

5. Fisenne, I. M.; Keller, H. W.; Keller, E. W. ^{210}Pb in Indoor Air by Total Alpha Measurement. *Health Phys.* **1996**, *71*, 723–726.

6. Thomas, J. W.; Hinchliffe, L. E. Filtration of 0.001 μm Particles by Wire Screens. *J. Aerosol Sci.* **1971**, *3*, 387–393.

7. Cheng, Y. S.; Yeh, H. C.; Brinsko, J. Use of Wire Screens as a Fan Filter. *Aerosol Sci. Technol.* **1985**, *4*, 165–174.

8. Reineking, A.; Porstendorfer, P. High Volume Screen Diffusion Batteries and Alpha Spectrometry for Measurement of the Radon Daughter Activity Size Distributions in the Environment. *J. Aerosol Sci.* **1986**, *17*, 873–880.

9. Holub, R. F.; Knutsen, E. O.; Soloman, S. Tests of the Graded Wire Screen Technique for Measuring the Amount and Size Distribution of Unattached Radon Progeny. *Radiat. Prot. Dosim.* **1988**, *24*, 265–268.

10. Hopke, P. K.; Ramamurthi, M.; Knutsen, E. O.; Tu, K. W.; Scofield, P.; Holub, R. F.; Cheng, Y. S.; Su, Y. F.; Winklmayr, W. The Measurement of Activity Weighted Size Distributions of Radon Progeny: Methods and Laboratory Intercomparison Studies. *Health Phys.* **1992**, *63*, 560–570.

11. Cheng, Y. S.; Yu, C. P.; Tu, K. W. Intercomparison of Activity Size Distributions of Thoron Progeny by Alpha and Gamma Counting Methods. *Health Phys.* **1994**, *66*, 72–79.

12. *International Intercalibration and Intercomparison Measurements of Radon Progeny Particle Size Distribution*; U.S. Department of Energy Environmental Measurement Laboratory Report, EML-589; EML: New York, July 1997.

13. Harley, N. H.; Cohen, B. S.; Robbins E. S. The Variability in Radon Decay Product Bronchial Dose. *Environ. Int.* **1996**, *22*, S959–964.

14. UNSCEAR. *Sources and Effects of Ionizing Radiation*. United Nations Scientific Committee on the Effects of Atomic Radiation, UNSCEAR Report to the General Assembly with Scientific Annexes; United Nations: New York, 2000.

Chapter 16

Visualization of DNA Double-Strand Break Repair at the Single-Molecule Level

William S. Dynan[1], Shuyi Li[1], Raymond Mernaugh[2], Stephanie Wragg[1], John Barrett[1], and Yoshihiko Takeda[1]

[1]Institute of Molecular Medicine and Genetics, Medical College of Georgia, Augusta, GA 30912
[2]Molecular Recognition Unit, Vanderbilt University, Nashville, TN 37232

Exposure to low doses of ionizing radiation is universal. The genotoxic effects of radiation are attributable, in large part, to DNA double-strand breaks (DSBs). This manuscript describes the development of reagents for visualizing individual DSBs, that is, with sensitivity at the single-molecule level. It also describes the use of these reagents to determine the pathway used for repair of rare DSBs induced at low radiation doses. To visualize DSBs in situ in living cells, we prepared antibodies to γ-H2AX, a modified histone isoform that accumulates at sites of unrepaired DSBs. We compared results with antibodies obtained using different technologies. Initially, we screened a phage display library to identify single chain antibody variable fragments (scFvs) that were specific for the characteristic C-terminal phosphopeptide of γ-H2AX. Although highly selective for this phosphopeptide in vitro, the antibodies were not able to detect γ-H2AX foci in situ, in irradiated cells. We found that conventional antibodies were superior to the scFvs for this purpose. Immunostaining with either affinity-purified rabbit antibodies or a commercial monoclonal antibody revealed distinct γ-H2AX foci in numbers consistent with the number of DSBs expected at a given dose. Separately, we developed a high-affinity scFv directed against a key DSB repair protein, the DNA-dependent protein kinase catalytic subunit. We introduced this scFv into human SK-MEL-28 melanoma cells by microinjection. The presence of the scFv caused radiation-induced γ-H2AX foci to persist to a much greater extent than in control cells. Similar results were obtained at 150 cGy and 10 cGy doses of gamma radiation. These studies provide the first direct evidence that the DNA-dependent protein kinase is required for the repair of rare DSBs induced at low radiation doses.

Introduction

The genotoxic effects of ionizing radiation arise from inhomogeneous deposition of energy within the cell. Ionization occurs along defined tracks. When a track intersects DNA, direct ionization, as well as reactive oxygen species generated from ionization of water, react to cause clusters of damage. When this damage affects both strands within a small region, double-strand breaks (DSBs) can occur. Because they interrupt chromosomal integrity, DSBs are far more lethal, on a per-event basis, than other forms of DNA damage. If not promptly repaired, genetic material on the distal side of the break, relative to the centromere, will not segregate properly at the next round of cell division, leading to loss of genes, and eventually, cell death. Moreover, when multiple breaks occur in the same cell, DNA ends from different chromosomes can form illegitimate unions, resulting in translocations and aberrations such as dicentric and ring chromosomes. These, in turn, result in ongoing genetic instability and, sometimes, neoplastic transformation. DSBs also induce unique signaling events, including the activation of the *ataxia telangiectasia mutated* (*ATM*) gene product, which initiates a cascade of regulatory phosphorylation (*1-3*).

To a large extent, the injurious effects of exposure to ionizing radiation are mitigated by DSB repair pathways. In eukaryotic cells, there are two principal repair pathways: nonhomologous end joining (NHEJ) and homologous recombination (HR) (Figure 1). NHEJ involves the ligase-catalyzed rejoining of broken DNA (reviewed in [*4*]). The ligation itself is carried out by DNA ligase IV (DNL IV), which forms a mixed tetramer with another gene product required for NHEJ, the XRCC4 protein (*5*). The DNA ligase IV/XRCC4 complex is present constitutively, and its sole function appears to be the rejoining of DSBs induced by DNA damaging agents and by certain recombination endonucleases. Surprisingly, purified DNL IV is not capable of joining the ends of double-stranded DNA fragments in vitro. It is, rather,

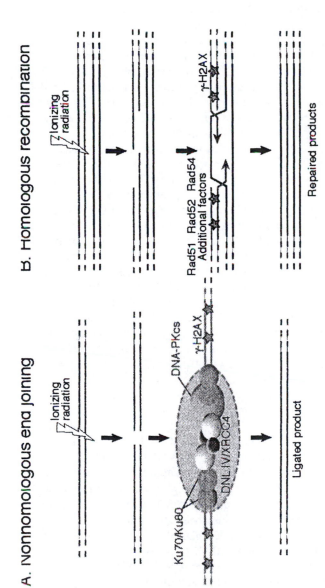

Figure 1. Pathways of DNA double-strand break repair. (A) Nonhomologous end joining pathway. Heterodimer Ku protein (composed of Ku70 and Ku80 subunits) binds to a DSB and recruits DNA-PKcs. This complex serves as a platform for recruitment of additional proteins, including DNA ligase IV, XRCC4, and at least one yet-to-be-identified factor. DNA-PKcs phosphorylates itself, XRCC4, and perhaps other targets. Other radiation-inducible kinases phosphorylate histone H2AX, causing a large scale change in chromatin structure in the vicinity of the DSB. (B) Homologous recombination pathway. A second, intact copy of a gene is used as a template for repair synthesis. A different set of proteins is required than for NHEJ.

absolutely dependent on a set of additional proteins that align the DNA ends and prepare them for rejoining. In mammalian cells, a minimum of fou. other polypeptides are required: the two subunits of Ku protein (Ku70 and Ku80), which carry out initial recognition of DNA ends, the DNA-dependent protein kinase catalytic subunit (DNA-PKcs), a serine/threonine kinase that phosphorylates one or more critical target sites within the repair complex, and at least one yet-to be-identified factor (4, 6). Loss of any of these factors through mutation results in severe radiosensitivity. NHEJ occurs rapidly, with a half-time of minutes (7, 8), but is intrinsically error-prone. Radiation-induced damage to DNA bases and deoxyribose moieties in the immediate vicinity of the broken end is believed to necessitate resection of at least a few residues at the extreme termini. When these ends are joined, a small deletion results.

The other pathway of DSB repair, homologous recombination, requires a second, intact copy of the broken chromosome, which is used as a template to synthesize DNA across the break (4, 9, 10). HR has been best studied in the yeast, *Saccharomyces cerevisiae*, where it is the predominant mechanism of repair. Like NHEJ, it requires the action of multiple proteins, including Rad 51, Rad 52, Rad 54, DNA polymerases, and others. In contrast to NHEJ, most of the proteins involved in HR have other functions in the cell, including mediating general genetic recombination. Most of the proteins involved in HR are required for cell viability in higher eukaryotes, making it impossible to isolate stable null alleles, and thus complicating genetic analysis (10). Because it involves templated DNA synthesis across the break, HR is potentially error-free.

Both NHEJ and HR are accompanied by large-scale chromatin changes in the vicinity of the DSB. These changes, which occur within about a megabase of the break site, involve phosphorylation of a histone H2A isoform, H2AX (11–13). The phosphorylated H2AX, known as γ-H2AX, can be detected with phosphospecific antibodies. These antibodies stain cells to form discrete foci. A variety of evidence suggests that these foci correspond on a direct, one-to-one basis, with DSBs. The number of foci correlates with the expected number of DSBs at a given dose (14, 15). Foci appear rapidly after irradiation and disappear on a time scale of 30-90 min, consistent with the kinetics of DSB repair. Moreover, in lymphocytes undergoing V(D)J recombination, a process that involves a DSB intermediate, very prominent foci are seen corresponding to recombination centers (16). Although γ-H2AX remains the prototypical marker for individual DSBs (that is, DSBs affecting a single DNA molecule), several other proteins have since been shown to accumulate in a similar distribution, including 53BP1, a phosphorylated isoform of DNA-PKcs, and a regulatory protein, NFBD1/MEC1 (17–20).

In mammalian cells, there is a balance between HR and NHEJ pathways. HR is believed to be predominant in the G2 phase of the cell cycle, after DNA replication, when a sister chromatid is available to use as a template (21, 22). An important caveat, however, is that virtually all studies of

DSB repair have been conducted at relatively high doses, sufficient to induce substantial cell killing. Conventional methods of detecting DSB repair in vivo, which are based on extraction of DNA and measurement of mobility in pulsed-field electrophoresis, require doses that induce 100 or more breaks per cell. For the human genome, these correspond to acute doses of approximately 3 Gray (Gy) or more of low-linear energy transfer (LET) radiation (8).

By contrast, environmentally relevant doses are far smaller. Environmental exposure is received from radioactive minerals and their decay products, cosmic rays and cosmogenic isotopes (such as tritium), and medical X-rays. There is also occupational exposure that is variable but, on the average, small. The average annual dose received by a U.S. resident is on the order of 3 mSv, equivalent to 3 mGy of low LET radiation (23). Thus, the *acute* dose in a typical laboratory experiment is about three orders of magnitude greater than the *annual* dose from background sources.

There is a paucity of information about how the ratio of NHEJ to HR is affected by radiation dose, especially in the environmentally relevant low dose range. The question is of considerable importance. A change in relative utilization of the two pathways as a function of dose could have a significant impact on risk estimates. For example, since NHEJ is error prone and HR is potentially error free, a shift from NHEJ to HR at low doses might reduce the risk of mutation per unit dose. The purpose of this project is to develop and assess reagents for investigation of the pathways used for DSB repair at low doses. We use these reagents to demonstrate that inhibition of the DNA-dependent protein kinase, a key enzyme in the NHEJ pathway, blocks repair of rare DSBs induced at low radiation doses.

Materials and Methods

ScFv Selection

Single chain antibody variable fragments (ScFvs) directed against γ-H2AX were obtained by phage antibody selection using biotinylated H2AX C-terminal peptides synthesized with or without site-specific phosphorylation (Figure 2). After two or three rounds of selection on peptides, phage were used to infect *Escherichia coli* TG1. Clones were picked and transferred to microtiter wells to produce soluble epitope-tagged scFvs, which were analyzed for γ-H2AX binding ability using peptides immobilized onto streptavidin-coated microtiter plates. Binding was detected using an anti-epitope tag monoclonal antibody conjugated to horseradish peroxidase, hydrogen peroxide, and the colorimetric horseradish peroxidase substrate, [2,2'-Azino-bis [3-ethylbenziazoline-6-sulfonic acid]]. ScFv 18-2, which is directed against the

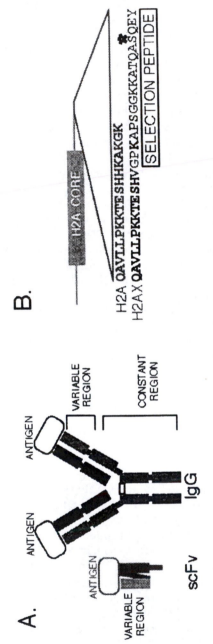

Figure 2. Strategy for production of scFvs against γ-H2AX. (A) General structure of an scFv compared with a naturally occurring IgG. (B) Peptide used for selection of scFvs from a combinatorial library. Asterisk denotes phosphoserine uniquely present in γ-H2AX.

DNA-dependent protein kinase catalytic subunit, was obtained by cDNA cloning from a monoclonal antibody 18-2-producing cell line (*24*). ScFvs were purified by conventional or anti-E tag affinity chromatography for use in enzyme-linked immunosorbent assay (ELISA), immunofluorescence assays, and microinjection experiments.

Rabbit Immunization

A peptide representing the C-terminal region of γ-H2AX was conjugated to keyhole limpet hemocyanin and used for immunization of rabbits. Immunizations were performed by Biosynthesis, Inc. (Louisville, TX). Primary immunization was performed using 200 μg of conjugated peptide in Freund's complete adjuvant. Booster injections were performed with the same amount of peptide in Freund's incomplete adjuvant at days 14, 28, 42, and 56. Production bleeds were performed at days 42, 51, and 86. Immunoaffinity purification was performed using serum collected at day 51. Serum was passed over a column of biotinylated γ-H2AX peptide bound to NeutrAvidin-agarose (Pierce, Rockford IL). Antibody was eluted from the column with Immunopure IgG elution buffer (Pierce).

ELISA and Surface Plasmon Resonance Assays

For anti γ-H2AX scFvs and sera, ELISA was performed using biotinylated H2AX peptides bound to a NeutrAvidin-coated multiwell plate (Pierce). Wells were incubated with scFv or rabbit antiserum as indicated in the legend to Figure 3. Binding of scFv was detected using anti-E tag secondary antibody (Amersham Biosciences) and alkaline-phosphatase conjugated goat anti-mouse tertiary antibody. Binding of rabbit antibodies was detected using alkaline phosphatase-conjugated goat anti-rabbit secondary antibody. Wells were developed with Blue Phos substrate (KPL, Gaithersburg, MD). Surface plasmon resonance was performed using a Biacore X instrument (Biacore, Inc., Piscataway, NJ) where the ligand was of a biotinylated γ-H2AX peptide bound to a streptavidin-coated sensor chip, and the analyte was scFv at the concentrations indicated in Figure 3. Data were analyzed assuming a 1:1 Langmuir binding model. For anti-DNA-PKcs scFv and mAb preparations, ELISA was performed by coating purified DNA-PKcs on a flat-bottom, high binding polystyrene plate essentially as described (*25*).

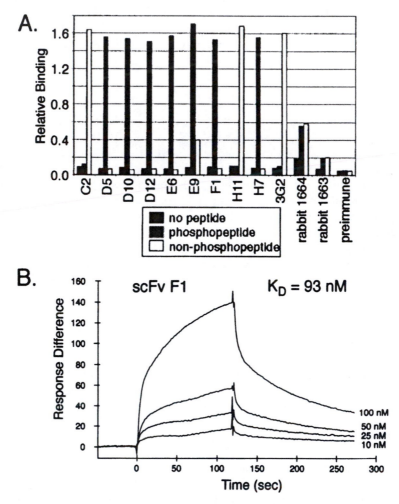

Figure 3. Characterization of scFvs selected against H2AX peptides. (A) ELISA assays performed as described in Materials/Methods. (B) Surface plasmon resonance sensorgram showing kinetics of association/dissociation for scFv F1 and phosphorylated peptide. (C) Immunofluorescence assays to test binding of scFv to nonirradiated and irradiated HeLa cells. Rare foci that do not increase in number upon irradiation are believed to reflect nonspecific staining. (D) Control experiment demonstrating ability to detect binding of scFv 18-2, directed against DNA-PKcs, under similar staining conditions. Human glioma cell line M059K contains DNA-PKcs, and M059J is a DNA-PKcs-negative variant.

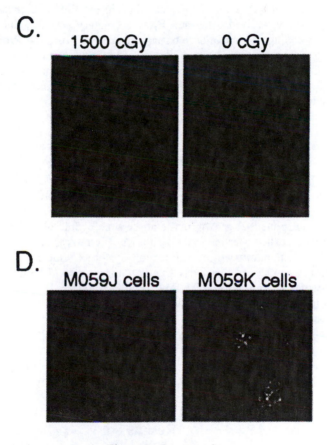

Figure 3. *Continued.*

Immunofluorescence Assays

SV40-transformed human fibroblasts (strain GM00637H, Coriell Cell Repository, Camden, NJ) were cultured on coverslips and exposed to ^{137}Cs irradiation (GammaCell 40 Exactor, MDS Nordion, Kanata, ON) at a rate of 1 Gy/min. Cells were allowed to recover for 30 min, fixed with 2% formaldehyde, permeabilized with 0.5% Triton X-100 and stained with scFv or rabbit anti-γ-H2AX antibody as described in the Figure Legends. Staining with scFv was visualized using anti-E tag secondary antibody and AlexaFluor 488 or 594-conjugated anti-mouse tertiary antibody (Molecular Probes, Eugene, OR). Staining with rabbit antibody was visualized using AlexaFluor 594-conjugated goat anti-rabbit secondary antibody (Molecular Probes). Images were collected with a Zeiss DeltaVision microscope. Staining of SK-MEL-28 human skin melanoma cells was as described (26).

Microinjection

The SK-MEL-28 human skin melanoma cell line (27) was grown in DMEM supplemented with 10% fetal bovine serum and antibiotics. Cells were seeded on 175 µm CELLocate coverslips (Eppendorf AG, Hamburg, Germany) and were in exponential growth phase at the time of microinjection. Cells were microinjected using sterile microcapillaries (Femtotips II, Eppendorf AG) mounted on an automated microinjection system (FemtoJet and InjectMan; Eppendorf AG) attached to a Zeiss Axiovert microscope. The injection mixture consisted of 1 mg/ml scFv, 15 µg/ml pEGFP-N1 DNA (Clontech, Palo Alto, CA), 10 mM KH_2PO_4, pH 7.4, and 75 mM KCl. Each injection was performed at a pressure of 50 hPa for 0.2 s. Cells were allowed to recover for 2-3 h at 37 °C and irradiated using a ^{137}Cs source at a rate of 1 Gy/min (for 150 cGy irradiation) or X-rays from a Varian 6 MV linear accelerator at the Georgia Radiation Therapy Center (for 10 cGy irradiation). Successfully injected cells were identified by GFP fluorescence after 12–24 hours.

Results

Experimental Design

Our studies required the development of antibody reagents for two purposes: (1) the identification of individual, unrepaired DSBs in situ in irradiated cells, and (2) the modulation of these repair pathways in living cells in order to establish which pathway is required for repair of rare DSBs induced at low doses of ionizing radiation. For the first of these purposes, we required antibodies against the γ-H2AX, a phosphorylated histone H2A isoform that has previously been shown to form discrete foci at sites of DNA double-strand rarebreaks (*11*). One of the advantages of γ-H2AX is that multiple phosphorylation events occur in response to each DSB event. This natural biological amplification facilitates detection of the individual DSB events, that is, detection at the single-molecule level. For the second purpose, modulation of repair pathways, we used a recombinant protein derived from an antibody that binds to the catalytic subunit of a key DSB repair enzyme, the DNA-dependent protein kinase. Our overall strategy was to inhibit DNA-PKcs function in living cells and measure the effect on persistence of unrepaired breaks, using the anti-γ-H2AX antibody.

Isolation of scFvs Directed Against γ-H2AX

The key technical requirement for detection of single-molecule events is optimization of signal-to-noise ratio. Although anti-peptide antibodies reactive with γ-H2AX have previously been described, we were interested in determining whether scFv technology might provide a better reagent. Figure 2A shows a comparison of the structure of an scFv and a naturally occurring immunoglobulin G (IgG) molecule (*28*). In the IgG, the antigen-binding surface is part of a domain formed by the interface of heavy chain variable (V_H) and light chain variable (V_L) regions. In the scFv, the V_H and V_L sequences are joined by a flexible linker to form a single molecule. ScFvs can be selected from a combinatorial library by phage display, without the need for immunization of an experimental animal. Unlike naturally occurring antibodies, scFvs are recombinant molecules that can be better standardized from batch to batch. Their sequences can be modified to enhance stability and antigen-binding affinity, or to introduce fluorophores to facilitate detection. Because scFvs are smaller than naturally occurring antibodies, they may be better able to access the relevant γ-H2AX epitope within the chromatin environment. In addition, they have the potential for expression as intrabodies within living cells.

We selected scFvs from a combinatorial phage display library using both phosphorylated and nonphosphorylated forms of a C-terminal γ-H2AX peptide. Figure 2B shows the peptide sequence used for selection. Ten separate bacteriophage isolates were chosen for further characterization. Soluble scFvs were produced in the periplasm of an *E. coli* host. Periplasmic extracts were either used directly or as a source of material for further purification.

Results with the ten chosen scFvs are shown in Figure 3A. ScFvs selected with phosphorylated peptide were highly specific for phosphorylated peptide over nonphosphorylated peptide (D5, D10, D12, E6, E9, F1, H7). ScFvs selected with the nonphosphorylated peptide showed reciprocal specificity (C2, H11, 3G2).

Binding of scFv F1 to the phosphorylated γ-H2AX peptide was further evaluated by surface plasmon resonance analysis. Binding was readily detected (Figure 3B. However, the scFv dissociated relatively rapidly, and the calculated binding affinity was weak (K_D approximately 100 nM). We next tested this scFv for their ability to detect γ-H2AX foci in situ in irradiated cells. Only scattered nonspecific staining was observed, and this was not radiation dependent (Figure 3C). As a control, we performed staining under similar conditions with scFv 18-2 (discussed subsequently), which binds with high affinity to DNA-PKcs. We saw robust staining, which was specific for DNA-PKcs-containing M059Kcells and absent from the matched, DNA-PKcs-deficient M059J cells (Figure 3D). Other γ-H2AX-specific scFv samples also gave negative results in the immunostaining assay. It may be that the affinity of the scFvs was too low to allow practical use in immunostaining. Alternatively, the scFvs may have been unable to recognize the γ-H2AX epitope in its native context. Although the scFv technology has promise, results suggested that considerable additional work might be required to obtain scFvs with optimal binding properties.

Alternative Approach Using Affinity-Purified Rabbit Anti-γ-H2AX

As an alternative source of anti-γ-H2AX antibodies for this project, we immunized rabbits with carrier-conjugated γ-H2AX phosphopeptide. ELISA and immunostaining assays were performed to assess the specificity of binding. Raw serum showed no specificity for the phosphopeptide, versus the non-phosphopeptide in the ELISA (Figure 3). Raw serum also showed only modest specificity for irradiated versus nonirradiated cells in the immunostaining assay, and this specificity declined with time after additional immunization (Figure 4, panel A). We hypothesized that the specificity of an initial immune response declined because of epitope spreading, a phenomenon that is commonly observed when an immune response is mounted against a self antigen (29). To enrich for phosphospecific antibodies, we applied the serum to a γ-H2AX phosphopeptide affinity column. After elution at low pH, affinity-purified antibodies were tested in an immunostaining assay. Staining was highly specific for radiation-exposed cells and was effectively competed by phosphopeptide (panel B).

To establish whether immunostaining was dose-responsive, an experiment was performed where human cells were exposed to doses of 10-100 cGy of gamma radiation (Figure 5). Cells were immunostained with affinity-purified antibody. We observed a dose-dependent increase in foci over the range 10-100 cGy. Quantitation showed that the number of foci corresponds well with

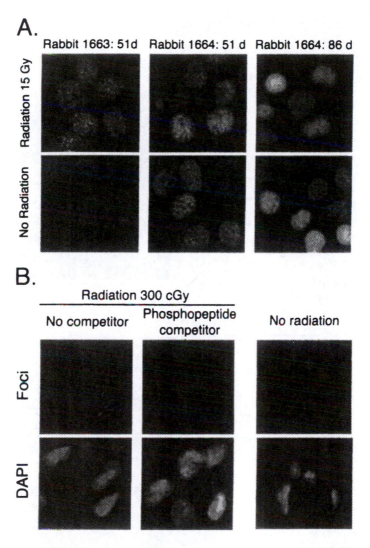

Figure 4. Development of affinity-purified rabbit anti-γ-H2AX antibodies.
Rabbits were immunized, sera were collected, and immunostaining was
performed using nonirradiated and irradiated HeLa cells as described in
Materials and Methods. (A) Raw serum collected at 51 or 86 days post-
immunization, as indicated. (B) Panels labeled "Foci" show immunostaining
with antibodies obtained by γ-H2AX phosphopeptide affinity chromatography of
serum from 51d bleed of rabbit 1663. Cells were treated with 300 cGy or no
radiation as indicated. Phosphopeptide competitor (1 μg/ml) was present where
indicated. Panels labeled "DAPI" show counterstaining for DNA.

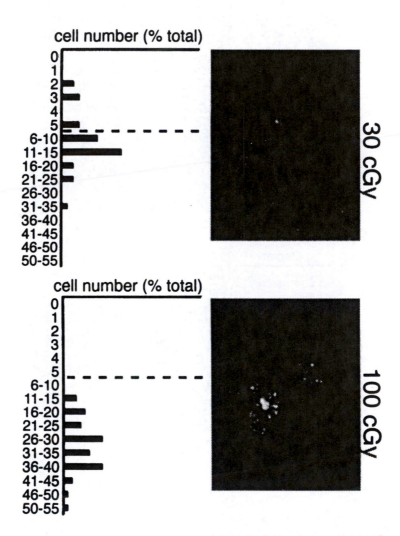

Figure 5. Detection and quantitation of γ-H2AX foci at low radiation doses. Human fibroblasts were immunostained with affinity-purified rabbit anti-γ-H2AX antibody and counterstained for DNA with DAPI. Gamma irradiation

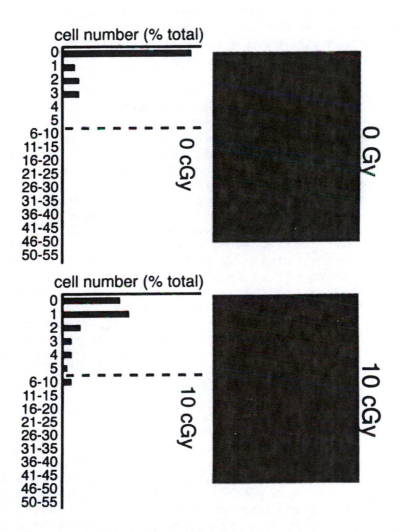

was performed as described in Materials and Methods. Approximately 30 cells were scored at each dose.

the number of anticipated double-strand breaks doses (ranging from roughly 3 breaks/cell at 10 cGy to 30 breaks/cell at 100 cGy). This suggests that each focus represents the response to an individual DSB repair event, even at very low doses that introduce only a few breaks per cell. Recent work in other laboratories, using independent anti-γ-H2AX preparations, has led to similar conclusions (*14, 15*).

Isolation of an ScFv Directed Against DNA-PKcs

The next phase of our experimental plan required the isolation of antibodies capable of modulating pathways of DSB repair in vivo, in living cells. To accomplish this, we developed a novel scFv, derived from a previously described monoclonal antibody (mAb) directed against DNA-PKcs, mAb 18-2 (*24*). A reverse transcriptase-polymerase chain reaction strategy was used to amplify the rearranged heavy and light chain variable region genes from mAb 18-2-expressing cells. Amplified genes were assembled into a scFv-encoding cDNA, which was subcloned for overexpression in the *E. coli* periplasm. Purified scFv selectively bound to purified DNA-PKcs in an ELISA (Figure 6). We have recently published elsewhere a more detailed characterization of the DNA-PKcs interaction properties of the scFv (*26*). Binding occurs with high affinity (K_D = 1-2 X 10^{-9} M) and is specific for a 25-residue linear sequence that occurs in almost the exact center of the 4128-residue DNA-PKcs polypeptide chain (residues 2000-2025). Like the parent mAb, the scFv only partially inhibits protein kinase activity (*24, 26*). However, it completely blocks DNA end-joining activity in a cell-free assay system (*26*).

Effect of scFv 18-2 on the Low Dose Radiation Response

We next tested scFv 18-2 for its effect on repair of rare DSBs induced at low radiation doses. Microinjection of antibodies is a well-established method to study intracellular protein function (*30, 31*). Although scFvs can, in principle, be expressed intracellularly by gene transfer, microinjection was chosen for the present study because it allows introduction of native, folded antibody directly into the nucleus. This eliminates concerns over disulfide bond formation and folding in the intracellular environment, which are common obstacles to use of scFvs for intracellular applications (*32*). Preliminary experiments showed that microinjected scFv 18-2 was stable inside cells and that it colocalized with DNA-PKcs in the nucleus (*26*).

Experiments were performed using human SK-MEL-28 cells, which are derived from a radioresistant melanoma (*27*). Figure 7, Panel A (a-g) shows images of cells that have been immunostained with anti-γ-H2AX and

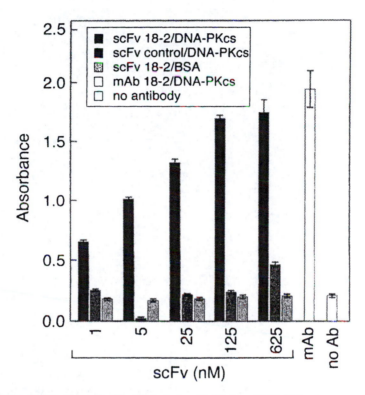

Figure 6. Binding of scFv 18-2 and mAb 18-2 to DNA-PKcs measured by ELISA. Standard ELISA assays were performed as described in Material and Methods. Wells were coated with DNA-PKcs or bovine serum albumin (BSA) as indicated. They contained the indicated amounts of scFv 18-2, control scFv (anti-γ-H2AX), mAb 18-2, or no antibody as indicated.

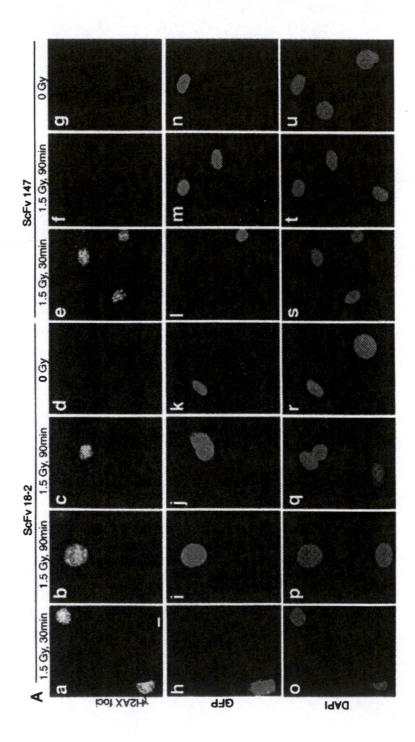

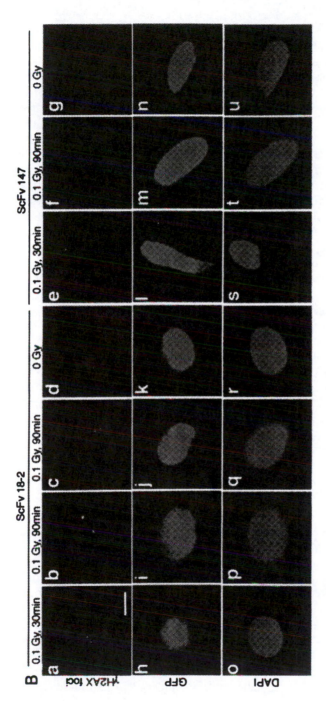

Figure 7. ScFv 18-2 prolongs the lifetime of γ-H2AX foci. (A) SK-MEL-28 cells were injected with scFv 18-2 or with scFv 147 and irradiated at 1.5 Gy. Cells were fixed at 30 min or 90 min after irradiation and stained with DAPI, anti-γ-H2AX antibody, and anti-GFP antibody as described in Materials and Methods. Cells were imaged using filters to detect each fluorophore separately. (B) As in panel A, except with 0.1 Gy of ionizing radiation. Scale bar indicates 10 μm in both panels. There were 50–100 successfully injected cells in each experimental group, and results shown are typical. These data have been published in a different format elsewhere (Reproduced with permission from reference 26. Copyright 2003 Oxford University Press.)

counterstained for DNA with DAPI. The staining protocol was similar to that in Figure 4 except that a commercial anti-γ-H2AX monoclonal antibody was used. This antibody became available late in the course of the project and gave essentially identical results to the affinity-purified rabbit anti-γ-H2AX rabbit serum. After treatment with 150 cGy of gamma radiation, which is calculated to produce ~50 DSBs per diploid human genome (*8*, *33*), prominent foci formed within 30 min (a, e). Foci were not observed in nonirradiated control cells (Panel A [d, g]).

Microinjected cells received either purified scFv 18-2, directed against DNA-PKcs, or a control scFv 147, directed against a small organic molecule not present in mammalian cells (*34*). Plasmid DNA vector encoding enhanced green fluorescent protein (EGFP) was co-injected to allow tracking of which cells were injected. Similar staining for γ-H2AX was seen at 30 min, regardless of which scFv the cells received, or if they were non-injected bystanders (Panel A (a, e). However, at 90 min post-irradiation, the γ-H2AX persisted at high levels only in cells receiving scFv 18-2 (b, c), and disappeared from non-injected cells in the same fields and from cells receiving control scFv (f).

Panel B shows the same experiment in SK-MEL-28 cells treated with 10 cGy of X-rays. Again, induction of γ-H2AX foci was similar in both irradiated groups (a, e). The foci persisted in cells receiving scFv 18-2 (b, c), but were quickly resolved in non-injected cells (not shown) and in cells receiving control scFv (f). Together, results suggest that scFv 18-2 blocks or delays repair of DSBs in vivo. Importantly, the magnitude of this effect was similar at both the high (150 cGy) and low (10 cGy) doses. These experiments provide the first direct evidence that DNA-PKcs is involved in the low dose radiation response.

Discussion

Here we have compared different approaches for measuring γ-H2AX repair foci. In a single round of selection, we readily obtained scFvs that were highly selective for the phosphorylated isoform of the H2AX histone tail. However, the practical utility of the scFvs was limited, as they failed to stain repair foci in an immunofluorescence assay. In principle, scFvs are a very powerful approach for identification of antibodies directed against almost any ligand. It is possible that the scFvs reported here could be improved with further rounds of mutagenesis and selection. However, affinity-purified rabbit antisera and the commercial monoclonal antibody were of more immediate value as a reagent for our studies.

The availability of methods to measure the response to small numbers of DSB events induced in single cells by environmentally relevant doses of radiation is useful for investigation of the mechanism of the response to low radiation doses that induce only one or a few breaks per cell. Here, we show that microinjection of a high-affinity scFv directed against DNA-PKcs leads to persistence of γ-H2AX repair foci and, by implication, persistence of unrepaired DSBs. These findings, portions of which have also been published elsewhere (26), provide the first direct evidence that DNA-PKcs is involved in the low dose radiation response.

The choice of repair pathways at low doses is a matter of considerable biological importance. The two main DSB repair pathways identified in high dose studies differ in their outcomes. NHEJ is rapid but intrinsically error-prone, because it often involves small deletions of damaged DNA segments. HR is potentially error-free, because it uses an intact copy of the gene as template. However, HR can introduce errors (for example, if recombination occurs between repeated sequences in different regions of the chromosome or genome) and, in mammalian cells, appears to be primarily active in the G2 phase of the cell cycle, a phase which is not populated in mature, non-dividing tissues. Information about the actual pathways of repair used at low doses will be helpful in modeling the effects of low-level radiation exposure and in understanding how polymorphisms in repair genes may impact on individual variation in risk.

Repair genes are highly conserved in vertebrate species. Thus, it will be of interest to extend the studies described here to other species, including laboratory model organisms or free-ranging wildlife. With support from the U.S. Department of Energy (DOE) Low Dose Radiation Research program, we have recently initiated experiments to apply some of the same approaches described here to the zebrafish, *Danio rerio* (D. Kozlowski and W.S. Dynan, unpublished results). The zebrafish has become a widely used model for biomedical research. The entire genome sequence is projected to be available by 2005, and hundreds of thousands of expressed sequence tag (EST) clones have been characterized. Embryos are optically transparent and develop outside the mother over a 48 h timescale, facilitating investigation of the effects of low dose radiation at the earliest and most vulnerable stages of development.

Acknowledgments

We thank Dr. Deborah Lewis for instruction in the microinjection technique, Dr. Alex Chiu for the use of the microinjection apparatus, Dr. Miyake Katsuya for assistance with image collection, Dr. David Munn for suggesting the use of the SK-MEL-28 cell line, Dr. Andrew Hayhurst for scFv 147, and Dr. David Bickel for statistical advice. We acknowledge the Medical College of Georgia Imaging Core Facility, Medical Illustration and Photography, and Office of Biostatistics and Bioinformatics for their services. This work was supported by an award from the DOE Low Dose Radiation Research program, DE-FG07-99ER62875 and by a U.S. Public Health Service grant, GM 35866.

References

1. Canman, C. E.; Lim, D. S.; Cimprich, K. A.; Taya, Y.; Tamai, K.; Sakaguchi, K.; Appella, E.; Kastan, M. B.; Siliciano, J. D. *Science* **1998**, *281*, 1677–1679.
2. Andegeko, Y.; Moyal, L.; Mittelman, L.; Tsarfaty, I.; Shiloh, Y.; Rotman, G. *J. Biol. Chem.* **2001**, *276*, 38224–38230.
3. Khanna, K. K.; Jackson, S. P. *Nat. Genet.* **2001**, *27*, 247–254.
4. Haber, J. E. *Trends Genet.* **2000**, *16*, 259–264.
5. Lee, K. J.; Huang, J.; Takeda, Y.; Dynan, W. S. *J. Biol. Chem.* **2000**, *275*, 34787–34796.
6. Featherstone, C.; Jackson, S. P. *Curr. Biol.* **1999**, *9*, R759–761.
7. Wang, H.; Zeng, Z. C.; Perrault, A. R.; Cheng, X.; Qin, W.; Iliakis, G. *Nucleic Acids Res.* **2001**, *29*, 1653–1660.
8. Metzger, L.; Iliakis, G. *Int. J. Radiat. Biol.* **1991**, *59*, 1325–1339.
9. Haber, J. E. *Curr. Opin. Cell Biol.* **1992**, *4*, 401–412.
10. Sonoda, E.; Takata, M.; Yamashita, Y. M.; Morrison, C.; Takeda, S. *Proc. Natl. Acad. Sci. U.S.A.* **2001**, *98*, 8388–8394.
11. Rogakou, E. P.; Boon, C.; Redon, C.; Bonner, W. M. *J. Cell Biol.* **1999**, *146*, 905–916.
12. Paull, T. T.; Rogakou, E. P.; Yamazaki, V.; Kirchgessner, C. U.; Gellert, M.; Bonner, W. M. *Curr. Biol.* **2000**, *10*, 886–895.
13. Rogakou, E. P.; Pilch, D. R.; Orr, A. H.; Ivanova, V. S.; Bonner, W. M. *J. Biol. Chem.* **1998**, *273*, 5858–5868.
14. Sedelnikova, O. A.; Rogakou, E. P.; Panyutin, I. G.; Bonner, W. M. *Radiat. Res.* **2002**, *158*, 486–492.
15. Rothkamm, K.; Löbrich, M. *Proc. Natl. Acad. Sci. U.S.A.* **2003**, *100*, 5057–5062.
16. Chen, H. T.; Bhandoola, A.; Difilippantonio, M. J.; Zhu, J.; Brown, M. J.; Tai, X.; Rogakou, E. P.; Brotz, T. M.; Bonner, W. M.; Ried, T.; Nussenzweig, A. *Science* **2000**, *290*, 1962–1965.
17. Shang, Y. L.; Bodero, A. J.; Chen, P. L. *J. Biol. Chem.* **2003**, *278*, 6323–6329.
18. Chan, D. W.; Chen, B. P.; Prithivirajsingh, S.; Kurimasa, A.; Story, M. D.; Qin, J.; Chen, D. J. *Genes Dev.* **2002**, *16*, 2333–2338.
19. Schultz, L. B.; Chehab, N. H.; Malikzay, A.; Halazonetis, T. D. *J. Cell Biol.* **2000**, *151*, 1381–1390.
20. Anderson, L.; Henderson, C.; Adachi, Y. *Mol. Cell. Biol.* **2001**, *21*, 1719–1729.

21. Takata, M.; Sasaki, M. S.; Sonoda, E.; Morrison, C.; Hashimoto, M.; Utsumi, H.; Yamaguchi-Iwai, Y.; Shinohara, A.; Takeda, S. *EMBO J.* **1998**, *17*, 5497–5508.
22. Lee, S. E.; Mitchell, R. A.; Cheng, A.; Hendrickson, E. A. *Mol. Cell. Biol.* **1997**, *17*, 1425–1433.
23. Hall, E. J. *Radiobiology for the Radiologist,* 5th ed.; Lippincott Williams & Wilkins: Philadelphia, PA, 2000.
24. Carter, T.; Vancurova, I.; Sun, I.; Lou, W.; DeLeon, S. *Mol. Cell. Biol.* **1990**, *10*, 6460–6471.
25. Jafri, F.; Hardin, J. A.; Dynan, W. S. *J. Immunol. Methods* **2001**, *251*, 53–61.
26. Li, S.; Takeda, Y.; Wragg, S.; Barrett, J.; Phillips, A. C.; Dynan, W. S. *Nucleic Acids Res.* **2003**, *31*, 5848–5857.
27. Carey, T. E.; Takahashi, T.; Resnick, L. A.; Oettgen, H.; Old, L. J. *Proc. Natl. Acad. Sci. U.S.A.* **1976**, *73*, 3278–3282.
28. Winter, G.; Milstein, C. *Nature* **1991**, *349*, 293–299.
29. Mamula, M. J.; Lin, R. H.; Janeway, C. A.; Hardin, J. A. *J. Immunol.* **1992**, *149*, 789–795.
30. Morgan, D. O.; Roth, R. A. *Immunol. Today* **1988**, *9*, 84–88.
31. McNeil, P. L. *Methods Cell. Biol.* **1989**, *29*, 153–173.
32. Cattaneo, A.; Biocca, S. *Trends Biotechnol.* **1999**, *17*, 115–121.
33. Ward, J. F. *Prog. Nucleic Acid Res. Mol. Biol.* **1988**, *35*, 95–125.
34. Hayhurst, A.; Harris, W. J. *Protein Expr. Purif.* **1999**, *15*, 336–343.

Indexes

Author Index

Subject Index

386

In-well softening, semivolatile organic compounds, 12–63
In-well vapor stripping principles, 16
In-well vapor stripping removal, dissolved TCE from groundwater, 23–24
India, arsenic-contaminated water, 100
Indium tin oxide electrodes
    bare electrode, voltammetry, pertechnetate ion, 312–314
    coating for pertechnetate determination, 309
    polymer-modified, voltammetry, pertechnetate ion, 314–320
Infiltration and outflow rates, infiltration test, Hell's Half Acre, 198–199, 200f–202f
Inorganic aqueous chemical systems, strontium behavior, 257–263
Interfacial properties in contaminated systems, compositional effects on DNAPL migration and retention, 160–182
Interfacial tension, dodecylamine and octanoic acid effects, 163–165
Intrafracture flow paths, coalescence, divergence, film flow, and dripping, 203, 209–215
Intrafracture flow processes, laboratory experiments, 203–216, 217f–219f
ITO. See Indium tin oxide

J

Japan, human mercury poisoning, 99

K

Kuramoto-Sivashinsky equation
    chaotic flow through fractures, 216, 220, 221f
    system feedback phenomena, 197

L

L-cysteine
    copper(II) ion detection, 300–304
    surface recognition agent, copper(II) ion, 291, 301f
Lachine shale, composite rate coefficient, cross-coupling reactivity, 76, 78t–80
Lead ion mobility dependence on soil humic substructures, 144–155
^{210}Lead tracer for radon decay product, aerosol size, 345, 347f–349
Lignin
    composite rate coefficient, cross-coupling reactivity, 76, 78t–80
    young humic substance, structure, 66, 69f
Lignocellulose/phenolics in pine shavings, organic amendment in ^{13}C tracer studies, heavy metal ion mobility, 144–155
Lignocellulose/silicates in wheat straw, organic amendment in ^{13}C tracer studies, heavy metal mobility, 144–155
Lithiophorite, structure, 85f
Low activity waste chemical pretreatment, process monitoring, 333, 337–340
Low dose radiation response, single chain antibody variable fragment (scFv) 18-2 effect, 366, 368f–370

M

M-VALOR, modified numerical simulator, organic liquid migration and entrapment predictions, 176–179f
Manganese oxide
    characterization and colloid coagulation, 86–88, 90–92f
    crystal structure, 88–91

388

Printed in the United States
137200LV00001B/28/A